Methods in Enzymology

Volume 412
AMYLOID, PRIONS, AND OTHER PROTEIN AGGREGATES
PART B

METHODS IN ENZYMOLOGY

EDITORS-IN-CHIEF

John N. Abelson   Melvin I. Simon

DIVISION OF BIOLOGY
CALIFORNIA INSTITUTE OF TECHNOLOGY
PASADENA, CALIFORNIA

FOUNDING EDITORS

Sidney P. Colowick and Nathan O. Kaplan

Methods in Enzymology

Volume 412

Amyloid, Prions, and Other Protein Aggregates Part B

EDITED BY

Indu Kheterpal

PENNINGTON BIOMEDICAL RESEARCH CENTER
LOUISIANA STATE UNIVERSITY SYSTEM
BATON ROUGE, LOUISIANA

Ronald Wetzel

GRADUATE SCHOOL OF MEDICINE
UNIVERSITY OF TENNESSEE
KNOXVILLE, TENNESSEE

AMSTERDAM • BOSTON • HEIDELBERG • LONDON
NEW YORK • OXFORD • PARIS • SAN DIEGO
SAN FRANCISCO • SINGAPORE • SYDNEY • TOKYO
Academic Press is an imprint of Elsevier

ELSEVIER

Academic Press is an imprint of Elsevier
525 B Street, Suite 1900, San Diego, California 92101-4495, USA
84 Theobald's Road, London WC1X 8RR, UK

This book is printed on acid-free paper.

For information on all Elsevier Academic Press publications
visit our Web site at www.books.elsevier.com

ISBN-13: 978-0-12-182817-2
ISBN-10: 0-12-182817-4

PRINTED IN THE UNITED STATES OF AMERICA
06 07 08 09 9 8 7 6 5 4 3 2 1

Table of Contents

Section I. Characterization of Protein Deposition *In Vivo* and *Ex Vivo*

Section II. Cell and Animal Models of Amyloid Formation and Toxicity

Section III. Computational Approaches and Theory

Contributors to Volume 412

Article numbers are in parentheses following the names of contributors.
Affiliations listed are current.

KARIM ABID (1), *University of Texas Medical Branch, Galveston, Texas 77555*

ERIC E. ABRAHAMSON (9), *Department of Neurology, University of Pittsburgh, Pittsburgh, Pennsylvania 15213*

SVIATOSLAV N. BAGRIANSTEV (3), *Laboratory for Molecular Biology, University of Illinois at Chicago, Chicago, Illinois 60607*

JORGE R. BARRIO (10), *Department of Molecular and Medical Pharmacology, UCLA School of Medicine, Los Angeles, California 90095*

VALERIE BERTHELIER (8), *Department of Medicine, University of Tenessee Graduate School of Medicine, Knoxville, Tennessee 37920*

JOSE M. BORREGUERO* (19), *Department of Physics, Center for Polymer Studies, Boston University, Boston, Massachusetts 02215*

HEATHER R. BRIGNULL (16), *Department of Biochemistry, Molecular Biology, and Cell Biology, Rice Institute for Biomedical Research, Northwestern University, Evanston, Illinois 60208*

JOAQUÌN CASTILLA (1), *University of Texas Medical Branch, Galveston, Texas 77555*

BYRON CAUGHEY (2, 14), *Laboratory of Persistent Viral Diseases, Rocky Mountain Laboratories, National Institute of Allergy and Infectious Diseases, National Institutes of Health, Hamilton, Montana 59840*

DHIANJALI CHANDRARATNA (15), *University of Cambridge, Department of Medicine, CIMR Wellcome Trust, Cambridge, United Kingdom CB2 2XY*

DONGMEI CHENG (6), *Center for Neurodegenerative Disease, School of Medicine, Emory University, Department of Human Genetics, Atlanta, Georgia 30322*

DAMIAN C. CROWTHER (15), *Department of Medicine, University of Cambridge, CIMR Wellcome Trust, Cambridge, United Kingdom CB2 2XY*

LUIS CRUZ (19), *Department of Physics, Center for Polymer Studies, Boston University, Boston, Massachusetts 02215*

MANIK L. DEBNATH (9), *Department of Psychiatry, University of Pittsburgh, Pittsburgh, Pennsylvania 15213*

STEVEN T. DEKOSKY (9), *Departments of Neurology and Psychiatry, University of Pittsburgh, Pittsburgh, Pennsylvania 15213*

HIROSHI DOI (5), *Laboratory for Structural Neuropathology, Riken Brain Science Institute, Saitama 351-0198, Japan*

DUC M. DUONG (6), *Departments of Human Genetics, Center for Neurodegenerative Disease, School of Medicine, Emory University, Atlanta, Georgia 30322*

Current affiliation: Department of Biology, Center for the Study of System Biology, Georgia Institute of Technology, Atlanta, Georgia 30318

FRANK A. FERRONE (17), *Department of Physics, Drexel University, Philadelphia, Pennsylvania 19104*

SUSANA M. GARCIA (16), *Department of Biochemistry, Molecular Biology, and Cell Biology, Rice Institute for Biomedical Research, Northwestern University, Evanston, Illinois 60208*

FLAVIANO GIORGINI (13), *Department of Genetics, University of Leicester, Leicester, United Kingdom LE1 7RH*

SHAUN GLEASON (11), *Siemens Medical Solutions USA, Inc., Molecular Imaging, Knoxville, Tennessee 37920*

YAIR M. GOZAL (6), *Department of Human Genetics, Center for Neurodegenerative Disease, School of Medicine, Emory University, Atlanta, Georgia 30322*

JENS GREGOR (11), *Department of Computer Science, University of Tennessee, Knoxville, Tennessee 37996*

JUN-TAO GUO (18), *Department of Biochemistry and Molecular Biology, and Institute of Bioinformatics, University of Georgia, Athens, Georgia 30602*

CAROL K. HALL (20), *Chemical Engineering Department, North Carolina State University, Raleigh, North Carolina 27695*

SUNG-CHENG HUANG (10), *Department of Molecular and Medical Pharmacology, UCLA School of Medicine, Los Angeles, California 90095*

ANDREW G. HUGHSON (2), *Laboratory of Persistent Viral Diseases, Rocky Mountain Laboratories, National Institute of Allergy and Infectious Diseases, National Institutes of Health, Hamilton, Montana 59840*

MILOS D. IKONOMOVIC (9), *Departments of Neurology and Psychiatry, University of Pittsburgh, Pittsburgh, Pennsylvania 15213*

BARBARA A. ISANSKI (9), *Department of Neurology, University of Pittsburgh, Pittsburgh, Pennsylvania 15213*

STEPHEN J. KENNEL (11), *Department of Medicine, University of Tennessee Graduate School of Medicine, Knoxville, Tennessee 37920*

VLADIMIR KEPE (10), *Department of Molecular and Medical Pharmacology, UCLA School of Medicine, Los Angeles, California 90095*

WILLIAM E. KLUNK (9), *Department of Psychiatry, University of Pittsburgh, Pittsburgh, Pennsylvania 15213*

RICHARD KNOCHENMUSS (7), *Novartis Institutes for Biomedical Research, 4002 Basel, Switzerland*

DAVID A. KOCISKO (14), *Laboratory of Persistent Viral Diseases, Rocky Mountain Laboratories, National Institute of Allergy and Infectious Diseases, National Institutes of Health, Hamilton, Montana 59840*

VITALY V. KUSHNIROV (3), *Cardiology Research Centre, Institute of Experimental Cardiology, Moscow, Russia 121552*

JAMES J. LAH (6), *Department of Human Genetics, Center for Neurodegenerative Disease, School of Medicine, Emory University, Atlanta, Georgia 30322*

ALLAN I. LEVEY (6), *Department of Human Genetics, Center for Neurodegenerative Disease, School of Medicine, Emory University, Atlanta, Georgia 30322*

SUSAN W. LIEBMAN (3), *Laboratory for Molecular Biology, University of Illinois at Chicago, Chicago, Illinois 60607*

DAVID A. LOMAS (15), *Department of Medicine, University of Cambridge,*

CIMR Wellcome Trust, Cambridge, United Kingdom CB2 2XY

CHESTER A. MATHIS (9), Department of Radiology, University of Pittsburgh, Pittsburgh, Pennsylvania 15213

KINSEY MAUNDRELL (1), Serono Pharmaceutical Research Institute, CH-1211 Geneva, Switzerland

GREGOR MCCOMBIE (7), Novartis Institute for Biomedical Research, Analytical and Imaging Sciences, 4002 Basel, Switzerland

KENICHI MITSUI (5), Research Resources Center, Riken Brain Science Institute, Saitama 351-0198, Japan

RODRIGO MORALES (1), University of Texas Medical Branch, Galveston, Texas 77555

RICHARD I. MORIMOTO (16), Department of Biochemistry, Molecular Biology, and Cell Biology, Rice Institute for Biomedical Research, Northwestern University, Evanston, Illinois 60208

JAMES F. MORLEY (16), Department of Biochemistry, Molecular Biology, and Cell Biology, Rice Institute for Biomedical Research, Northwestern University, Evanston, Illinois 60208

PAUL J. MUCHOWSKI (13), University of California, San Francisco, Gladstone Institute of Neurological Disease, San Francisco, California 94158

DIETER MUELLER (7), Novartis Institutes for Biomedical Research, 4002 Basel, Switzerland

CHARLES L. MURPHY (4), Human Immunology and Cancer Program, University of Tennessee Graduate School of Medicine, Knoxville, Tennessee 37920

NOBUYUKI NUKINA (5), Laboratory for Structrural Neuropathology, Riken Brain Science Institute, Saitama 351-0198, Japan

ALEXANDER P. OSMAND (8), Department of Medicine, University of Tenessee Graduate School of Medicine, Knoxville, Tennessee 37920

RICHARD PAGE (15), Department of Medicine, University of Cambridge, CIMR Wellcome Trust, Cambridge, United Kingdom CB2 2XY

MICHAEL J. PAULUS (11), Preclinical Product Management, Siemens Medical Solutions USA Inc., Molecular Imaging, Knoxville, Tennessee 37920

JUNMIN PENG (6), Department of Human Genetics, Center for Neurodegenerative Disease, School of Medicine, Emory University, Atlanta, Georgia 30322

TATIANA ROHNER (7), Novartis Institutes for Biomedical Research, 4002 Basel, Switzerland

PAULA SAÀ (1), University of Texas Medical Branch, Galveston, Texas 77555

NAGICHETTIAR SATYAMURTHY (10), Department of Molecular and Medical Pharmacology, UCLA School of Medicine, Los Angeles, California 90095

JAY R. SILVEIRA (2), Laboratory of Persistent Viral Diseases, Rocky Mountain Laboratories, National Institute of Allergy and Infectious Diseases, National Institutes of Health, Hamilton, Montana 59840

GARY W. SMALL (10), Department of Psychiatry and Biobehavioral Sciences, UCLA School of Medicine, Los Angeles, California 90095

ALAN SOLOMON (4, 11), Human Immunology and Cancer Program, University of Tennessee Graduate School of Medicine, Knoxville, Tennessee 37920

CLAUDIO SOTO (1), University of Texas Medical Branch, Galveston, Texas 77555

DIETER STAAB (7), *Novartis Institutes for Biomedical Research, 4002 Basel, Switzerland*

H. EUGENE STANLEY (19), *Department of Physics, Center for Polymer Studies, Boston University, Boston, Massachusetts 02215*

MARKUS STOECKLI (7), *Novartis Institute for Biomedical Research, 4002 Basel, Switzerland*

MOTOMASA TANAKA (12), *University of California San Francisco, Department of Cellular and Molecular Pharmacology, Howard Hughes Medical Institute, San Francisco, California 94143; PRESTO, Japan Science and Technology Agency, Saitama 332-0012, Japan*

BRIGITA URBANC (19) *Department of Physics, Center for Polymer Studies, Boston University, Boston, Massachusetts 02215*

VICTORIA A. WAGONER (20), *Chemical Engineering Department, North Carolina State University, Raleigh, North Carolina 27695*

JONATHAN S. WALL (11), *Human Immunology and Cancer Program, University of Tennessee Graduate School of Medicine, Knoxville, Tennessee 37920*

SHUCHING WANG (4), *Human Immunology and Cancer Program, University of Tennessee Graduate School of Medicine, Knoxville, Tennessee 37920*

DEBORAH T. WEISS (4), *Human Immunology and Cancer Program, University of Tennessee Graduate School of Medicine, Knoxville, Tennessee 37920*

JONATHAN S. WEISSMAN (12), *Department of Cellular and Molecular Pharmacology, Howard Hughes Medical Institute, University of California San Francisco, San Francisco, California 94143*

RONALD WETZEL[†] (8), *Department of Medicine, University of Tenessee Graduate School of Medicine, Knoxville, Tennessee 37920*

KARL-HIENZ WIEDERHOLD (7), *Novartis Institutes for Biomedical Research, 4002 Basel, Switzerland*

TERESA WILLIAMS (4), *Human Immunology and Cancer Program, University of Tennessee Graduate School of Medicine, Knoxville, Tennessee 37920*

YING XU (18), *Department of Biochemistry and Molecular Biology, and Institute of Bioinformatics, University of Georgia, Athens, Georgia 30602*

[†]*Current affiliation: Department of Structural Biology, University of Pittsburgh School of Medicine, Pittsburgh, Pennsylvania 15260*

Preface

After one of us suffered through a series of unsuccessful attempts to obtain funding for protein aggregation projects in the early 1990s, a biophysicist colleague pointed out that a major problem with such applications was that there simply weren't many good tools available for obtaining useful information about protein aggregate structures and how they are formed. While this feedback was hard to hear, accept, and constructively respond to, it was largely accurate. The state-of-the-art methods commonly available and in use for following aggregation kinetics in those years were turbidity assays, or perhaps, with luck, ThT fluorescence. What was considered to be high-end structural information on aggregates, meanwhile, came from electron microscopy, FTIR, and X-ray fiber diffraction. The biophysicist colleague was too kind to point out another major problem with these applications: very few scientists really cared about protein aggregation.

Times have changed, and by 2006, misfolding and aggregation studies have moved to center stage in the protein science and biomedical funding theaters. Protein aggregates are found in many of the common neurodegenerative diseases, in addition to a well-established group of peripheral amyloidoses and other deposition diseases. At the same time, some protein aggregates are known to play useful roles in a variety of life forms, possibly even in humans. The fossilized footprints of protein aggregation in molecular and cellular evolution are revealed by the existence of a growing list of cellular systems involved in preventing, managing, and/or limiting the consequences of aggregation. It is now recognized that disordered polypeptides, whether exiting the ribosome or a membrane channel, or emerging from denaturant into native buffer, often undergo kinetic partitioning between a native state maturation pathway and one or more alternative aggregation pathways. Protein aggregation also plays important roles in biotechnology, and our accumulated experience in working with amyloid fibrils and other ordered aggregates may still prepare the way for exploiting these structures as nanomaterials.

Not so very long ago, the conventional wisdom in the protein folding community was that all proteins know how to fold properly, and if one has the misfortune to encounter aggregation in a folding experiment, it is not because there is anything interesting or fundamental going on in the experiment, it is only because one doesn't know the "proper conditions" for avoiding aggregation. In the past decade, as aggregation studies have become not only acceptable but almost glamorous, this wisdom has been turned on its head, such that now if one's protein can't be induced to form amyloid fibrils, it must be because one hasn't found the correct protocol. Aggregation is now thought of as a central feature of the polypeptide molecules that do Nature's heavy lifting, and, as such, protein aggregation has a central place in the biophysics of life processes. As a major constraint in the "other genetic code" that controls how and whether the linear sequence of amino acids encoded in a stretch of DNA expresses its intrinsic functionality, aggregation is also important as a mode by which biological information transfer can ultimately be corrupted, and as such plays a continuing role in molecular evolution.

Not only has the funding environment improved over the past 15 years, so have the tools. Thanks to the foresight, commitment, and ingenuity of many research groups working on amyloid and other protein aggregation, the quantity, quality, and breadth of the technology available to study protein aggregation has greatly expanded. This volume (Part B, Volume 412) and its companion volume (Part C, Volume 413), which along with Volume 309 constitute what amounts to a *Methods in Enzymology* amyloid trilogy, celebrate this achievement of the field by highlighting a variety of new methods now being applied to protein aggregation and aggregates in a variety of settings and from a variety of points of view. Part B focuses on: (a) methods for working with, detecting, and characterizing protein aggregates derived from cells or tissues; (b) cell and animal models of amyloid formation and toxicity; and (c) computational and theoretical analyses of the aggregation process.

Chapter 1 by Soto and colleagues undertakes a detailed description of the authors' method for amplifying prion particles as a means of sensitive detection. Chapters 2 and 3 by the Caughey and Liebman groups describe methods for the extremely difficult task of fractionating heterogeneous mixtures of aggregates, a task that grows in importance as we recognize the variety of aggregate morphologies that can be generated even by the same polypeptide. These methods should be useful for fractionation of aggregates generated both *in vivo* and *in vitro*.

Chapters 4 through 11 deal with various levels of analysis of aggregates in tissue. In Chapter 4, Solomon and colleagues discuss chemical confirmation of the major molecular constituent(s) of amyloid deposits by mass spectrometric analysis of the purified components. Chapters 5, 6, and 7 also focus on the characterization of the components of protein deposits in tissue using mass spectrometric approaches. Chapters 7 through 11 deal with imaging of amyloid deposits in tissue either *ex vivo* (Chapters 7 through 9) or in the living organism (Chapters 10 through 11).

Chapter 12 by Weissman and colleagues describes their important methods for introducing *in vitro*-generated aggregates into yeast cells, which allowed them to confirm the protein-only nature of yeast prions and the conformational nature of prion strains. Chapter 13 by Giorgini and Muchowski describes an approach to screening for genetic modifiers of amyloid toxicity in a yeast model. Kocisko and Caughey in Chapter 14 describe a cell based assay for identifying compounds that compromise prion toxicity. Chapters 15 and 16 describe two examples of amyloid disease models in multicellular organisms.

As we struggle to understand the molecular details of the structures of aggregate-prone polypeptides and aggregates and the assembly mechanisms that connect the two, the value of computational approaches to these issues is becoming increasingly clear. Chapter 17 deals with an analysis of some of the nuances of nucleation theory. Chapters 18 through 20 describe different methods for approaching an understanding of aggregate structure and the aggregation process.

For the editors, the preparation of these volumes was a labor of love. The labor, along with a love of the subject matter, was shared by the many colleagues who participated in a variety of ways. We are grateful to all of the authors, who responded to our urgings to make their methods both available and accessible, and to the anonymous reviewers, who helped the authors to meet that challenge by reading these chapters critically and carefully and providing frank feedback. We are also especially grateful for the administrative and organizational skills of Pam Trentham at the University of Tennessee and for our editor at Elsevier, Cindy Minor, who both did their best to keep us on track. We gladly acknowledge our sources of support that helped make these volumes possible: the National Institutes of Health (RW), the National Science Foundation (IK), and the Louisiana Board of Regents (IK).

INDU KHETERPAL
RONALD WETZEL

METHODS IN ENZYMOLOGY

Section I

Characterization of Protein Deposition
In Vivo and *Ex Vivo*

[1] Protein Misfolding Cyclic Amplification for Diagnosis and Prion Propagation Studies

By JOAQUÍN CASTILLA, PAULA SAÁ, RODRIGO MORALES, KARIM ABID, KINSEY MAUNDRELL, and CLAUDIO SOTO

Abstract

Diverse human disorders are thought to arise from the misfolding and aggregation of an underlying protein. Among them, prion diseases are some of the most intriguing disorders that can be transmitted by an unprecedented infectious agent, termed prion, composed mainly (if not exclusively) of the misfolded prion protein. The hallmark event in the disease is the conversion of the native prion protein into the disease-associated misfolded protein. We have recently described a novel technology to mimic the prion conversion process *in vitro*. This procedure, named protein misfolding cyclic amplification (PMCA), conceptually analogous to DNA amplification by polymerase chain reaction (PCR), has important applications for research and diagnosis. In this chapter we describe the rational behind PMCA and some of the many potential applications of this novel technology. We also describe in detail the technical and methodological aspects of PMCA, as well as its application in automatic and serial modes that have been developed with a view to improving disease diagnosis.

Introduction

Prion diseases or transmissible spongiform encephalopathies (TSEs) are neurodegenerative disorders of humans and animals usually characterized by the presence of PrP^{res}, an abnormal, protease-resistant isoform of a cellular protein called PrP^C. Historically, scrapie has been the most common TSE in animals, affecting sheep for more than 200 years (Collinge, 2001). TSEs have also been identified in mink and mule deer since the 1960s. The most recent and worrisome outbreak of an animal TSE disease is bovine spongiform encephalopathy (BSE) in cattle, which originated in Britain in the 1980s (Prusiner, 1997). BSE has important implications for human health, because the infectious agent can be transmitted to humans producing a new disease, termed variant Creutzfeldt-Jakob disease (vCJD) (Collinge, 1999; Will *et al.*, 1996). TSEs are characterized by an extremely long incubation period, followed by a brief and invariably fatal clinical disease (Roos *et al.*, 1973). To date, no therapy or early diagnosis is available.

METHODS IN ENZYMOLOGY, VOL. 412   0076-6879/06 $35.00
          DOI: 10.1016/S0076-6879(06)12001-7

The pathogen responsible for TSEs, called "prion" (Prusiner, 1982), is composed mainly of a misfolded protein named PrP^{Sc}, which is a post-translationally modified version of the normal protein, PrP^C (Cohen and Prusiner, 1998). The conversion seems to involve a conformational change during which the α-helical content of the normal protein diminishes and the amount of β-sheet increases (Caughey *et al.*, 1991; Pan *et al.*, 1993). The structural changes are accompanied by alterations in the biochemical properties: PrP^C is soluble in nondenaturing detergents, PrP^{Sc} is insoluble; PrP^C is readily digested by proteases, whereas PrP^{Sc} is partially resistant, resulting in the formation of a N-terminally truncated fragment known as PrP^{res} (Baldwin *et al.*, 1995; Cohen and Prusiner, 1998).

At present, there is no accurate diagnosis for TSEs (Budka *et al.*, 1995; Weber *et al.*, 1997). In the case of sporadic CJD (sCJD) or vCJD, diagnosis is currently based almost entirely on clinical observation, because even though different molecules, such as protein S-100 or the 14-3-3 protein, have been proposed as markers of the disease, none of them are pathognomonic of the syndrome. For this reason, according to the operational diagnosis currently in use by the European Surveillance of CJD, definitive diagnosis can only be established by postmortem neuropathological examination and detection of PrP^{res} by immunohistochemistry, histoblot, or Western blot (Budka *et al.*, 1995; Weber *et al.*, 1997). Presymptomatic detection of sCJD or vCJD in living people is currently not possible.

To minimize the propagation of the bovine disease, several tests have been developed to diagnose BSE in postmortem brain tissue (Moynagh and Schimmer, 1999; Soto, 2004). However, in cattle, as in humans, there is no reliable way to identify affected animals early after infection (Schiermeier, 2001), because the problem of a diagnosis on the basis of PrP^{res} detection is that this form of the protein is abundant only in the brain at advanced stages of the disease.

Infectivity studies have been used to show that prions are also present in low amounts in peripheral tissues, such as lymphoid organs and blood (Aguzzi, 2000; Brown *et al.*, 2001; Collinge, 2001; Wadsworth *et al.*, 2001), and on the basis of these observations, different bioassays showing high sensitivity have been developed. In these methods, animals are injected with very low quantities of PrP^{Sc}, and the clinical signs indicating the presence of infectious material are monitored (Brown *et al.*, 2001). The biggest practical problem for using the infectivity assay in routine diagnosis, however, is that prion replication during the incubation phase progresses very slowly, and several months or even years may elapse before a detectable quantity of PrP^{Sc} has accumulated in the brain.

In vivo, prion replication is an extraordinary phenomenon that still remains not entirely understood. Although it is known that conversion of

PrP^C to PrP^{Sc} is an essential element in the etiology of the disease, the intrinsic mechanism by which this occurs, and whether other factors are involved, are crucial questions that remain to be answered.

The Birth of PMCA

To understand the mechanism of prion conversion, the nature of the infectious agent, and to attempt sensitive diagnosis, we have recently developed a technique referred to as protein misfolding cyclic amplification (PMCA) in which it is possible to simulate prion replication in the test tube in an accelerated mode (Saborio *et al.*, 2001). PMCA is a cyclic process leading to accelerated prion replication (Saborio *et al.*, 2001; Soto *et al.*, 2002). Each cycle is composed of two phases (Fig. 1). During the first phase, the sample, containing minute amounts of PrP^{Sc} and a large excess of PrP^C, is incubated to induce formation of PrP^{Sc} polymers. In the second phase, the sample is sonicated to break down the polymers, thus multiplying the number of growth sites for subsequent conversion. With each successive cycle, there is an exponential increase in the number of "seeds,"

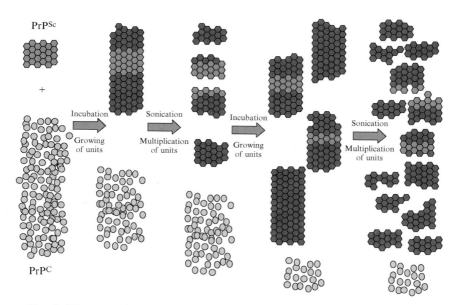

FIG. 1. Diagrammatic representation of principle behind PMCA. Cyclic amplification consists of subjecting a sample containing minute quantities of PrP^{Sc} and a large excess of PrP^C to cycles consisting of phases of growing of polymers and multiplication of converting units.

and thus the conversion process is dramatically accelerated (Fig. 1) (Saborio *et al.*, 2001). The cyclic nature of the system permits the use of as many cycles as is required to reach the amplification state needed for the detection of PrP^{Sc} in a given sample. Recently, we have shown that the *in vitro*–generated forms of PrP^{res} share similar biochemical and structural properties compared with PrP^{res} derived from sick brains (Castilla *et al.*, 2005a). Furthermore, inoculation of wild-type hamsters with *in vitro*–amplified PrP^{res} led to a scrapie-like disease identical to the illness produced using infectious material from diseased brains (Castilla *et al.*, 2005a). The technology has been automated, leading to a dramatic increase in efficiency of amplification and its application to detect PrP^{Sc} in blood of hamsters experimentally infected with scrapie (Castilla *et al.*, 2005b).

Applications of PMCA

Simulation of the process of prion conversion using PMCA represents a novel platform technology, which is likely to have a sustained impact in the field of prion biology. By using PMCA, we have been able to demonstrate definitively that the *in vitro*–generated PrP^{res} is fully infectious when injected into wild-type animals (Castilla *et al.*, 2005a). This provides the crucial demonstration that the *in vitro* conversion that occurs during PMCA closely mimics the events that take place over a protracted period *in vivo*, leading to disease and, ultimately, to death of the organism. The ability to simulate this process in accelerated mode, under controlled conditions *in vitro*, thus provides an opportunity to examine many aspects of prion biology that hitherto have been inaccessible to experimentation. Following is a brief description of the multiple areas in which PMCA may contribute.

The Molecular Mechanism of Species Barrier and
   Prion Strains Phenomena

As a consequence of the transmission of BSE to humans, a great concern has arisen regarding interspecies infectivity and tissues having a high enough quantity of prions to transmit the disease (Hill *et al.*, 2000; Wadsworth *et al.*, 2001). The molecular aspects that underlie the species barrier and the strain phenomena are still not understood (Bruce, 2003; Clarke *et al.*, 2001; Kascsak *et al.*, 1991). It has been shown that the sequence identity between infectious PrP^{Sc} and host prion protein plays a crucial role in determining species barrier (Telling *et al.*, 1996). It is clear that a few amino acid differences between both proteins can modify dramatically the incubation time and the course of the disease (Asante and

Collinge, 2001; DeArmond and Prusiner, 1996). So far, the investigation of the species barrier, prion strains, and the tissues carrying infectivity has been done mostly using the biological assay of infectivity (Clarke *et al.*, 2001; Wadsworth *et al.*, 2001). However, these studies are time consuming, because it is necessary to wait for several months or even years until the animals develop the clinical symptoms. In addition, the assessment of the species barrier for transmission of prions to humans is compromised by the use of animal models. PMCA can provide a complement to the *in vivo* studies of the species barrier and prion strains phenomenon by combining PrP^{Sc} and PrP^{C} from different sources in distinct quantities and evaluating quantitatively the efficiency of the conversion. In this sense, it has to be noted that the cell-free conversion system developed by Caughey and colleagues (Kocisko *et al.*, 1994) has been used successfully to compare and predict species barrier effects and the pertinent underlying mechanisms (Kocisko *et al.*, 1995; Raymond *et al.*, 1997).

Investigation of Factors Involved in the PrP^{C} to PrP^{Sc} Conversion

Another important issue in prion propagation is to know whether other factors have any role in the PrP^{C} to PrP^{Sc} conversion. We reported previously that the conversion procedure does not occur using highly purified prion proteins (PrP^{Sc} and PrP^{C}) under our experimental conditions (Saborio *et al.*, 1999). However, the activity is recovered when the bulk of cellular proteins is reincorporated into the sample (Saborio *et al.*, 1999). This finding provides direct evidence that other factors present in the brain are essential to catalyze prion propagation. In this direction, PMCA could also contribute to a better understanding of the mechanism of prion conversion and the identification of additional factors involved. Indeed, Supattapone and coworkers have used PMCA to show that metal cations, such as copper and zinc, and polyanions including diverse types of RNA molecules can modulate PrP conversion *in vitro* (Deleaut *et al.*, 2003, 2005; Nishina *et al.*, 2004).

Screening for Inhibitors of Prion Propagation

In the same manner that prion propagation can be used to discover novel drug targets for TSEs in culture cells, PMCA also shows a great advantage in these types of studies. Inhibitors and promoters could be tested quickly in different contexts using human and bovine prions, for which no prion-permissive culture cells have been generated. One of the best targets for TSE therapy is the inhibition and reversal of PrP^{C} to PrP^{res} conversion (Head and Ironside, 2000; Soto and Saborio, 2001). In drug development, it is crucial to have a relevant and robust *in vitro* assay to screen compounds for activity before testing them in more time-consuming and expensive *in vivo* assays. PMCA represents a convenient biochemical tool to identify and evaluate the activity of drug candidates for TSE treatment, because it mimics *in vitro* the central pathogenic process of the disease. Also the simplicity of the method and the relatively rapid outcome are important features of these types of studies. Moreover, the fact that PMCA can be applied to prion conversion in different species provides the opportunity to validate the use in humans of drugs that have been evaluated in experimental animal models of the disease.

Diagnosis

One of the most valuable applications of PMCA is in TSE diagnosis. As stated previously, the biggest problem facing a biochemical test to detect PrP^{res} presymptomatically in tissues other than brain is the very low amount of PrP^{res} existing in them. Most of the efforts to develop a diagnostic system for prion diseases have been focused on the increase of sensitivity of the current detection methods. PMCA offers the opportunity to enhance existing methods by amplifying the amount of PrP^{res} in the sample. Combining the strategy of reproducing prions *in vitro* with any of the high-sensitive detection methods, the early diagnosis of TSE may be achieved. The aim would be not only to detect prions in the brain in early presymptomatic cases but also to generate a test to diagnose living animals and humans. For this purpose, a tissue other than brain is required and, to have an easier noninvasive method, detection of prions in body fluids such as urine or blood are the best options. A blood test for CJD can have many applications, including screening of blood banks, identification of populations at risk, reduction of iatrogenic transmission of CJD, and early diagnosis of the disease (Soto, 2004).

Extension to Other Protein Misfolding Diseases

Besides TSEs, several other diseases involve changes in the conformation of a natural protein to an altered structure with toxic properties capable of inducing tissue damage and organ dysfunction (Carrell and Lomas, 1997; Dobson, 2004; Kelly, 1998; Soto, 2001). This group of diseases called protein misfolding disorders includes several forms of neurodegenerative diseases such as Alzheimer's, Parkinson's, and Huntington's diseases, as well as a group of more than 15 distinct disorders involving amyloid deposition in diverse organs (Soto, 2001). In a similar way to PrP^{res} in TSE, the protein conformational changes associated with the pathogenesis of these diseases result in the formation of abnormal proteins rich in β-sheet structure, partially resistant to proteolysis and with a high tendency to aggregate (Soto,

2001). The process of misfolding and aggregation also follows a seeding-nucleation mechanism, and hence the principles of PMCA might be applied to amplify the abnormal folding of these proteins as well. Therefore, PMCA may have a broader application for research and diagnosis of diseases in which misfolding and aggregation of a protein are hallmark events.

Method and Technical Details

Protein misfolding cyclic amplification in its original mode was done by manual operation (Saborio *et al.*, 2001), but we have recently developed an automated mode (aPMCA), which has been developed to increase sensitivity, specificity, and throughput. The increased throughput of aPMCA has allowed us to evaluate the importance of numerous variables including temperature, pH, substrate concentration, type, and concentration of the detergents, power and length of sonication, and so on. In addition, the sensitivity has been increased further by the introduction of a new concept involving serial rounds of amplification. This procedure is named serial automated PMCA (saPMCA) and is similar to application of multiple rounds of PCR amplification to reach high sensitivity detection of DNA.

Buffer

Conversion Buffer. Composition of the conversion buffer (CB) has been established and optimized after exhaustive studies, and we have found that even small changes may dramatically affect the efficiency of the amplification process. Thus, we highly recommend using the following conversion buffer: PBS; NaCl, 0.15 *M*; Triton X-100, 1%; and complete protease inhibitor cocktail 1× (Roche, cat#: 1836145). A pH of between 7.0 and 7.3 is necessary to obtain the best results. Low concentrations of SDS may also be included in CB but are not usually necessary. When used, the SDS concentration should be optimized depending on the type of the PrP^{Sc} species to be amplified.

Equipment

Sonicator. In the original PMCA protocol, the proof of concept was established using a manual sonicator using a single microtip (Saborio *et al.*, 2001). However, with the increased need for high throughput and automation, we have implemented a programmable sonicator that uses a 96-well plate format (Misonix, USA, model S3000MP sonicator) and satisfies the principal requirements for PMCA even though the machine was originally designed for other purposes. Improvements to the equipment planned for the near future should lead to a full adaptation to the needs of PMCA.

Homogenizer. A principal component in the PMCA reaction is the PrP^C used as substrate. At this point we consider that a normal brain homogenate (NBH) is the best substrate for high-efficiency amplification. For brain homogenization, we recommend using the high-viscosity mixer Eurostar PWR BSC S1 (IKA, USA), whereas for manual homogenization, the Potter homogenizer is a perfectly satisfactory option.

Preparation of Samples for Amplification

The correct preparation of the inoculum (PrP^{Sc} used as starting material) and substrate (material used as source of PrP^C) samples is critical to achieve a good efficiency of amplification (Fig. 2). To prepare the best samples for PMCA it is important to know several critical parameters concerning the material to be amplified, in particular, (1) the animal species; (2) the type of tissue in which PrP^{Sc} is to be detected; (3) an estimation of the amount of PrP^{Sc} in the sample to be amplified; (4) the storage conditions of the sample; and (5) possible inhibitors that may interfere with the amplification.

Whenever possible, it is preferable to use a substrate from the same species as the PrP^{Sc} to be amplified (see "Preparation of Substrate"). In this way, any potential problems caused by species barrier can be avoided. On the other hand, the use of substrates from different species can be useful, for example, in studies to understand the nature of the species barrier.

Another important parameter is the condition in which the sample to be amplified has been stored. Although PrP^{Sc} is resistant to high temperature (Castilla *et al.*, 2005a), treatment at $>100°$ can promote the formation of large aggregates in the samples, which interfere with efficient amplification. The use of samples previously denatured using chaotropic agents or ionic detergents at high concentrations is also incompatible with PMCA and should be avoided. Although not many studies have been done using formalin-fixed samples, it is not recommended to use this type of sample for amplification.

Finally, either the samples to be amplified or the substrates prepared for the amplification may contain potential inhibitory molecules such as plasminogen, cations, or other, as yet, unidentified blood compounds that have been found to interfere with PMCA (data not shown). Because it is extremely important to eliminate such molecules, some samples will require pretreatment before the amplification process (see "Pretreatment in Preparation of Samples from Peripheral Sources"). In addition, special precautions need to be taken when using blood, CSF, saliva, milk, urine, or feces (see "Preparation of Samples from Peripheral Sources").

Preparation of Prion-Infected Samples from CNS. Infectious brain material should be homogenized in conversion buffer (CB) at 10% (w/v)

I. Substrate preparation

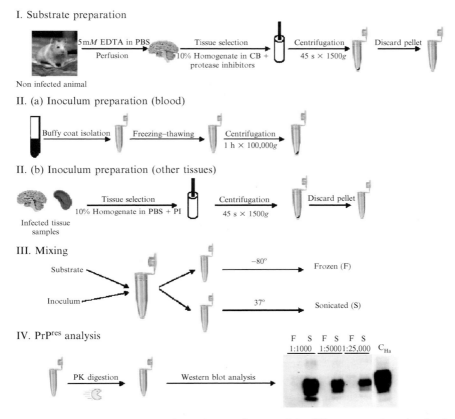

FIG. 2. PMCA method. The scheme shows a diagram of the different procedures involved in PMCA, including the preparation of the substrate (I), inoculum (II), the mixing and amplification process (III), and the PrP^res detection (IV).

at 4° using a high-viscosity mixer (see "Equipment"). The homogenate should be centrifuged at 1500g for 30 sec and the supernatant retained for further use. Because much of the PrP^Sc is present in the membrane fractions, centrifugation at a higher speed should be avoided to reduce the risk of losing infectious material in the pellet. Storage of the homogenate in aliquots at −80° is recommended.

Preparation of Prion-Infected Samples from Peripheral Sources. Although the most frequent samples for testing will be brain homogenates, other tissues could also be used as a source of infectious material. In most cases, the samples should be prepared in the same manner, with the exception of some special tissues that may require some form of pretreatment. We have separated these latter samples into: (1) blood samples, (2) tissue samples containing high amount of blood, and (3) other fluids such as CSF, saliva, milk, urine, and feces.

BLOOD SAMPLES. Blood is probably the most interesting sample to use in prion diagnosis. However, it is also considered one of the most complicated tissues for these purposes. The extremely low amount of PrP^{Sc} present in blood and the presence of inhibitors of prion replication make it necessary to take special precautions. The blood should be collected using a syringe containing EDTA to avoid clotting and placed in tubes containing sodium citrate. For standard infectious blood material, it is necessary to use at least 1 ml of whole blood. One milliliter of PBS should be added to the total blood and buffy coat should be prepared by centrifugation on a Ficoll gradient using standard procedures. The isolated buffy coat fraction should be subjected to three consecutive freezing–thawing cycles to break the cells, and then centrifuged for $100,000g$ for 1 h at $4°$ to pellet PrP^{Sc}. For PMCA, the pellet should be resuspended directly in 100 μl of normal brain homogenate. The amount of PrP^{Sc} present in 1 ml of infectious blood is usually not sufficient for standard aPMCA, and in this case serial rounds of aPMCA are needed (Castilla *et al.*, 2005b) as described later (see "Serial Automatic PMCA Procedure").

TISSUE SAMPLES CONTAINING HIGH AMOUNT OF BLOOD. We have observed during development of PMCA that small amounts of plasma or serum can inhibit the PMCA reaction. In addition, the proteinase K treatment, which is performed after amplification to detect PrP^{Sc}, can be inhibited by protease inhibitors present in blood. For this reason, whenever possible, it is recommended that animals be perfused with PBS + 5 mM EDTA before dissection of the tissues. When this is not possible, freshly dissected tissues should be washed carefully in PBS + 5 mM EDTA before the preparation of the samples.

OTHER SOURCES OF PrP^{Sc} SUCH AS CSF, SALIVA, MILK, URINE, AND FECES. Although we have not yet attempted to use PMCA to amplify samples from these biological fluids, later we provide some advice based on our experience with similar samples. We would recommend diluting these samples with PBS + 5 mM EDTA followed by centrifugation at $100,000g$ for 1 h. This procedure requires exhaustive washing of the sample with CB and a further centrifugation at $100,000g$ for 1 h. The pellet, most probably invisible, should be resuspended directly in substrate to avoid more dilution. It is important to note that these special samples could contain enzymes, urea, and other molecules that may coprecipitate with the infectious material, thereby further complicating the amplification step.

Once prepared, samples for amplification should be divided into aliquots and frozen at $-80°$. It is known that PrP^{Sc} has a strong tendency to aggregate, and this seems to increase with repeated freezing and thawing. Although higher levels of aggregation seem not to affect the ability of PrP^{Sc} to act as template for amplification (samples have been frozen and retested more than 20 times without any significant difference), the size and number of the large aggregates can lead to errors in sample dilution, which thus may produce variability in the level of amplification.

In vivo studies have shown that the PK-treated PrP^{Sc} is still infectious, even though its infectivity is diminished compared with that of nontreated PrP^{Sc}. The latter may be due to an increased propensity to aggregation. Samples digested with PK treatment can also be amplified by PMCA even though, as with *in vivo* infectivity, the level of amplification is slightly diminished. Despite this, the use of PK to remove proteinaceous contaminants might be a good option for certain samples.

Preparation of Substrate. We consider the preparation of the substrates to be the most critical step in achieving successful PMCA. In our hands, the best and most convenient substrate is normal brain homogenate from the same species as the prion sample to be amplified, although other substrates have also been used successfully (see "Other Substrates").

Normal Brain Homogenate. As mentioned previously, the presence of cations and certain blood components can seriously affect the amplification process. For this reason, we consider it highly beneficial to perfuse the animals with PBS + 5 mM EDTA before the brain extraction. After perfusion, a totally white brain can be obtained from the animal. We recommend the use of a CO_2 chamber for euthanizing, to avoid using anesthesia, which may also interfere with the subsequent amplification.

Whenever possible, we recommend preparing substrate from animals of the same species as the infectious sample to be amplified. In the case of larger animals such as cattle, sheep, goats, and so on in which perfusion before tissue dissection is not possible, we would recommend removing the entire brain as quickly as possible to reduce postmortem lysis, and then washing the fresh tissue immediately with cold PBS + 5 mM EDTA to remove as much blood as possible. In case obtaining brains from the same species is a problem (e.g., humans), we have successfully used transgenic mice brain overexpressing PrP^C.

We have not experimented extensively at this point to determine which part of the brain is the most suitable for PMCA studies; however, currently we would recommend using the entire encephalic area including brain stem. On the basis of *in vivo* experiments, we also recommend using animals as young as possible, although we avoid using fetal tissue.

After removal, the brain should be placed into conversion buffer at 4° and immediately homogenized at 10% (w/v) using a high-viscosity mixer (see "Equipment"). In our hands, the highest amplification is obtained using

7.5–10% of substrate, but it is also possible to use 2.5–5%, although with lower yield of amplification. After homogenization, large pieces of tissue and unbroken cells should be removed by a low-speed centrifugation. The low speed (1500g for 30 sec) is important to avoid losing or destabilizing membrane components that seem to be essential for conversion. The final substrate preparation should be turbid with visible membrane fragments still present. If the homogenate is transparent, the efficiency of conversion will not be good.

Homogenates, once prepared, should be stored at −80° and can be thawed and refrozen approximately 10 times without significant loss of efficiency. Storage at −20° is not recommended; however, for short-term use, homogenates can be kept at 4° for up to 7 days. Following the studies from Supattapone's group (Deleault *et al.*, 2003), it is recommended that the work be done in RNAse-free conditions.

Other Substrates. The normal brain homogenate is considered, at this moment, to be the most efficient substrate for amplification, particularly for TSE diagnosis purposes. However, the use of alternative substrates could be beneficial for other studies aimed, for example, at understanding the tissues capable to propagate prions or to localize other factors involved in PrP^C to PrP^{Sc} conversion.

CELLS. Protein extracts from whole cells can provide a good substrate for specific applications of PMCA. For these studies, transient or stably transfected knock out-PrP (PrP-KO) cells overexpressing different PrP^C transgenes can provide useful substrates for subsequent experimentation. Our experience at this point has been focused on PrP knock out N2a cells overexpressing hamster or mouse PrP as substrates. Cells should be resuspended in a small volume of PBS (0.5–1 ml) and centrifuged in a 1.5-ml Eppendorf tube for 5 min at 2000g. The supernatant must be completely removed, and the pellet containing cells should be resuspended in 50–100 μl of CB and the centrifugation step repeated. To enhance the level of amplification using cells as substrate, we have found that it is useful to supplement the PMCA reaction with a "PrP inert substrate" such as PrP-KO brain homogenate or normal brain homogenate from a species resistant to prion propagation such as rabbit. This material provides additional quantities of a yet unknown "conversion factor," which is highly expressed in brain.

LIPID RAFTS. The detergent-resistant membrane (DRM) or lipid–raft fraction, should also be considered as a good alternative substrate to the whole brain homogenate. PrP^C is attached to the outer cell membrane by a glycosyl phosphatidylinositol (GPI) anchor, and, like other GPI-anchored proteins, PrP^C is found in the cholesterol, glycosphingolipid, sphingomyelin-rich membrane subdomains known as lipid rafts (Vey *et al.*, 1996). This membrane fraction seems to contain all elements required for prion conversion (unpublished data). Various methods have been described to isolate lipid rafts both from brain homogenates and from neuronal cell lines. We routinely use Optiprep (Axis-Shield) density gradients to isolate lipid raft from N2a neuroblastoma cells or from brain.

PURIFIED PrP^C. Currently, one of the most intriguing issues in the prion field is the identity of the factors required for PrP^C to PrP^{Sc} conversion. With this in mind, we have designed a purification technique to obtain PrP^C free of other components but suitable for conversion to PrP^{Sc}. This purified substrate can be used to study the effects of well-characterized biological fractions on the conversion process. Mixing PrP^C with a sample that is able to complement the conversion enables us to have a better understanding of the requirements of the conversion process (e.g., nucleic acids, lipids, proteins). The ability to purify PrP^C will allow us to mix it with PrP^{Sc} and screen for cellular components capable of reconstituting a conversion competent environment. This is currently being worked out and will be reported at a later time.

Automated PMCA Procedure

In the original PMCA procedure (Saborio *et al.*, 2001), sonication was performed manually using a single probe sonicator. More recently, an automated version of PMCA has been developed (Castilla *et al.*, 2005a) that shows improved efficiency and reproducibility. This procedure now referred to as automatic PMCA (aPMCA) uses an inverted 96-well sonicator that can be programmed for automatic operation (see "Equipment"). This technique has proved to be of great value for diagnosis and for other prion propagation studies. aPMCA overcomes one of the major drawbacks of manual PMCA, namely cross-contamination, because there is no direct contact between the sonicator probe and the sample. The following recommendations for standard use of this procedure should be observed:

1. Samples to be amplified are placed at different dilutions into 0.2-ml PCR tubes and mixed with 10% substrate (see "Preparation of Substrate"). The final volume should be between 60 and 100 μl. For each condition, three tubes are prepared. One is frozen immediately (frozen control), and the second is subjected to multiple cycles of incubation/sonication (PMCA samples) (Fig. 2).

2. Samples are incubated for 30–60 min at 37° in the reservoir of the automatic sonicator. The duration of the incubation phase needs to be optimized for each sample, because factors such as the prion strain and the amount of PrP^{Sc} in the sample will require different incubation periods. There are numerous parameters that can be modified (including time,

temperature, and agitation rate) to reach highest efficiency for a particular sample; however, in this chapter we will limit the description to our standard procedure. As further knowledge about prion replication *in vitro* accumulates, and as more sophisticated equipment becomes available, additional modifications to the technique will be implemented.

3. Samples are sonicated for a single pulse of 40 sec. The sonication is the most critical step in this technique, and variation in the level of sonication can generate huge differences in the results. Using the optimal level of sonication is crucial to break down and multiply the PrP^{Sc} polymers without affecting their capacity to act as "seed" for further PrP^{C} conversion. It is also important to note that the ultrasound strength needed to amplify PrP^{Sc} of distinct strains and from diverse species can be different (Soto *et al.*, 2005), and hence low or even no amplification at all may be obtained for new samples under conditions that work very well for others. These findings are probably related to the specific conformation/ aggregation state of each strain of prion, which has been proposed to explain the differences in clinical, pathological, and biochemical features of distinct strains.

The sonication step is the most difficult to monitor adequately, and many factors can influence the final amplification observed. These factors, which we describe later, need to be taken into account to achieve maximal amplification.

   a. *Power of sonication:* The power of sonication for the 263K hamster prion strain should be set to the maximum potency of this sonicator (level 8–10). For other species/strains, sonication power should be optimized experimentally and is in general lower than for 263K strain.
   b. *Wavelength:* At present, we do not know how wavelength affects the effectiveness of sonication; however, we should be able to determine this once other equipment becomes available.
   c. *Water in the sonication reservoir:* The reservoir has to be filled with 140 ml of water (see "Equipment"), which decreases at a rate of around 2.5 ml/h at 37°. Tubes should be incubated without touching the sonication plate.
   d. *Tubes for sonication:* It is very important to use thin-walled 0.2-ml tubes to obtain the most effective penetration of ultrasound waves.
   e. *Number of tubes:* The rack used in this sonicator is designed to hold 96 tubes. However, our experiments have shown that the effective power of the sonicator diminishes when the rack is completely full. This is probably because each tube attenuates to some extent the effect of the ultrasound waves. If all positions need to be used, we would recommend increasing the power of sonication. In our

standard procedure, only 60% of the rack is used, with tubes being placed at random positions across the plate.

4. The incubation/sonication cycle (steps 2–3) should be repeated as many times as needed to reach the desired level of amplification. For the standard reaction, we recommend approximately 24 h of cyclic amplification.

It is preferable to complete the entire amplification experiment without freezing–thawing the samples. If it is necessary to interrupt the amplification, samples should always be frozen at $-80°$. It is not necessary to use a quick freezing procedure.

Although the theoretical limit of amplification will be the amount of PrP^C substrate present in the tube, we regularly observe that the efficiency of amplification starts to decrease after approximately 150 cycles (75 h of incubation). This problem is most likely the result of a deleterious effect of prolonged incubation at $37°$. Under these conditions, the PrP^C substrate or other brain-derived cofactors necessary to promote the conversion of PrP^C into PrP^{Sc} might be altered or consumed. In view of this, and to increase further the level of achievable amplification, we have extended the technique of aPMCA to include serial rounds, in which at each new round, the amplified samples are rediluted into fresh substrate. This new approach, termed serial automated PMCA (saPMCA), will be described in the following.

To maintain a good reproducibility of the aPMCA technique some points need to be carefully considered, especially when small amounts of PrP^{Sc} are used for amplification. In particular, the following situations should be avoided:

a. Low sample volumes ($<50\ \mu$l).
b. Low water level (<100 ml) in the reservoir of the automatic sonicator.
c. Bubble formation in the sample that could prevent a good transmission of sonication waves.

5. After the last pulse of sonication is completed, the samples are ready for protease K digestion (see "Detection of Amplified Product"). If digestion is not performed immediately, the amplified samples should be stored at $-80°$.

Serial Automatic PMCA (saPMCA) Procedure

Serial automatic PMCA consists of successive rounds of aPMCA in which at each round the amplified sample is diluted into fresh substrate. This approach is highly recommended for experiments requiring elevated levels of amplification, especially when working with samples containing minute initial amounts of PrP^{Sc}, such as blood, CSF, or peripheral nonlymphoid tissues.

As mentioned previously, the efficiency of PMCA decreases after approximately 75 h of constant incubation at 37°. However, efficient conversion is restored when the amplified samples are diluted into fresh substrate. Therefore, after a first round of standard aPMCA, the amplified samples are diluted into NBH and amplified in a second round. The dilution factor depends on the purpose of the study and the original dilution of the PrP^{Sc}. For experiments in which the aim is to simply eliminate the original inoculum (e.g., for comparative studies, infectivity experiments), and where the initial amount of PrP^{Sc} is relatively high (4 log LD_{50}), a 100- or 1000-fold dilution can be performed at each stage of saPMCA. However, in studies where undetectable amounts or PrP^{Sc} are present, even after a first round of PMCA, a 10-fold dilution is enough to refresh the substrate. The rounds of saPMCA can be repeated as many times as is needed to reach the detection threshold of Western blotting. Samples remaining negative after eight rounds of saPMCA can be considered negative, because according to our experience, approximately six rounds of saPMCA can amplify the minimum amount of material required for amplification (approximately a few hundred molecules of PrP^{Sc} monomers) (unpublished data).

Because of the PCR-like nature of PMCA, special care should be taken during the manipulation of the samples when performing serial dilutions of the amplified material. Thus, after each round of aPMCA, the samples should be gently spun down to remove material present in the lid, which arises during sonication or because of condensation of the sample. Given the power of this procedure, inclusion of negative control samples (NBH without PrP^{Sc}) that are amplified and serially diluted in parallel with the experimental samples is highly recommended.

For safety conditions, filter tips should be used for liquid handling, and sonication should be performed in a closed container inside a BSL-2 hood to avoid the spread of infectious material.

Detection of Amplified Product

After amplification, the two samples (amplified and frozen) are digested with proteinase K (PK), and PrP^{res} is detected by immunological methods. Because distinct species/strains of prion show a different extent of resistance to proteolytic degradation, the optimal PK treatment condition should be determined beforehand. The critical issue is to make sure that no PrP^{C} remains undigested after PK treatment, because it is a common mistake to confuse incomplete digestion of PrP^{C} with false-positive PrP^{res} formation. When PrP is detected by Western blotting, it is easy to distinguish incomplete PrP^{C} digestion from bona-fide PrP^{res}, because the latter exhibit a switch on molecular weight because of the removal of the first $\sim$90 amino acids. To ensure complete digestion, especially after extended incubations, a higher concentration of PK may be required to digest increasingly larger aggregates. Addition of up to 0.05% SDS in the buffer used for the PK treatment may also help. Digestions using temperatures between 42° and 64° and with shaking at 350–450 rpm are also recommended. Our standard procedure, which can be taken as a basis for further optimization, is as follows.

1. Prion-containing samples (20 μl) are incubated with standard concentrations of PK (50 μg/ml for hamster, 25 μg/ml for mouse, 20 μg/ml for cattle, or 40–50 μg/ml for CJD) for 1 h at 45° with shaking at 350–450 rpm. Aliquots of PK (10 mg/ml) are stored frozen at $-20°$, and in the interests of reproducibility, any thawed, unused enzyme is discarded at the end of the experiment. Note: Blood contains protease inhibitors that can interfere with the PK digestion, and in those samples where the presence of blood is unavoidable, the PK concentration should be adjusted accordingly.

2. The PK digestion is stopped by addition of phenyl-methyl-sulfonyl-fluoride or SDS-PAGE loading buffer. Samples can be analyzed for the presence of PrP^{res} using any of the established immunological methods, such as Western blotting or ELISA (Soto, 2004).

Acknowledgments

This work is partially supported by NIH grants AG0224642 and NS049173. Dr. Soto is part of the European Community project TSELAB.

References

Aguzzi, A. (2000). Prion diseases, blood and the immune system: Concerns and reality. *Haematologica* **85,** 3–10.

Asante, E. A., and Collinge, J. (2001). Transgenic studies of the influence of the PrP structure on TSE diseases. *Adv. Protein Chem.* **57,** 273–311.

Baldwin, M. A., Cohen, F. E., and Prusiner, S. B. (1995). Prion protein isoforms, a convergence of biological and structural investigations. *J. Biol. Chem.* **270,** 19197–19200.

Brown, P., Cervenakova, L., and Diringer, H. (2001). Blood infectivity and the prospects for a diagnostic screening test in Creutzfeldt-Jakob disease. *J. Lab. Clin. Invest.* **137,** 5–13.

Bruce, M. E. (2003). TSE strain variation. *Br. Med. Bull.* **66,** 99–108.

Budka, H., Aguzzi, A., Brown, P., Brucher, J. M., Bugiani, O., Gullotta, F., Haltia, M., Hauw, J. J., Ironside, J. W., Jellinger, K., Kretzschmar, H. A., Lantos, P. L., Masullo, C., Schlote, W., Tateishi, J., and Weller, R. O. (1995). Neuropathological diagnostic criteria for Creutzfeldt-Jakob disease (CJD) and other human spongiform encephalopathies (Prion diseases). *Brain Pathol.* **5,** 459–466.

Carrell, R. W., and Lomas, D. A. (1997). Conformational disease. *Lancet* **350,** 134–138.

Castilla, J., Saá, P., Hetz, C., and Soto, C. (2005a). *In vitro* generation of infectious scrapie prions. *Cell* **121,** 195–206.

Castilla, J., Saá, P., and Soto, C. (2005b). Detection of prions in blood. *Nat. Med.* **11,** 982–985.

Caughey, B. W., Dong, A., Bhat, K. S., Ernst, D., Hayes, S. F., and Caughey, W. S. (1991). Secondary structure analysis of the scrapie-associated protein PrP 27–30 in water by infrared spectroscopy. *Biochemistry* **30,** 7672–7680.

Clarke, A. R., Jackson, G. S., and Collinge, J. (2001). The molecular biology of prion propagation. *Philos. Trans. R. Soc. Lond. B Biol. Sci.* **356**, 185–195.

Cohen, F. E., and Prusiner, S. B. (1998). Pathologic conformations of prion proteins. *Ann. Rev. Biochem.* **67**, 793–819.

Collinge, J. (1999). Variant Creutzfeldt-Jakob disease. *Lancet* **354**, 317–323.

Collinge, J. (2001). Prion diseases of humans and animals: Their causes and molecular basis. *Annu. Rev. Neurosci.* **24**, 519–550.

DeArmond, S. J., and Prusiner, S. B. (1996). Transgenetics and neuropathology of prion diseases. *Curr. Top. Microbiol. Immunol.* **207**, 125–146.

Deleault, N. R., Lucassen, R. W., and Supattapone, S. (2003). RNA molecules stimulate prion protein conversion. *Nature* **425**, 717–720.

Deleault, N. R., Lucassen, R. W., and Supattapone, S. (2005). PrPres amplification reconstituted with purified prion proteins and synthetic polyanions. *J. Biol. Chem.* **280**, 26873–26879.

Dobson, C. M. (2004). Protein chemistry. In the footsteps of alchemists. *Science* **304**, 1259–1262.

Head, M. W., and Ironside, J. W. (2000). Inhibition of prion protein conversion: A therapeutic tool? *Trends Microbiol.* **6**, 6–8.

Hill, A. F., Joiner, S., Linehan, J., Desbruslais, M., Lantos, P. L., and Collinge, J. (2000). Species-barrier-independent prion replication in apparently resistant species. *Proc. Natl. Acad. Sci. USA* **97**, 10248–10253.

Kascsak, R. J., Rubenstein, R., and Carp, R. I. (1991). Evidence for biological and structural diversity among scrapie strains. *Curr. Top. Microbiol. Immunol.* **172**, 139–152.

Kelly, J. W. (1998). The alternative conformations of amyloidogenic proteins and their multi-step assembly pathways. *Curr. Opin. Struct. Biol.* **8**, 101–106.

Kocisko, D. A., Come, J. H., Priola, S. A., Chesebro, B., Raymond, G. J., Lansbury, P. T., and Caughey, B. (1994). Cell-free formation of protease-resistant prion protein. *Nature* **370**, 471–474.

Kocisko, D. A., Priola, S. A., Raymond, G. J., Chesebro, B., Lansbury, P. T., Jr., and Caughey, B. (1995). Species specificity in the cell-free conversion of prion protein to protease-resistant forms: A model for the scrapie species barrier. *Proc. Natl. Acad. Sci. USA* **92**, 3923–3927.

Moynagh, J., and Schimmer, H. (1999). Test for BSE evaluated. *Nature* **400**, 105.

Nishina, K., Jenks, S., and Supattapone, S. (2004). Ionic strength and transition metals control PrPSc protease resistance and conversion-inducing activity. *J. Biol. Chem.* **43**, 2613–2621.

Pan, K. M., Baldwin, M., Njuyen, J., Gassett, M., Serban, A., Groth, D., Mehlhorn, I., and Prusiner, S. B. (1993). Conversion of alpha-helices into b-sheets features in the formation of scrapie prion proteins. *Proc. Natl. Acad. Sci. USA* **90**, 10962–10966.

Prusiner, S. B. (1982). Novel proteinaceous infectious particles cause scrapie. *Science* **216**, 136–144.

Prusiner, S. B. (1997). Prion diseases and the BSE crisis. *Science* **278**, 245–251.

Raymond, G. J., Hope, J., Kocisko, D. A., Priola, S. A., Raymond, L. D., Bossers, A., Ironside, J., Will, R. G., Chen, S. G., Petersen, R. B., Gambetti, P., Rubenstein, R., Smits, M. A., Lansbury, P. T., Jr., and Caughey, B. (1997). Molecular assessment of the potential transmissibilities of BSE and scrapie to humans. *Nature* **388**, 285–288.

Roos, R., Gajdusek, D. C., and Gibbs, C. J., Jr. (1973). The clinical characteristics of transmissible Creutzfeldt-Jakob disease. *Brain* **96**, 1–20.

Saborio, G. P., Permanne, B., and Soto, C. (2001). Sensitive detection of pathological prion protein by cyclic amplification of protein misfolding. *Nature* **411**, 810–813.

Saborio, G. P., Soto, C., Kascsak, R. J., Levy, E., Kascsak, R., Harris, D. A., and Frangione, B. (1999). Cell-lysate conversion of prion protein into its protease-resistant isoform suggests the participation of a cellular chaperone. *Biochem. Biophys. Res. Commun.* **258**, 470–475.

Schiermeier, Q. (2001). Testing times for BSE. *Nature* **409**, 658–659.

Soto, C. (2001). Protein misfolding and disease; protein refolding and therapy. *FEBS Lett.* **498**, 204–207.

Soto, C. (2004). Diagnosing prion diseases: Needs, challenges and hopes. *Nat. Rev. Microbiol.* **2**, 809–819.

Soto, C., and Saborio, G. P. (2001). Prions: Disease propagation and disease therapy by conformational transmission. *Trends Mol. Med.* **7**, 109–114.

Soto, C., Saborio, G. P., and Anderes, L. (2002). Cyclic amplification of protein misfolding: Application to prion-related disorders and beyond. *Trends Neurosci.* **25**, 390–394.

Soto, C., Anderes, L., Suardi, S., Cardone, F., Castilla, J., Frossard, M. J., Peano, S., Saá, P., Limido, L., Carbonatto, M., Ironside, J., Torres, J. M., Pocchiari, M., and Tagliavini, F. (2005). Pre-symptomatic detection of prions by cyclic amplification of protein misfolding. *FEBS Lett.* **579**, 638–642.

Telling, G. C., Parchi, P., DeArmond, S. J., Cortelli, P., Montagna, P., Gabizon, R., Mastrianni, J., Lugaresi, E., Gambetti, P., and Prusiner, S. B. (1996). Evidence for the conformation of the pathologic isoform of the prion protein enciphering and propagating prion diversity. *Science* **274**, 2079–2082.

Vey, M., Pilkuhn, S., Wille, H., Nixon, R., DeArmond, S. J., Smart, E. J., Anderson, R. G., Taraboulos, A., and Prusiner, S. B. (1996). Subcellular colocalization of the cellular and scrapie prion proteins in caveolae-like membranous domains. *Proc. Natl. Acad. Sci. USA* **93**, 14945–14949.

Wadsworth, J. D., Joiner, S., Hill, A. F., Campbell, T. A., Desbruslais, M., Luthert, P. J., and Collinge, J. (2001). Tissue distribution of protease resistant prion protein in variant Creutzfeldt-Jakob disease using a highly sensitive immunoblotting assay. *Lancet* **358**, 171–180.

Weber, T., Otto, M., Bodemer, M., and Zerr, I. (1997). Diagnosis of Creutzfeldt-Jakob disease and related human spongiform encephalopathies. *Biomed. Pharmacother.* **51**, 381–387.

Will, R. G., Ironside, J. W., Zeidler, M., Cousens, S. N., Estibeiro, K., Alperovitch, A., Poser, S., Pocchiari, M., Hofman, A., and Smith, P. G. (1996). A new variant of Creutzfeldt-Jakob disease in the UK. *Lancet* **347**, 921–925.

[2] Fractionation of Prion Protein Aggregates by Asymmetrical Flow Field-Flow Fractionation

By JAY R. SILVEIRA, ANDREW G. HUGHSON, and BYRON CAUGHEY

Abstract

Achieving the successful separation and analysis of amyloid and other large protein aggregates can be a difficult proposition. Field-flow fractionation (FFF) is a flow-based separation method like chromatography; however, FFF is capable of high-resolution separations in the absence of a stationary matrix. Thus, FFF is a relatively gentle technique and is well suited to the task of separating large macromolecules and macromolecular

METHODS IN ENZYMOLOGY, VOL. 412   0076-6879/06 $35.00
                                                                   DOI: 10.1016/S0076-6879(06)12002-9

complexes. Flow field-flow fractionation (FlFFF), one of the techniques in the FFF family, has been used to successfully fractionate a wide size range of prion protein aggregates, allowing their subsequent characterization by several biophysical and biochemical methods. The ability to easily adjust the strength of the field used during separation means that FlFFF could be applied to particles ranging from 1 nm to nearly 100 μm in size. This flexibility, coupled with the ability to produce fast, high-resolution separations, makes FFF a potentially valuable tool in the field of amyloid research.

Introduction

Several neurodegenerative diseases, including the transmissible spongiform encephalopathies (TSEs or prion diseases), are associated with the accumulation of protein aggregates that can range in size from large amyloid fibrils to small subfibrillar oligomers (Caughey and Lansbury, 2003). Understanding how the size of these aggregates relates to their potential neurotoxic or infectious properties is essential in developing strategies aimed at the treatment and prevention of these diseases. However, the relatively large size and limited solubility of many aggregates represents a challenge to the traditional techniques of protein separation and analysis. Size exclusion chromatography (SEC) (Safar *et al.*, 1990), polyacrylamide gel electrophoresis (PAGE) (Brown *et al.*, 1990; Hope, 1994), ultracentrifugation (Caughey *et al.*, 1995, 1997), and filtration (Caughey *et al.*, 1995) have all been used to separate TSE-associated protein aggregates for characterization, but each technique has notable drawbacks. In those techniques that use a stationary phase or matrix (SEC, PAGE, filtration), large aggregates may be caught up and/or subjected to shearing forces, and some techniques (ultracentrifugation, filtration) provide relatively limited resolution. In addition, some of these techniques provide relatively narrow fractionation ranges. Field-flow fractionation techniques can separate infectious TSE particles over an extremely broad range of sizes (Silveira *et al.*, 2005; Sklaviadis *et al.*, 1992), while avoiding many of the pitfalls associated with other separation techniques.

Field-Flow Fractionation

Field-flow fractionation (FFF) is an extremely varied and versatile family of separation techniques that has existed for several decades (Giddings, 1993; Schimpf *et al.*, 2000). Although FFF's gentle method of separation is applicable to biological samples, its adoption has been slow in the field of biotechnology and it remains perhaps "the best kept secret in bioanalysis" (Reschiglian *et al.*, 2005). Like chromatography, FFF is a flow-based method in which a narrow band of sample is injected into a stream flowing through an elongated chamber. However, unlike chromatography, in which separation typically takes place when molecules interact with a "stationary phase" distributed throughout the flow chamber, FFF requires no stationary phase. FFF separations take place in an empty and inherently inert chamber called a *channel*, where sample retention is caused by the action of an externally generated field (sedimentation, thermal, electrical, magnetic, crossflow, etc.) applied at right angles to the direction of flow.

Because the open channel design produces minimal shear forces (Giddings, 1993), and the surface area available for adsorption is relatively small, FFF systems are attractive candidates for the separation of macromolecules and large macromolecular complexes (Ratanathanawongs-Williams and Giddings, 2000). Unlike the interfaces used in chromatography, where sample interaction is required for separation, the surfaces of the FFF channel are intended to be inert toward samples. Thus, although adsorption can occur, surface chemistries or carrier compositions can often be altered to minimize or remove interactions without compromising separation. Using FFF, Li and colleagues obtained absolute recoveries of 95–98% for several proteins (cytochrome C, bovine serum albumin, γ-globulin, thyroglobulin, high-density lipoprotein, and low-density lipoprotein) using regenerated cellulose membranes and physiological phosphate buffer as a carrier (Li *et al.*, 1997; Ratanathanawongs-Williams and Giddings, 2000).

In flow field-flow fractionation (FlFFF), the field is generated by a second independent stream (called the crossflow stream) that runs across the channel at right angles to the primary channel flow stream (Giddings *et al.*, 1976). Symmetrical FlFFF uses a crossflow stream that is generated by flow passing through porous walls at the top (depletion wall) and bottom (accumulation wall) of the channel (Fig. 1A), whereas asymmetrical FlFFF uses a nonpermeable depletion wall, and crossflow enters through the same inlet as the channel flow (Fig. 1B) (Wahlund and Giddings, 1987). The combined flows are typically split by regulating the channel outflow and forcing the remainder of the flow to pass through the membrane. In this process, a portion of the channel flow is gradually lost to crossflow as liquid travels down the channel, and in a rectangular channel, this leads to a continuous decrease in the velocity of the channel flow. However, the channel flow velocity can be maintained if an appropriate trapezoidal channel is used, which has the added benefit of reducing peak dilution (Litzen and Wahlund, 1991).

Before separation, samples injected into the FFF channel must typically undergo a "relaxation" process, in which channel flow is stopped, and the

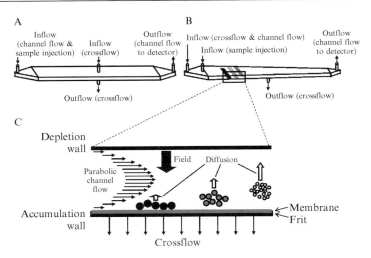

FIG. 1. Schematic of FlFFF channel features. Geometry and flow pathways of typical (A) symmetrical and (B) asymmetrical FlFFF channels. (C) Exploded view of the asymmetrical FlFFF channel, where smaller particles diffuse further into the parabolic channel flow and travel faster than larger particles.

sample is allowed to reach an equilibrium distribution near the accumulation wall under the influence of the field. Because the flow in asymmetrical FlFFF cannot be stopped independently of the field, stop-flow relaxation is not an option, and another technique must be used. Asymmetrical FlFFF samples are relaxed by temporarily pumping fluid in through both the inlet and outlet ports at specific rates and allowing all the fluid to exit through the membrane. This process will focus the sample at a predetermined point slightly downstream of the sample injection port. This technique, known as downstream central injection, reduces focusing/relaxation time, permits loading of larger sample volumes, and allows for rapid, high-resolution separations (Litzen and Wahlund, 1989; Wahlund and Litzen, 1989).

FlFFF separation is based solely on diffusion coefficients/hydrodynamic size, and because moving fluid is capable of acting on all unattached objects in its path, FlFFF is considered the most universal FFF technique (Giddings *et al.*, 1976). FlFFF has been applied successfully to particles with molar masses as low as 300 g/mol (Beckett *et al.*, 1987) and is currently useful for separating particles from approximately 1 nm to nearly 100 μm in diameter (Giddings *et al.*, 1992; Ratanathanawongs-Williams, 2000). The lower size limit is due to the high pressures resulting from the combination of low molecular weight cutoff membranes and strong crossflow rates necessary to counteract diffusion in small particles, whereas the upper limit is roughly 20% of the channel thickness (Ratanathanawongs-Williams, 2000).

As with other FFF subtypes, FlFFF uses a field (liquid flow) to drive particles toward the accumulation wall. This force is opposed by diffusion of particles away from the wall, and the balance of these two forces results in an equilibrium distribution of particles at some distance from the accumulation wall (Fig. 1C). Thus, under the normal (or Brownian) mode of FlFFF that applies to submicrometer particles, smaller particles with higher diffusion rates tend to reach an equilibrium position farther away from the accumulation wall. Because the flow profile in the channel is parabolic, smaller particles will penetrate further into the fast streamlines of the flow and will migrate faster than larger particles. The end result is a separation based on size, in which smaller particles elute earlier than larger ones.

With the ability to alter channel flow and/or crossflow rates, one FlFFF instrument can be used to separate a wide range of particle sizes without the need to change system components. Although a single FlFFF separation with constant channel flow and crossflow might be capable of resolving particles over an approximately sevenfold range in diameter, "programmed elution," in which channel flow and/or crossflow rates are varied during the run, increases this range to roughly 50-fold (Ratanathanawongs and Giddings, 1992). Figure 2A shows the light-scattering traces of a group of size standards subjected to asymmetrical FlFFF with constant channel flow and various crossflow rates on an Eclipse F separation system (Wyatt Technology Europe, Woldert). At 1 ml/min channel flow, a doubling of the crossflow rate from 2 to 4 ml/min produces only a modest shift in elution time; however, further changes in elution time can be achieved by simultaneously adjusting channel flow (Ratanathanawongs and Giddings, 1992). The use of programmed elution to gradually reduce the strength of the field throughout a single run can promote the elution of large, well-retained particles while still maintaining the resolution of smaller particles (Botana et al., 1995; Wahlund et al., 1986). Shown in Fig. 2B are the elution positions for seven different particles. Each particle sample was individually loaded and eluted from the Eclipse F using a programmed elution of decreasing crossflow intensity, and the seven elution profiles were superimposed in the figure for comparison. Although baseline resolution is not achieved for all particles, separation of a nearly 14-fold range in particle diameter is achieved by field programming alone.

Separation of PrP-res Aggregates by Asymmetrical FlFFF

Although the following method has been optimized to fractionate PrP-res aggregates isolated from the brains of hamsters infected with the 263K

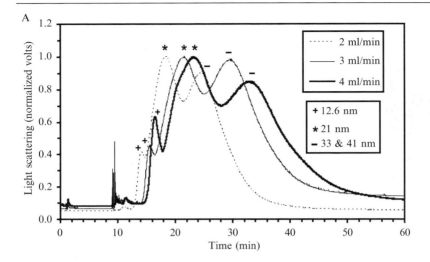

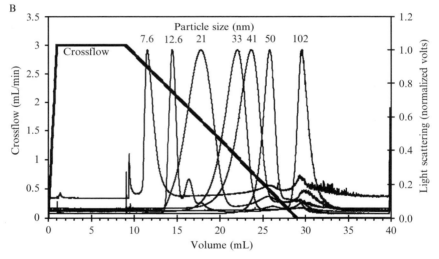

FIG. 2. Asymmetrical FlFFF of size standards under various crossflow conditions. Bovine carbonic anhydrase (7.6-nm diameter), horse spleen apoferritin (12.6-nm diameter), and five sizes of polystyrene size standards (21-, 33-, 41-, 50-, and 102-nm diameters; Duke Scientific, Palo Alto) were separated on the Eclipse F system. (A) A mixture of four size standards was eluted under constant cross-flow of 2, 3, or 4 ml/min. (B) Each of the seven size standards was individually loaded onto the Eclipse F system and eluted with programmed crossflow decreasing linearly from 3 ml/min to 0 ml/min over the course of 20 min. The seven elution profiles were then superimposed onto one plot for comparison. All samples were run in 0.1% SUS, 20 m*M* Tris, pH 7.0, with a constant channel flow of 1 ml/min, and a cellulose triacetate membrane was used at the channel's accumulation wall. Sample elution was detected by right angle light scattering.

strain of scrapie, the inherent flexibility of FlFFF in terms of membrane and channel composition, channel size, as well as channel flow and cross-flow rates, makes the general technique amenable to a wide variety of protein aggregates. The starting material for the procedure is purified, proteinase K–treated PrP-res (Raymond and Chabry, 2004), which is brought to 0.1 mg/ml in 20 mM Tris, pH 7.0, containing 1% sodium n-undecyl sulfate (SUS) detergent, and subjected to freezing, heating, and sonication to promote partial disaggregation of aggregates as described previously (Silveira et al., 2005). A typical sample contains 20–50 μg PrP-res protein, because this is the minimum with which we have been able to obtain useful light scattering and concentration data for the small minority of highly infectious, intermediate-size particles. Before injection, the sample is filtered through a 0.2-μm Nanosep centrifugal device (Pall Life Sciences, Ann Arbor) by centrifugation (500g, 20 min, 25°) to remove dust or any excessively large aggregates that resisted disaggregation. Larger samples (>200 μl) are generally split between several filters to reduce the chances of clogging the membranes. Typically, little material is lost during this step, but quantitative analyses of PrP indicate that as much as one third of the sample may be retained on the filter, depending on the source of the PrP-res and the effectiveness of the disaggregation procedure.

For all our asymmetrical FlFFF work, we have used the Eclipse F separation system with a trapezoidal channel that is 26.5 cm in length, 350 μm in height, and 21-mm wide at the sample inlet. The depletion wall is composed of polymethylmethacrylate, and the accumulation wall is lined with a 10-kDa cutoff polyethersulfone (PES) membrane. We have observed that the composition of the membrane used in the system can have a dramatic impact on the success of a separation. Although PES membranes in combination with 0.1% SUS in the mobile phase have worked well for fractionating samples of PrP-res, polystyrene size standards required a more hydrophilic membrane (cellulose triacetate) to prevent irreversible binding. Additional considerations for minimizing sample loss include avoiding excessively high field strengths, which have been shown to promote sample loss (Giddings and Caldwell, 1984; Li et al., 1990), and in the case of FlFFF, choosing a membrane with an appropriate molecular weight cutoff to avoid sample loss through the membrane wall (Li and Giddings, 1996; Li et al., 1997; Ratanathanawongs-Williams and Giddings, 2000). Although the typical sample volume of our PrP-res samples (200–500 μl) is larger than the standard 100-μl injection loop in our system, these samples are loaded by filling the loop, injecting, focusing, and repeating the process (Wahlund and Litzen, 1989). Although this method may result in slightly longer FlFFF runs, it alleviates the need to change injection loops within the instrument to accommodate different sample volumes. Once focus flow is established in the channel during a 1-min "prefocus" step, each 100-μl aliquot of sample is injected at 0.2 ml/min for 2 min and focused for an additional 3 min. To ensure that the entire sample is delivered to the channel, the total injection volume is at least three times the volume of the loop (Wahlund, 2000), and the inject flow is kept relatively low to minimize disruption of the focus flow (Wahlund and Giddings, 1987). After the last round of injection and focusing, an additional 12 min of focusing is applied to the sample to ensure that particles of all sizes are localized into a tight band before elution. Focusing has the added benefit of buffer exchange, which equilibrates the sample in the mobile phase (20 mM Tris, pH 7.0, containing 0.1% SUS). However, one should be aware of the potential risk of aggregation because of excessive focusing. Detergent is present throughout the run to reduce the chances of sample reaggregation and losses because of binding, but it is kept below the SUS critical micelle concentration (CMC) of approximately 0.4% (Huisman, 1964; Ranganathan *et al.*, 2000). Attempts to use mobile phases containing 1% SUS or SDS (above the CMC for these detergents) have resulted in overpressure issues because of the increased viscosity and problems with light scattering analyses because of the presence of micelles. The typical elution profile uses a constant channel flow of 1 ml/min and a programmed crossflow decreasing from 3 ml/min to 0 ml/min over 20 min. The crossflow gradient is followed by an additional 10 min without crossflow to elute the largest aggregates.

The fractionated sample is monitored in line by static light scattering, refractive index, and dynamic light-scattering analyses on DAWN EOS, Optilab DSP, and WyattQELS instruments respectively (Wyatt Technologies, Santa Barbara), and 1-ml fractions are collected for additional analyses. Figure 3A shows raw data from both the refractive index detector used to determine solute concentration and the right-angle detector of the 18-angle DAWN EOS static light-scattering instrument. The first major peak (fractions 4–8) contains a substantial amount of PrP protein in the form of monomers and small oligomers (visualized in Fig. 4A), which scatter relatively little light. In contrast, the second major peak (fractions 18–24) contains larger PrP fibrils and aggregates that are easily detected by light scattering. The combined data from the detectors were processed using Wyatt's ASTRA analysis software (version 4.90.07) to determine the weight-average molar mass (M_W), z-average radius of gyration ((r_g)$_z$), and z-average hydrodynamic radius ((r_h)$_z$) for the fractionated particles (Fig. 3B). The fractionated aggregates contained a wide range of M_W values (less than 10^5 to greater than 10^7 Da), r_g values ranging from less than 10 nm to approximately 250 nm, and r_h values from roughly 5 nm to more than 50 nm. This separation used the same crossflow gradient

A

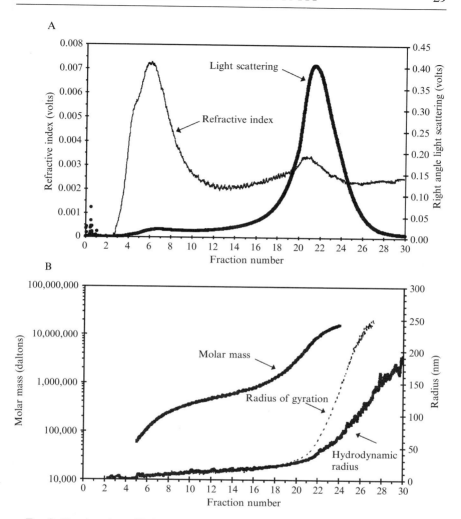

FIG. 3. Fractionation of PrP-res aggregates by asymmetrical FlFFF. Samples of partially disaggregated 263K PrP-res were fractionated on an Eclipse F separation system and analyzed in-line by light scattering. (A) Mean refractive index and right-angle light-scattering signals from four independent PrP-res fractionations. (B) The raw data shown in panel A were processed in combination with additional static and dynamic light scattering measurements to determine the mean values for molar mass ($n = 3$), hydrodynamic radius ($n = 4$), and radius of gyration ($n = 4$) of the fractionated particles using Wyatt's ASTRA analysis software (version 4.90.07). Adapted from Silveira et al. (2005). Nature 437, 257–261.

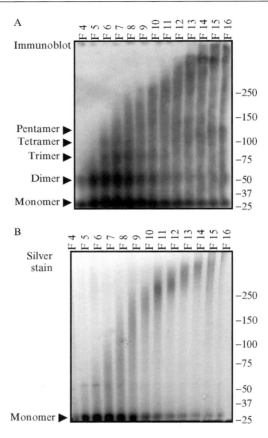

FIG. 4. Nondenaturing polyacrylamide gel electrophoresis of fractionated PrP-res. Samples from FlFFF fractions were brought to 5% glycerol and 0.02% bromophenol blue, subjected to PAGE on 3–8% Tris-acetate gels, and analyzed by either (A) immunoblotting with anti-PrP monoclonal antibody 3F4 after transfer to PVDF or (B) silver stain. PrP monomer and small oligomers are best visualized on the immunoblot, whereas silver staining more clearly shows the distribution of larger aggregates. Fraction numbers are shown at the top, PrP oligomers are indicated on the left, and molecular weight standards (kDa) are shown on the right. Adapted from Silveira *et al.* (2005). *Nature* **437**, 257–261.

outlined in Fig. 2B, which permits resolution of a wide range of aggregates in a single FlFFF run. With this method, particles in the 10–30-nm range were the most highly resolved (see Fig. 3B and Fig. 2B), whereas lower levels of resolution are observed for particles outside this range. As seen in Fig. 4A, PrP monomers and small oligomers containing less than six PrP molecules were not resolved from each other, eluting together in the early fractions of the gradient. However, larger oligomers in fractions 10–16, which are associated with the highest levels of specific TSE infectivity (Silveira *et al.*, 2005), are more highly resolved (Fig. 4B). Despite lower levels of resolution at the extremes, the overall distribution of particles analyzed in this single run represents a nearly 30-fold difference in hydrodynamic radii, which is substantially more than could be resolved without the use of programmed elution. Changes in the starting and ending crossflow rates, as well as the slope of the crossflow gradient, can easily be made to enhance resolution of smaller or larger particles.

Conclusion

The FFF family of techniques is capable of high-resolution separations over an extremely large range of particle sizes (1 nm–100 μm), all in the absence of a stationary phase. Thus, FFF is uniquely suited to accomplish separations of amyloid and other large bioparticles that have typically presented a challenge to some of the more traditional sizing methods found in biomedical laboratories. Sklaviadis and colleagues have used sedimentation field-flow fractionation (SdFFF) to characterize ~60-nm infectious particles from the brains of hamsters infected with Creutzfeldt-Jakob disease (Sklaviadis *et al.*, 1992), and in our laboratory, FlFFF has been invaluable for the separation and characterization of prion protein particles ranging from monomers to large infectious amyloid fibrils (Silveira *et al.*, 2005). In this era in which questions abound regarding the relationship between particle size and pathological activities of various abnormal protein aggregates, we suspect that FFF would also be highly applicable to studies of many important protein-misfolding diseases.

Acknowledgments

This research was supported by the Intramural Research Program of the NIH, NIAID.

References

Beckett, R., Jue, Z., and Giddings, J. C. (1987). Determination of molecular weight distributions of fulvic and humic acids using flow field-flow fractionation. *Environ. Sci. Technol.* **21,** 289–295.

Botana, A. M., Ratanathanawongs, S. K., and Giddings, J. C. (1995). Field-programmed flow field-flow fractionation. *J. Microcolumn Sep.* **7,** 395–402.

Brown, P., Liberski, P. P., Wolff, A., and Gajdusek, D. C. (1990). Conservation of infectivity in purified fibrillary extracts of scrapie-infected hamster brain after sequential enzymatic digestion or polyacrylamide gel electrophoresis. *Proc. Natl. Acad. Sci. USA* **87,** 7240–7244.

Caughey, B., Kocisko, D. A., Raymond, G. J., and Lansbury, P. T., Jr. (1995). Aggregates of scrapie-associated prion protein induce the cell-free conversion of protease-sensitive prion protein to the protease-resistant state. *Chem. Biol.* **2,** 807–817.

Caughey, B., and Lansbury, P. T. (2003). Protofibrils, pores, fibrils, and neurodegeneration: Separating the responsible protein aggregates from the innocent bystanders. *Annu. Rev. Neurosci.* **26,** 267–298.

Caughey, B., Raymond, G. J., Kocisko, D. A., and Lansbury, P. T., Jr. (1997). Scrapie infectivity correlates with converting activity, protease resistance, and aggregation of scrapie-associated prion protein in guanidine denaturation studies. *J. Virol.* **71,** 4107–4110.

Giddings, J. C. (1993). Field-flow fractionation: Analysis of macromolecular, colloidal, and particulate materials. *Science* **260,** 1456–1465.

Giddings, J. C., Benincasa, M. A., Liu, M.-K., and Li, P. (1992). Separation of water-soluble synthetic and biological macromolecules by flow field-flow fractionation. *J. Liquid Chromatogr.* **15,** 1729–1747.

Giddings, J. C., and Caldwell, K. D. (1984). Field-flow fractionation: Choices in programmed and nonprogrammed operation. *Anal. Chem.* **56,** 2093–2099.

Giddings, J. C., Yang, F. J., and Myers, M. N. (1976). Flow-field-flow fractionation: A versatile new separation method. *Science* **193,** 1244–1245.

Hope, J. (1994). The nature of the scrapie agent: The evolution of the virino. *Ann. N. Y. Acad. Sci.* **724,** 282–289.

Huisman, H. F. (1964). Light scattering of solutions of ionic detergents III. *Proceedings of the Koninklijke Nederlandse Akademie van Wetenschappen* **B67,** 388–406.

Li, J. M., Caldwell, K. D., and Machtle, W. (1990). Particle characterization in centrifugal fields. Comparison between ultracentrifugation and sedimentation field-flow fractionation. *J. Chromatogr.* **517,** 361–376.

Li, P., and Giddings, J. C. (1996). Isolation and measurement of colloids in human plasma by membrane-selective flow field-flow fractionation: Lipoproteins and pharmaceutical colloids. *J. Pharm. Sci.* **85,** 895–898.

Li, P., Hansen, M., and Giddings, J. C. (1997). Separation of lipoproteins from human plasma by flow field-flow fractionation. *J. Liquid Chromotogr. Rel. Technol.* **20,** 2777–2802.

Litzen, A., and Wahlund, K. G. (1989). Improved separation speed and efficiency for proteins, nucleic acids and viruses in asymmetrical flow field flow fractionation. *J. Chromatogr.* **476,** 413–421.

Litzen, A., and Wahlund, K.-G. (1991). Zone broadening and dilution in rectangular and trapezoidal asymmetrical flow field-flow fractionation channels. *Anal. Chem.* **63,** 1001–1007.

Ranganathan, R., Tran, L., and Bales, B. L. (2000). Surfactant- and salt-induced growth of normal sodium alkyl sulfate micelles well above their critical micelle concentrations. *J. Phys. Chem.* **104,** 2260–2264.

Ratanathanawongs, S. K., and Giddings, J. C. (1992). Dual-field and flow-programmed lift hyperlayer field-flow fractionation. *Anal. Chem.* **64,** 6–15.

Ratanathanawongs-Williams, S. K. (2000). Flow field-flow fractionation. *In* "Field-flow Fractionation Handbook" (M. E. Schimpf, K. Caldwell, and J. C. Giddings, eds.), pp. 257–277. John Wiley & Sons, Inc., New York.

Ratanathanawongs-Williams, S. K., and Giddings, J. C. (2000). Sample recovery. *In* "Field-flow Fractionation Handbook" (M. E. Schimpf, K. Caldwell, and J. C. Giddings, eds.), pp. 325–343. John Wiley & Sons, Inc., New York.

Raymond, G. J., and Chabry, J. (2004). Purification of the pathological isoform of prion protein (PrPSc or PrP-res) from transmissible spongiform encephalopathy-affected brain tissue. *In* "Techniques in Prion Research" (S. Lehmann and J. Grassi, eds.), pp. 16–26. Birkhauser Verlag, Basel.

Reschiglian, P., Zattoni, A., Roda, B., Michelini, E., and Roda, A. (2005). Field-flow fractionation and biotechnology. *Trends Biotechnol.* **23**, 475–483.

Safar, J., Wang, W., Padgett, M. P., Ceroni, M., Piccardo, P., Zopf, D., Gajdusek, D. C., and Gibbs, C. J., Jr. (1990). Molecular mass, biochemical composition, and physicochemical behavior of the infectious form of the scrapie precursor protein monomer. *Proc. Natl. Acad. Sci. USA* **87**, 6373–6377.

Schimpf, M. E., Caldwell, K., and Giddings, J. C. (2000). "Field-flow Fractionation Handbook." John Wiley & Sons, Inc., New York.

Silveira, J. R., Raymond, G. J., Hughson, A. G., Race, R. E., Sim, V. L., Hayes, S. F., and Caughey, B. (2005). The most infectious prion protein particles. *Nature* **437**, 257–261.

Sklaviadis, T., Dreyer, R., and Manuelidis, L. (1992). Analysis of Creutzfeldt-Jakob disease infectious fractions by gel permeation chromatography and sedimentation field flow fractionation. *Virus Res.* **26**, 241–254.

Wahlund, K. G., and Giddings, J. C. (1987). Properties of an asymmetrical flow field-flow fractionation channel having one permeable wall. *Anal. Chem.* **59**, 1332–1339.

Wahlund, K. G., and Litzen, A. (1989). Application of an asymmetrical flow field-flow fractionation channel to the separation and characterization of proteins, plasmids, plasmid fragments, polysaccharides and unicellular algae. *J. Chromatogr.* **461**, 73–87.

Wahlund, K. G., Winegarner, H. S., Caldwell, K. D., and Giddings, J. C. (1986). Improved flow field-flow fractionation system applied to water-soluble polymers: Programming, outlet stream splitting, and flow optimization. *Anal. Chem.* **58**, 573–578.

Wahlund, K.-G. (2000). Asymmetrical flow field-flow fractionation. *In* "Field-flow Fractionation Handbook" (M. E. Schimpf, K. Caldwell, and J. C. Giddings, eds.), pp. 279–294. John Wiley & Sons, Inc., New York.

[3] Analysis of Amyloid Aggregates Using Agarose Gel Electrophoresis

By Sviatoslav N. Bagriantsev, Vitaly V. Kushnirov, and Susan W. Liebman

Abstract

Amyloid aggregates are associated with a number of mammalian neuro-degenerative diseases. Infectious aggregates of the mammalian prion protein PrP^{sc} are hallmarks of transmissible spongiform encephalopathies in humans and cattle (Griffith, 1967; Legname *et al.*, 2004; Prusiner, 1982; Silveira *et al.*, 2004). Likewise, SDS-stable aggregates and low-n oligomers of the Aβ peptide (Selkoe *et al.*, 1982; Walsh *et al.*, 2002) cause toxic effects associated with Alzheimer's disease (Selkoe, 2004). The discovery of prions in lower eukaryotes, for example, yeast prions [PSI$^+$], [PIN$^+$], and [URE3] suggested that prion phenomena may represent a fundamental process that is widespread among living organisms (Chernoff, 2004; Uptain and Lindquist, 2002; Wickner, 1994; Wickner *et al.*, 2004). These protein structures are more

METHODS IN ENZYMOLOGY, VOL. 412
Copyright 2006, Elsevier Inc. All rights reserved.

0076-6879/06 $35.00
DOI: 10.1016/S0076-6879(06)12003-0

stable than other cellular protein complexes, which generally dissolve in SDS at room temperature. In contrast, the prion polymers withstand these conditions, while losing their association with their non-prion partners. These bulky protein particles cannot be analyzed in polyacrylamide gels, because their pores are too small to allow the passage and acceptable resolution of the large complexes. This problem was first circumvented by Kryndushkin *et al.* (2003), who used Western blots of protein complexes separated on agarose gels to analyze the sizes of SDS-resistant protein complexes associated with the yeast prion [*PSI*⁺]. Further studies have used this approach to characterize [*PSI*⁺] (Allen *et al.*, 2005; Bagriantsev and Liebman, 2004; Salnikova *et al.*, 2005), and another yeast prion [*PIN*⁺] (Bagriantsev and Liebman, 2004).

In this chapter, we use this method to assay amyloid aggregates of recombinant proteins Sup35NM and Aβ42 and present protocols for Western blot analysis of high molecular weight (>5 MDa) amyloid aggregates resolved in agarose gels. The technique is suitable for the analysis of any large proteins or SDS-stable high molecular weight complexes.

Yeast Prions [*PSI*⁺] and [*PIN*⁺]

Yeast prions are protein structures that serve as epigenetic elements of inheritance. [*PSI*⁺] is a prion form of the yeast translational termination factor Sup35p (for reviews see Chernoff, 2001; True and Lindquist, 2000; Tuite and Cox, 2003; Uptain and Lindquist, 2002; Wickner *et al.*, 2004). The glutamine- and asparagine-rich N-terminal domain of Sup35p (Sup35NM) (Ter-Avanesyan *et al.*, 1993) is dispensable for Sup35p's activity in translation termination but is required for Sup35p's prion properties (Ter-Avanesyan *et al.*, 1994; Bradley and Liebman, 2004). Recombinant full-length Sup35p or its prion domain alone can quickly polymerize into amyloid fibers *in vitro* (Glover *et al.*, 1997; King *et al.*, 1997). This process is greatly accelerated by the addition of preformed polymers of Sup35NM (a process called self-seeding) (Glover *et al.*, 1997). Another yeast prion, [*PIN*⁺] (also called [*RNQ*⁺]) is a form of Rnq1p, a yeast protein with unknown function (Derkatch *et al.*, 2001; Sondheimer and Lindquist, 2000). Like the N-terminal domain of Sup35p, Rnq1p's C-terminal domain has a high asparagine and glutamine content and seems to be responsible for the prion properties of Rnq1p (Sondheimer and Lindquist, 2000). It was also shown that prionized Rnq1p molecules could form self-seeding amyloid polymers (Sondheimer and Lindquist, 2000). The presence of [*PIN*⁺] dramatically enhances the rate of [*PSI*⁺] appearance (Derkatch *et al.*, 1997, 2001) by a cross-seeding mechanism (Derkatch *et al.*, 2004).

The mammalian prion protein PrP^sc was proposed to have a number of stably inherited prion conformations, or "strains," which accounted for the variety of clinically different prion phenotypes (Bruce, 2003; Collinge, 2001; Prusiner et al., 1998). Likewise, [PSI⁺] and [PIN⁺] have "strains" of their own, referred to herein as "variants" (Bradley and Liebman, 2003; Bradley et al., 2002; Derkatch et al., 1996). A growing body of evidence suggests that prion variants, indeed, correspond to various distinct, heritable aggregate types of the same prion (Chien and Weissman, 2001; DePace and Weissman, 2002; King and Diaz-Avalos, 2004; Tanaka et al., 2004; Uptain et al., 2001). Variants of the [PSI⁺] prion are usually distinguished by their rates of nonsense codon readthrough (nonsense suppression). For example, [psi⁻] cells carrying ade1-14, a premature nonsense codon in ADE1, cannot produce adenine, accumulate a red pigment, and fail to grow on adenine-deficient media. [PSI⁺] cells can read through the ade1-14 nonsense codon, remain white, and grow without adenine. On the basis of how effectively the ade1-14 nonsense codon is suppressed, [PSI⁺] variants are referred to as "strong" or "weak" (Derkatch et al., 1996). Variants of [PIN⁺] were first characterized as having different [PSI⁺]-induction efficiency (Derkatch et al., 1997, 2000) from "low" to "very high" (Bradley et al., 2002). These variants were further subdivided by their Rnq1p-GFP fluorescence pattern defined as "single-dot" or "multiple-dot" (Bradley and Liebman, 2003).

Intracellularly, both [PSI⁺] (Patino et al., 1996; Paushkin et al., 1996) and [PIN⁺] (Sondheimer and Lindquist, 2000) form large, possibly amyloid aggregates (Kimura et al., 2003). Gel filtration experiments revealed that Sup35p from [PSI⁺] cells eluted in fractions corresponding to structures with molecular weight higher than 1000 kDa (Paushkin et al., 1997). The growth of the aggregates is controlled by the yeast chaperone Hsp104p, which is required for successful propagation of [PSI⁺] and [PIN⁺] (Chernoff et al., 1995; Derkatch et al., 1996, 1997; Sondheimer and Lindquist, 2000). Hsp104p shears prion aggregates ensuring they are in sufficient numbers to be divided between mother and daughter yeast cells (Kushnirov and Ter-Avanesyan, 1998). When this function of Hsp104p is abolished (e.g., by inhibition with millimolar concentrations of guanidine hydrochloride), [PSI⁺] and [PIN⁺] aggregates continue to grow but do not multiply, which eventually leads to the complete loss of the prion aggregates in the daughter cells (Ness et al., 2002; Wegrzyn et al., 2001).

The idea of analyzing yeast prions by using agarose gels followed by transfer onto a membrane and immunodetection revealed that aggregates of [PSI⁺] and [PIN⁺] represent bulky structures that disassemble upon SDS treatment into smaller, yet SDS stable, structures called herein "subparticles." Using the agarose gel technique described herein, the subparticles were discovered and analyzed first for [PSI⁺] (Kryndushkin et al., 2003), and then for [PIN⁺] (Bagriantsev and Liebman, 2004). It was shown that the size range of the subparticles was variant-specific and corresponded to the size of ~9–50 monomers of Sup35p (Kryndushkin et al.,

2003) or ~20–100 monomers of Rnq1p (Bagriantsev and Liebman, 2004). The subparticles are probably polymers of Sup35p or Rnq1p. It was also demonstrated that subparticles of different prion variants have specific size distributions and thermal stability. They increased in size as a result of inhibition of Hsp104p's activity by guanidine (Bagriantsev and Liebman, 2004; Kryndushkin *et al.*, 2003). Variant-specific thermal stability of [*PIN*⁺] subparticles was used to establish that two variants of [*PIN*⁺] cannot simultaneously propagate within one cell. When yeast bearing different [*PIN*⁺] variants were crossed, only one type of [*PIN*⁺] subparticle was present in the diploid. Importantly, evidence for a highly specific mechanism of prion assembly *in vivo* was presented: the heterologous prions [*PSI*⁺] and [*PIN*⁺] do not form heterogeneous subparticles when they coexist in one cell but retain "purity" and lack any appreciably tight interactions with each other (Bagriantsev and Liebman, 2004). However, in [*psi*⁻][*PIN*⁺] cells overexpressing Sup35p, a noticeable proportion of Rnq1p was found in Sup35p polymers (Salnikova *et al.*, 2005).

Buffers, Electrophoresis, and Transfer Conditions

Buffers

Protein Extraction Buffer. 50 mM Tris-HCl, pH 7.5, containing 50 mM KCl and 10 mM MgCl$_2$. An antiprotease cocktail with a high anti-serine protease content, for example, 10 mM PMSF, in combination with 8 μg/ml aprotinin and 80 μM TLCK should be used to prevent proteolytic degradation. To perform electrophoresis and transfer, two buffer systems may be used with equal efficacy.

Buffer A. 20 mM Tris, 200 mM glycine (Laemmli, 1970). *Sample buffer A*: 50 mM Tris, pH 6.8, 2% (w/v) SDS, 0.025% bromophenol blue, 5% glycerol. *Running buffer A*: Buffer A, 0.1% SDS. *Electrophoretic transfer buffer A*: Buffer A, 0.1% SDS, 15% (v/v) methanol.

Buffer B. 40 mM Tris-acetate, 1 mM EDTA (Sambrook *et al.*, 1989). *Sample buffer B*: 20 mM Tris-acetate, 0.5 mM EDTA, 2% (w/v) SDS, 0.025% bromophenol blue, 5% glycerol. *Running, electrophoretic and vacuum transfer buffer B*: Buffer B, 0.1% SDS.

Protein Electrophoresis in Agarose

Low melting point agarose should be avoided, because it may melt during transfer. Agarose (1.2–1.5% w/v) is melted in *running buffer A* or *B* without SDS. While stirring the melted agarose with a Teflon-coated stirring bar, 10% (w/v) SDS solution is added (drop by drop to avoid local solidification of the agarose) to a 0.1% final SDS concentration. Addition of SDS before melting may result in the formation of thick foam and undesirable bubbles in the gel. Regular high melting point agarose normally used for DNA gel electrophoresis and a regular DNA electrophoresis casting chamber is used to cast the gel.

For analysis of $[PSI^+]$ or $[PIN^+]$ subparticles, 60–80 μg of lysate protein per lane is sufficient to obtain a good image; for Sup35NM or Aβ42, 1 μg of fibers is sufficient (see later). Before loading the samples onto the gel, they should be pretreated in SDS-containing *sample buffer*. At room temperature the detergent does not dissolve any of the amyloid polymers or prion subparticles that we tested, but disaggregates their higher-order complexes and most of the other protein complexes in the lysate. Importantly, the detergent molecules apply a negative charge to the polymers, rendering them mobile in an electric field. The final concentration of SDS in the sample, as well as the time and temperature of incubation, may differ for different amyloid samples (see later). A 7-min incubation at room temperature in the presence of 1% SDS was found to be suitable for all samples described here. A preparation of chicken pectoralis extract (Kim and Keller, 2002) pretreated at room temperature in the same *sample buffer* represents a convenient molecular weight ladder. Coomassie staining of the extract reveals several abundant muscular proteins: a doublet titin (~3 MDa), nebulin (~750 kDa), and myosin heavy chain (~200 kDa) (Bagriantsev and Liebman, 2004). Virtually the same protein pattern is obtained with a preparation from rabbit muscle (Kryndushkin *et al.*, 2003). Although neither ladder can be used for a precise determination of the molecular masses of the amyloid polymers, they do provide acceptable estimates. Samples are resolved using a regular horizontal gel chamber designed for DNA electrophoresis in *running buffer A* (Bagriantsev and Liebman, 2004) or *B* (Allen *et al.*, 2005; Kryndushkin *et al.*, 2003; Salnikova *et al.*, 2005) at a constant 125 V until the bromophenol touches the edge of the gel. Lower voltage leads to increased lateral diffusion of the sample, whereas higher voltage may cause undesirable heat build-up. If transfer onto a membrane is not required, the proteins can be stained in the gel. To diminish diffusion of the sample during the staining and washing steps, the gel should be partially dried with a vacuum dryer (Vacuum Slab Dryer Model 483, Bio-Rad Laboratories) until 1.5–2 mm thick, followed by staining with the water-soluble protein dye Coomassie G-250 (Bio-Safe Coomassie, Bio-Rad) (see Fig. 5A).

Transfer of Proteins from Agarose Gels and Immunostaining

The aggregates can be transferred by using a modified Mini-Protean 2 transfer apparatus (Bio-Rad) (Bagriantsev and Liebman, 2004) originally designed for the transfer of proteins from acrylamide gels. Modifications are required, because the cassette supplied with the device, which holds the gel/membrane sandwich, cannot accommodate the thick agarose gel without damaging it and causing a distorted image. Because we could not find a cassette of the proper thickness commercially available, we widened the Mini-Protean 2 standard cassette by the easy manipulations shown in Fig. 1. The inner part of the cassette (A) is cut out using a jigsaw. This procedure yields the frame (B) and the insert (C). Four pieces of 1-mm-thick plastic spacers (we used teeth of a gel comb) are glued to the corners of frame (B) as shown. The spacers should be positioned so they do not obstruct the cassette from sliding back into the transfer chamber. The insert (C) is then glued onto the spacers. The entire procedure is repeated for the second half of the cassette. Cassettes made this way are thick enough to accommodate the agarose gel sandwich and still fit into the Mini-Protean 2 transfer chamber. Note that the Mini-Protean 2 transfer chamber can accommodate only one cassette customized in this way. The transfer sandwich should be assembled as for regular protein blotting: a sponge (cut to fit into the space freed from the insert), a layer of filter paper, the gel, polyvinylidene fluoride (PVDF), or nitrocellulose membrane (preequilibrated in *transfer buffer*); a layer of filter paper; and a cut sponge. PVDF should be used if detection of monomers is required, because it has a higher protein binding capacity. This is important because more than 90% of proteins run in the "monomer" region of the membrane. Negatively

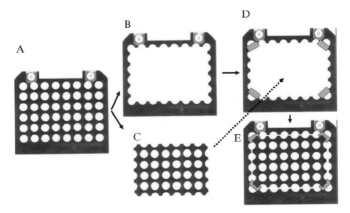

FIG. 1. Customization of the transfer cassette from Mini-Protean 2 transfer apparatus. Using a jigsaw, cut out the insert (C) from the cassette (A). Glue plastic spacers onto the frame (B) as shown (D). Glue the insert (C) onto D as shown. Repeat the procedure for the second half of the cassette (see text).

charged protein–SDS complexes will migrate to the positive pole. It is important to ensure that the sandwich is not loose when the cassette is closed. If gels that are too thin are used, especially in combination with worn-out sponges, insufficient or uneven tightness of the gel–membrane contact surface may lead to undirected transfer of the proteins and result in a faulty image, as shown in Fig. 2. An additional layer of filter paper or a new pair of sponges should be used to ensure proper tightness. We perform the transfer in *transfer buffer* for 45 min in the presence of an ice bar at a constant 100 V with continuous stirring. We then replace the melted ice bar and continue the transfer for another 45 min. It is important to keep the transfer chamber cold, because unlike acrylamide gels, the integrity of the agarose gels is highly sensitive to heat.

Alternately, a semi-dry transfer method may be used. Semi-dry blotting unit FB-SDB-2020 (Fisher Scientific) was used to transfer Aβ42 polymers from agarose gels 18-cm long (see "Analysis of Sup35NM and Aβ42 Amyloid Aggregates by Electrophoresis in Agarose"), which could not be accommodated by the Mini-Protean 2 transfer chamber. In this case, no customization procedures are necessary. As mentioned previously, relatively firm contact

FIG. 2. Distorted image caused by insufficient tightness of the gel–membrane contact. SDS-treated lysates were resolved in 1.5% agarose gel and transferred onto a polyvinylidene fluoride membrane using a customized Mini-Protean 2 chamber. Insufficient tightness of the gel–membrane contact led to an undirected transfer of the proteins and resulted in a distorted image.

between the gel surface and the membrane should be ensured. This can be achieved by adding additional layers of filter paper (soaked in the *transfer buffer*) and/or adjusting the lid pressure. The transfer is performed at 4.5 mA per square centimeter of the gel for 1.5 h. Although the semi-dry transfer is easier to handle, it seems to give a distorted image more frequently. Two spacers slightly thinner than the gel should be placed on the sides of the gel to prevent the upper electrode from squeezing the gel.

The other option is to perform transfer using vacuum blotter (Allen *et al.*, 2005; Kryndushkin *et al.*, 2003). Proteins were transferred from gels using the VacuGene XL system (Amersham Biosciences) at 30 mBar pressure overnight or for a minimum of 4 h. Running buffer B was constantly supplied to the top of the gel using a filter paper wick.

After the transfer is complete, the membrane is blocked and probed with desired antibodies in the normal way. The membrane strip containing the molecular weight ladder may be cut out before antibody staining (but after blocking) and the bands revealed with Coomassie Brilliant Blue R-250 or Ponceau S. Alternately, the entire membrane may be stained with the dyes after all immunostaining procedures are complete.

Analysis of [*PSI*$^+$] and [*PIN*$^+$] Prion Subparticles from Yeast Lysates by Electrophoresis in Agarose

This method requires a relatively high concentration of protein in lysate. This can be obtained from 250 ml of YPD media (2% dextrose, 2% peptone, 1% yeast extract) culture at A_{600} ~1 or 50 ml of culture at A_{600} 4–5, which each yield ~350 mg of wet yeast pellet. We have not detected any changes in the size distribution of the [*PSI*$^+$] or [*PIN*$^+$] prion subparticles from cultures with A_{600} in the range of 1–6. The lysis of cells resuspended in *lysis buffer* is performed through disruption of the cell wall by vortexing with glass beads. The size distribution of [*PSI*$^+$] subparticles is not changed if a more gentle method of protein extraction, such as spheroplasting, is used (Kryndushkin *et al.*, 2003), indicating that glass beads do not shatter prion aggregates. Importantly, intensive preclearing of the crude lysate from the cell debris may result in cosedimentation of the prion aggregates with the debris. For example, centrifugation of certain weak [*PSI*$^+$] variants in L1845 yeast strain for 10 min at 10,000g (a procedure that is often used as a preclearing step in yeast lysate preparations) leads to an ~90% decline in the amount of the prionized Sup35p in the supernatant. Therefore, we recommend removing cell debris by 1 min centrifugation at 600g (4°) for trial experiments. This procedure yields ~500 μl of turbid lysate from ~350 mg of wet yeast pellet, with a total protein concentration of 3–10 mg/ml (as detected by Bradford reagent, Bradford, 1976). Lysates with a total protein content of less than 3 mg/ml may require concentration, because ~60–80 μg of total protein per lane is needed to obtain a good image. The lysates can be stored at $-70°$ for up to 1 wk without detectably affecting the image. Lysates stored longer may give a low-intensity distorted image. Before loading onto the gel, lysates should be incubated with *sample buffer*, and then loaded on and analyzed in a 1.2–1.5% (w/v) agarose gel. The final SDS concentration in the sample may vary from 0.5–7.5% (w/v) without detectably affecting the mobility of the subparticles (we usually use 2% SDS). Although even brief incubation at room temperature or on ice is enough to attach SDS onto the protein molecules, we prefer a 7-min incubation at room temperature, because it does not affect the mobility of the subparticles and makes it easier to handle several samples simultaneously. More importantly, the full 70-min incubation at $60°$ or $95°$ is required for the partial or complete breakage of some subparticles (Bagriantsev and Liebman, 2004) (Fig. 3). Electrophoresis and transfer using the customized Mini-Protean 2 or semi-dry methods are performed as described previously. To immunostain [PSI⁺] subparticles, we use monoclonal antibodies against Sup35p's C-terminal domain (Fig. 3). As seen on the image, subparticles from a weak [PSI⁺] variant are usually larger on average than those from a strong [PSI⁺] (Bagriantsev and Liebman, 2004; Kryndushkin *et al.*, 2003). Monoclonal antibodies against Sup35p's N terminal domain can also be used, but they were reported to recognize [PSI⁺] subparticles less efficiently (Allen *et al.*, 2005). Possibly, the prion conformation of the Sup35p's N-terminal region makes it less accessible to antibody. To stain [PIN⁺] subparticles, we use anti-Rnq1p antibodies (a kind gift from S. Lindquist) (Fig. 3). As described elsewhere (Bagriantsev and Liebman, 2004), subparticles derived from so-called multiple-dot and single-dot [PIN⁺] variants (Bradley *et al.*, 2002) show a clearly distinguishable size distribution and thermal stability (Bagriantsev and Liebman, 2004). Assuming that [PSI⁺] and [PIN⁺] subparticles are homogeneous polymers of Sup35p and Rnq1p, respectively, they should contain approximately 10–90 monomers of corresponding proteins. The number of monomers differs for a given prion variant (Bagriantsev and Liebman, 2004; Kryndushkin *et al.*, 2003).

Analysis of Sup35NM and Aβ42 Amyloid Aggregates by Electrophoresis in Agarose

Analysis of Sup35NM Amyloid Fibers

The N-terminal domain of Sup35p that is rich in asparagines and glutamines has been shown to be responsible for its prion properties (Derkatch *et al.*, 1996; Ter-Avanesyan *et al.*, 1994). Purification of the

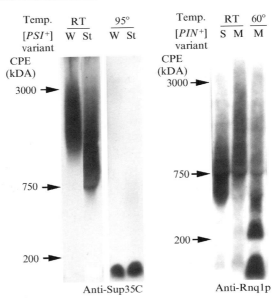

FIG. 3. Subparticles of yeast prions [*PSI*⁺] and [*PIN*⁺] in agarose gels. Yeast lysates (80 μg of protein) were treated with SDS for 7 min at various temperatures, resolved in 1.5% agarose gels, transferred onto a polyvinylidene fluoride membrane using a customized Mini-Protean 2 chamber, and detected with antibodies. Chicken pectoralis extract (CPE) was used as a molecular weight ladder. (Left panel) Shown are subparticles of a weak (W) [*PSI*⁺] variant from a derivative of strain L1845 and a strong (St) [*PSI*⁺] variant from a derivative of strain 74D-694. Lysates were pretreated with 7.5% SDS at room temperature (RT) or at 95°. The membrane was stained with monoclonal antibodies against Sup35p's C-terminal domain (BE4). As seen on the image, subparticles from the weak [*PSI*⁺] variant are larger on average than those from the strong [*PSI*⁺]. Both types of subparticles can withstand 2% SDS treatment at up to 60° (not shown) but disaggregate into monomers at 95°. (Right panel) Shown are subparticles of single-dot (S) [*PIN*⁺] and multiple-dot (M) [*PIN*⁺] variants from derivatives of strain 74D-694. Lysates were pretreated with 2% SDS at room temperature (RT) or at 60°. The membrane was stained with polyclonal antibodies against Rnq1p (a kind gift from S. Lindquist). Subparticles from single- and multiple-dot [*PIN*⁺] variants have distinct size distributions and dissociate at 95° into monomers (not shown). Although subparticles from the single-dot [*PIN*⁺] variant retain structural integrity at 60° in the presence of 2% SDS (not shown), subparticles from the multiple-dot [*PIN*⁺] variant disaggregate in these conditions into subfragments corresponding to the size of quartermers and octamers of Rnq1p.

His-Tagged N-terminal domain of Sup35p (Sup35NM) was described else-where (Serio *et al.*, 1999). The following modifications were introduced in the procedure. Sup35NM is stored unconcentrated, or its concentration is adjusted to ~0.5–1 mg/ml (~17–35 μM) with the urea-containing denaturing buffer. To trigger the polymerization reaction, we replace the denaturing buffer with the phosphate-buffered saline solution using Micro Bio Spin 6 desalting columns (Bio-Rad). Polymerization occurs spontaneously during rotation of the sample (60 rpm) at various temperatures for 48 h. To obtain a strong signal on the immunoblot, 1 μg of Sup35NM per lane is enough. The final concentration of SDS in the sample may vary from 0.5–2% (w/v) without affecting mobility of the polymers. Heat treatment is a much more important factor here. Although all Sup35NM fibers we analyzed were able to withstand at least a 7-min incubation at room temperature, treatment at 60° partially or completely disaggregated some fibers, depending on the conditions under which they were originally polymerized. For example, Sup35NM fibers that were polymerized at constant rotation were stable at 60° with 2% SDS, whereas those polymerized with occasional shaking disassembled under these conditions into monomers (Fig. 4). This is consistent

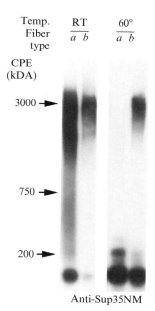

FIG. 4. Polymers of Sup35NM resolved in agarose gels. Samples containing 1 μg of His-Tagged Sup35NM polymerized in phosphate-buffered saline at occasional shaking (fiber type *a*) or at constant rotation (fiber type *b*) were treated with 1% SDS for 7 min at room temperature (RT) or at 60°, resolved in 1.5% agarose gel, transferred onto a polyvinylidene fluoride membrane using a customized Mini-Protean 2 chamber, and detected with anti-Sup35NM monoclonal antibodies (Ab0332). Chicken pectoralis extract (CPE) was used as a molecular weight ladder. As seen from the image, fibers of Sup35NM made during constant rotation (type *b*) exhibit a higher stability to SDS and heat treatment than those made with occasional shaking (type *a*).

with previously reported data suggesting that polymerization conditions affect physical properties of Sup35NM fibers (Tanaka *et al.*, 2004). We suggest using a 1–2% (w/v) final SDS concentration and a 7-min incubation at room temperature as the starting conditions to analyze polymers. A 7-min treatment at 95° completely disaggregated all Sup35NM fibers we tested into monomers (not shown). We perform electrophoresis and transfer using the customized Mini-Protean 2 chamber as described previously (the semi-dry transfer was not tested for Sup35NM) and detect Sup35NM polymers with the anti-Sup35NM monoclonal antibodies (Fig. 4). The size distribution of the Sup35NM polymers is within the range of ~20–100 monomers of Sup35NM.

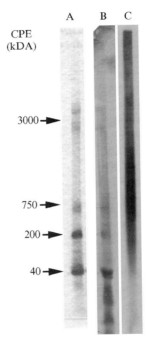

FIG. 5. Chicken pectoralis extract and Aβ42 polymers resolved in long agarose gels. (A) A preparation of CPE was resolved in a 1.4% agarose gel (18 cm). The gel was partially dried using a Vacuum Slab Dryer and was stained with Coomassie G-250. (B, C) An 18-cm-long 1.4% agarose gel was used to separate a preparation of CPE (B) and a sample containing 10 μg of Aβ42 (C) polymerized at room temperature and pretreated with 1% SDS for 7 min at room temperature. Incubation with 2% SDS did not affect the mobility of the samples (not shown). Proteins were transferred onto a polyvinylidene fluoride membrane using a semi-dry blotting unit and stained with monoclonal antibodies against Aβ's residues 1–17 (6E10 from Signet Laboratories, Inc., Dedham, MA). CPE was revealed with Coomassie Brilliant Blue R-250.

Analysis of Aβ42 Amyloid Fibers. The amyloidogenic peptide Aβ42, associated with Alzheimer's disease, is known to assemble into highly stable amyloid polymers under various conditions. Recombinant Aβ42 (powdered Aβ42-acetate from Rpeptide, Athens, GA) was polymerized according to the manufacturer's suggestions. A 1-mg/ml solution of Aβ42 was made by suspending 0.5 mg of Aβ42 powder in 100 μl of 2.5 mM NaOH and adding 400 μl of phosphate-buffered saline solution. Polymerization proceeded at room temperature with constant rotation (60 rpm). We perform electrophoresis of Aβ42 using long (18 cm) agarose gels, because it allows for better visualization of the peptide's largest polymers. Each sample contained 10 μg of Aβ42 in *sample buffer A* with 1% (w/v) SDS (2% SDS did not detectably affect the size distribution of the polymers). Although ultrasonication did not disassemble Aβ42, temperature treatment at 60° or at 95° for 7 min caused partial disaggregation of Aβ42 polymers (not shown). Aβ42 polymers were transferred onto a membrane using the semi-dry blotting unit and were detected with monoclonal antibodies reactive to Aβ42 (Fig. 5C). The ladder was visualized by Coomassie Brilliant Blue R-250 (Fig. 5B). As seen from the image, Aβ42 polymerized under conditions mentioned previously, forms polymers ranging in size from ~40 kDa–~10 MDa. Most polymerized Aβ is within the range of ~750 kDa–~5 MDa (~180–1200 monomers of Aβ42). Conditions of the polymerization reaction can substantially affect morphology of the amyloid fibers. It was shown that even light disturbances occurring during polymerization (e.g., agitation versus quiescence) lead to formation of Aβ40 fibers with different heritable morphology (Petkova *et al.*, 2005).

Acknowledgments

The work was supported by National Institute of Health Grant GM56350 (to S. W. L.) and the Welcome Trust and Howard Hughes Medical Institute (to V. V. K.). We thank Yakov Vitrenko for providing Sup35NM fibers *type a* used in Fig. 4.

References

Allen, K. D., Wegrzyn, R. D., Chernova, T. A., Muller, S., Newnam, G. P., Winslett, P. A., Wittich, K. B., Wilkinson, K. D., and Chernoff, Y. O. (2005). Hsp70 chaperones as modulators of prion life cycle: Novel effects of Ssa and Ssb on the *Saccharomyces cerevisiae* prion [*PSI*$^+$]. *Genetics* **169**, 1227–1242.

Bagriantsev, S., and Liebman, S. W. (2004). Specificity of Prion Assembly *in vivo*: [*PSI*$^+$] and [*PIN*$^+$] form separate structures in yeast. *J. Biol. Chem.* **279**, 51042–51048.

Bradford, M. M. (1976). A rapid and sensitive method for the quantitation of microgram quantities of protein utilizing the principle of protein-dye binding. *Anal. Biochem.* **72**, 248–254.

Bradley, M. E., Edskes, H. K., Hong, J. Y., Wickner, R. B., and Liebman, S. W. (2002). Interactions among prions and prion "strains" in yeast. *Proc. Natl. Acad. Sci. USA* **99** (Suppl. 4), 16392–16399.

Bradley, M. E., and Liebman, S. W. (2003). Destabilizing interactions among [*PSI*+] and [*PIN*+] yeast prion variants. *Genetics* **165**, 1675–1685.

Bradley, M. E., and Liebman, S. W. (2004). The Sup35 domains required for maintenance of weak, strong or undifferentiated yeast [*PSI*+] prions. *Mol. Microbiol.* **51**, 1649–1659.

Bruce, M. E. (2003). TSE strain variation. *Br. Med. Bull.* **66**, 99–108.

Chernoff, Y. O. (2001). Mutation processes at the protein level: Is Lamarck back? *Mutat. Res.* **488**, 39–64.

Chernoff, Y. O. (2004). Amyloidogenic domains, prions and structural inheritance: Rudiments of early life or recent acquisition? *Curr. Opin. Chem. Biol.* **8**, 665–671.

Chernoff, Y. O., Lindquist, S. L., Ono, B., Inge-Vechtomov, S. G., and Liebman, S. W. (1995). Role of the chaperone protein Hsp104 in propagation of the yeast prion-like factor [*PSI*+]. *Science* **268**, 880–884.

Chien, P., and Weissman, J. S. (2001). Conformational diversity in a yeast prion dictates its seeding specificity. *Nature* **410**, 223–227.

Collinge, J. (2001). Prion diseases of humans and animals: Their causes and molecular basis. *Annu. Rev. Neurosci.* **24**, 519–550.

DePace, A. H., and Weissman, J. S. (2002). Origins and kinetic consequences of diversity in Sup35 yeast prion fibers. *Nat. Struct. Biol.* **9**, 389–396.

Derkatch, I. L., Bradley, M. E., Hong, J. Y., and Liebman, S. W. (2001). Prions affect the appearance of other prions: The story of [*PIN*+]. *Cell* **106**, 171–182.

Derkatch, I. L., Bradley, M. E., Masse, S. V., Zadorsky, S. P., Polozkov, G. V., Inge-Vechtomov, S. G., and Liebman, S. W. (2000). Dependence and independence of [*PSI*+] and [*PIN*+]: A two-prion system in yeast? *EMBO J.* **19**, 1942–1952.

Derkatch, I. L., Bradley, M. E., Zhou, P., Chernoff, Y. O., and Liebman, S. W. (1997). Genetic and environmental factors affecting the *de novo* appearance of the [*PSI*+] prion in *Saccharomyces cerevisiae*. *Genetics* **147**, 507–519.

Derkatch, I. L., Chernoff, Y. O., Kushnirov, V. V., Inge-Vechtomov, S. G., and Liebman, S. W. (1996). Genesis and variability of [*PSI*+] prion factors in *Saccharomyces cerevisiae*. *Genetics* **144**, 1375–1386.

Derkatch, I. L., Uptain, S. M., Outeiro, T. F., Krishnan, R., Lindquist, S. L., and Liebman, S. W. (2004). Effects of Q/N-rich, polyQ, and non-polyQ amyloids on the *de novo* formation of the [*PSI*+] prion in yeast and aggregation of Sup35 *in vitro*. *Proc. Natl. Acad. Sci. USA* **101**, 12934–12939.

Glover, J. R., Kowal, A. S., Schirmer, E. C., Patino, M. M., Liu, J. J., and Lindquist, S. (1997). Self-seeded fibers formed by Sup35, the protein determinant of [*PSI*+], a heritable prion-like factor of *S. cerevisiae*. *Cell* **89**, 811–819.

Griffith, J. S. (1967). Self-replication and scrapie. *Nature* **215**, 1043–1044.

Kim, K., and Keller, T. C., 3rd. (2002). Smitin, a novel smooth muscle titin-like protein, interacts with myosin filaments *in vivo* and *in vitro*. *J. Cell Biol.* **156**, 101–111.

Kimura, Y., Koitabashi, S., and Fujita, T. (2003). Analysis of yeast prion aggregates with amyloid-staining compound *in vivo*. *Cell Struct. Funct.* **28**, 187–193.

King, C. Y., and Diaz-Avalos, R. (2004). Protein-only transmission of three yeast prion strains. *Nature* **428**, 319–323.

King, C. Y., Tittmann, P., Gross, H., Gebert, R., Aebi, M., and Wuthrich, K. (1997). Prion-inducing domain 2–114 of yeast Sup35 protein transforms *in vitro* into amyloid-like filaments. *Proc. Natl. Acad. Sci. USA* **94**, 6618–6622.

Kryndushkin, D. S., Alexandrov, I. M., Ter-Avanesyan, M. D., and Kushnirov, V. V. (2003). Yeast [*PSI*⁺] prion aggregates are formed by small Sup35 polymers fragmented by Hsp104. *J. Biol. Chem.* **278,** 49636–49643.

Kushnirov, V. V., and Ter-Avanesyan, M. D. (1998). Structure and replication of yeast prions. *Cell* **94,** 13–16.

Laemmli, U. K. (1970). Cleavage of structural proteins during the assembly of the head of bacteriophage T4. *Nature* **227,** 680–685.

Legname, G., Baskakov, I. V., Nguyen, H. O., Riesner, D., Cohen, F. E., DeArmond, S. J., and Prusiner, S. B. (2004). Synthetic mammalian prions. *Science* **305,** 673–676.

Ness, F., Ferreira, P., Cox, B. S., and Tuite, M. F. (2002). Guanidine hydrochloride inhibits the generation of prion "seeds" but not prion protein aggregation in yeast. *Mol. Cell. Biol.* **22,** 5593–5605.

Patino, M. M., Liu, J. J., Glover, J. R., and Lindquist, S. (1996). Support for the prion hypothesis for inheritance of a phenotypic trait in yeast. *Science* **273,** 622–626.

Paushkin, S. V., Kushnirov, V. V., Smirnov, V. N., and Ter-Avanesyan, M. D. (1996). Propagation of the yeast prion-like [*PSI*⁺] determinant is mediated by oligomerization of the SUP35-encoded polypeptide chain release factor. *EMBO J.* **15,** 3127–3134.

Paushkin, S. V., Kushnirov, V. V., Smirnov, V. N., and Ter-Avanesyan, M. D. (1997). *In vitro* propagation of the prion-like state of yeast Sup35 protein. *Science* **277,** 381–383.

Petkova, A. T., Leapman, R. D., Guo, Z., Yau, W. M., Mattson, M. P., and Tycko, R. (2005). Self-propagating, molecular-level polymorphism in Alzheimer's {beta}-amyloid fibrils. *Science* **307,** 262–265.

Prusiner, S. B. (1982). Novel proteinaceous infectious particles cause scrapie. *Science* **216,** 136–144.

Prusiner, S. B., Scott, M. R., DeArmond, S. J., and Cohen, F. E. (1998). Prion protein biology. *Cell* **93,** 337–348.

Salnikova, A. B., Kryndushkin, D. S., Smirnov, V. N., Kushnirov, V. V., and Ter-Avanesyan, M. D. (2005). Nonsense suppression in yeast cells overproducing Sup35 (eRF3) is caused by its non-heritable amyloids. *J. Biol. Chem.* **280,** 8808–8812.

Sambrook, J., Fritsch, E. E., and Maniatis, T. (1989). "Molecular Cloning: A laboratory manual." Cold Spring Harbor Laboratory Press, Cold Spring Harbor, NY.

Selkoe, D. J., Ihara, Y., and Salazar, F. J. (1982). Alzheimer's disease: Insolubility of partially purified paired helical filaments in sodium dodecyl sulfate and urea. *Science* **215,** 1243–1245.

Selkoe, D. J. (2004). Cell biology of protein misfolding: The examples of Alzheimer's and Parkinson's diseases. *Nat. Cell Biol.* **6,** 1054–1061.

Serio, T. R., Cashikar, A. G., Moslehi, J. J., Kowal, A. S., and Lindquist, S. L. (1999). Yeast prion [*PSI*⁺] and its determinant, Sup35p. *Methods Enzymol.* **309,** 649–673.

Silveira, J. R., Caughey, B., and Baron, G. S. (2004). Prion protein and the molecular features of transmissible spongiform encephalopathy agents. *Curr. Top. Microbiol. Immunol.* **284,** 1–50.

Sondheimer, N., and Lindquist, S. (2000). Rnq1: An epigenetic modifier of protein function in yeast. *Mol. Cell* **5,** 163–172.

Tanaka, M., Chien, P., Naber, N., Cooke, R., and Weissman, J. S. (2004). Conformational variations in an infectious protein determine prion strain differences. *Nature* **428,** 323–328.

Ter-Avanesyan, M. D., Dagkesamanskaya, A. R., Kushnirov, V. V., and Smirnov, V. N. (1994). The SUP35 omnipotent suppressor gene is involved in the maintenance of the non-Mendelian determinant [*PSI*⁺] in the yeast *Saccharomyces cerevisiae*. *Genetics* **137,** 671–676.

Ter-Avanesyan, M. D., Kushnirov, V. V., Dagkesamanskaya, A. R., Didichenko, S. A., Chernoff, Y. O., Inge-Vechtomov, S. G., and Smirnov, V. N. (1993). Deletion analysis of the SUP35 gene of the yeast *Saccharomyces cerevisiae* reveals two non-overlapping functional regions in the encoded protein. *Mol. Microbiol.* **7,** 683–692.

True, H. L., and Lindquist, S. L. (2000). A yeast prion provides a mechanism for genetic variation and phenotypic diversity. *Nature* **407,** 477–483.

Tuite, M. F., and Cox, B. S. (2003). Propagation of yeast prions. *Nat. Rev. Mol. Cell. Biol.* **4,** 878–890.

Uptain, S. M., and Lindquist, S. (2002). Prions as protein-based genetic elements. *Annu. Rev. Microbiol.* **56,** 703–741.

Uptain, S. M., Sawicki, G. J., Caughey, B., and Lindquist, S. (2001). Strains of [*PSI*$^+$] are distinguished by their efficiencies of prion-mediated conformational conversion. *EMBO J.* **20,** 6236–6245.

Walsh, D. M., Klyubin, I., Fadeeva, J. V., Rowan, M. J., and Selkoe, D. J. (2002). Amyloid-beta oligomers: Their production, toxicity and therapeutic inhibition. *Biochem. Soc. Trans.* **30,** 552–557.

Wegrzyn, R. D., Bapat, K., Newnam, G. P., Zink, A. D., and Chernoff, Y. O. (2001). Mechanism of prion loss after Hsp104 inactivation in yeast. *Mol. Cell. Biol.* **21,** 4656–4669.

Wickner, R. B. (1994). [URE3] as an altered URE2 protein: Evidence for a prion analog in *Saccharomyces cerevisiae*. *Science* **264,** 566–569.

Wickner, R. B., Liebman, S. W., and Saupe, S. J. (2004). Prions of yeast and filamentous fungi: [URE3], [*PSI*$^+$], [*PIN*$^+$], and [Het-s]. *In* "Prion Biology and Diseases" (S. B. Prusiner, ed.), pp. 305–372. Cold Spring Harbor Laboratory Press, Cold Spring Harbor, NY.

[4] Characterization of Systemic Amyloid Deposits by Mass Spectrometry

By Charles L. Murphy, Shuching Wang, Teresa Williams, Deborah T. Weiss, and Alan Solomon

Abstract

The human systemic (noncerebral) amyloidoses represent a heterogeneous group of disorders characterized by the widespread deposition of proteins as fibrils in organs or tissues throughout the body. The unequivocal identification of the type of amyloid deposited is critical to the correct diagnosis and treatment of patients with these illnesses. Heretofore, this information was inferred from clinical data, ancillary laboratory tests, and results of immunohistochemical, as well as genetic, analyses. However, to establish definitively the type of amyloid present, the chemical composition of the fibrillar components must be determined. For this purpose, we have developed micro-methods, whereby this information can be obtained

METHODS IN ENZYMOLOGY, VOL. 412   0076-6879/06 $35.00
DOI: 10.1016/S0076-6879(06)12004-2

by tandem mass spectrometry (MS/MS) using material extracted from formalin-fixed, amyloid-containing tissue biopsy specimens or subcutaneous fat aspirates. The ability to identify precisely the protein nature of the pathologic deposits has diagnostic, therapeutic, and prognostic implications for patients with amyloid-associated disease.

Introduction

To date, at least 23 different wild-type or mutated proteins have been identified as constituting the fibrils found in body tissues or organs from individuals with systemic (noncerebral) amyloidosis (Westermark *et al.*, 2005) (Table I). Despite the differing chemical nature of these molecules, all forms of amyloid have virtually identical tinctorial and ultrastructural features and, as a result, cannot be differentiated microscopically. Furthermore, although the various disorders associated with amyloid deposition

TABLE I
AMYLOIDOGENIC PRECURSOR PROTEINS ASSOCIATED WITH THE HUMAN SYSTEMIC
(NON-CEREBRAL) AMYLOIDOSES[a]

Amyloidogenic precursor	Amyloid designation	Familial
Immunoglobulin light chain	AL	
Immunoglobulin heavy chain	AH	
Transthyretin (mutated)	ATTR	+
Transthyretin (wild-type)	ATTR	
Serum amyloid A	AA	
Apolipoprotein AI (mutated)	AApoAI	+
Apolipoprotein AI (wild-type)	AApoAI	
Apolipoprotein AII (mutated)	AApoAII	+
Apolipoprotein AIV (wild-type)	AApoAIV	
Beta 2-microglobulin	$A\beta_2M$	
Lysozyme (mutated)	ALys	+
Fibrinogen (mutated)	AFib	+
Gelsolin (mutated)	AGel	+
Calcitonin	ACal	
Islet amyloid polypeptide	AIAPP	
Atrial natriuretic factor	AANF	
Prolactin	APro	
Insulin	AIns	
Lactadherin	AMed	
Kerato-epithelin	AKer	
Lactoferrin	ALac	
FLJ20513	APin	
Semenogelin I	ASgI	

[a] Modified from Westermark *et al.*, 2005.

(acquired or inherited) may be distinguished from one another on the basis of particular medical criteria, there can be considerable overlap in disease manifestations (Benson, 2001) and other confounding factors (Lachmann *et al.*, 2002), thus making a precise diagnosis impossible. Because the cause, treatment, and prognosis of each of these illnesses differ, it is essential that the exact nature of the fibrils be established.

Typically, the diagnosis of amyloidosis is made by pathologists and is based on the presence in tissue or subcutaneous fat aspirates of green birefringent congophilic deposits, as evidenced by polarizing microscopy. In addition, immunological procedures (e.g., immunohistochemistry) that use antisera specific for the most common types of amyloid proteins (Arbustini *et al*, 2002; Gallo *et al.*, 1986; Kaplan *et al.*, 1997, 1999a; Linke, 1985; Olsen *et al.*, 1999; Röcken *et al.*, 1996) have been used to determine the protein composition of such material. Unfortunately, these methods have several technical limitations, including the lack of appropriate antisera with the requisite specificity. For example, commercially available reagents are prepared against the native form of the amyloidogenic precursor molecule and, thus, may fail to react with the amyloid because of either its fragmentary nature or the conformational changes that result from fibrillogenesis. Furthermore, the presence of normal tissue components may lead to false-positive reactions. Therefore, to establish definitively the nature of the amyloid protein, it must be extracted and analyzed chemically (Solomon and Westermark, 2002). Because such studies require relatively large amounts of fresh tissue that, most often, are not readily available, micromethods have been developed to isolate, purify, and sequence amyloid protein extracted from specimens (Kaplan *et al.*, 1999b, 2001, 2004), even after formalin fixation (Ikeda *et al.*, 1998; Layfield *et al.*, 1996, 1997). These approaches, however, also are restricted by sample size, dependence on antisera with appropriate reactivity, and the extracted protein's susceptibility to Edman degradation.

To circumvent these difficulties, we have used a more sensitive technique—tandem mass spectrometry (MS/MS)—to identify the nature of the amyloid found in needle biopsy specimens, as well as subcutaneous fat aspirates (Murphy *et al.*, 2001) and here describe the procedures used in our laboratory for these analyses.

Sample Preparation

Tissue Specimens

A 4-μm-thick section cut from formalin-fixed, paraffin-embedded tissue is stained with a freshly prepared solution of alkaline Congo red (Westermark *et al.*, 1999) and examined under polarizing microscopy at a magnification of

×100. A qualitative assessment of amyloid deposition is made on the basis of the relative extent of green birefringence seen in the entire specimen. Depending on the volume of tissue available and amount of congophilia present, approximately 6–30 sections are placed on poly-L-lysine–coated glass microscopic slides. The sections are first deparaffinized by immersion in AmeriClear (Baxter, Deerfield, IL) for 2–4 h and then rehydrated by exposure for 5-min periods to the following graded series of solvents: 100% ethanol, 95% ethanol/water, 80% ethanol/water and, finally, distilled water. Excess liquid is drained onto a paper towel and the slides stored in a suitable container until completely dry. The sections are loosened with a No. 11 scalpel blade, scraped from the slide into a 1.8-ml capacity Eppendorf tube to which is added 200 μl of a 0.25 M Tris (hydroxymethyl) aminomethane-hydrochloride, pH 8.0, 1.0 M disodium-EDTA solution, and 800 μl of 8 M guanidine hydrochloride (Gd HCl) (Pierce, Rockford, IL). The tube is then incubated at 37° and sonicated (Tekmar Sonic Disrupter, Tekmar Control Systems, Inc., Spokane, WA) once or twice daily in an ice bath (15–20 sec pulse) until maximum clarity of the preparation occurs (usually between 8 and 10 days). The protein is reduced by addition of 10 μl of 2-mercaptoethanol (Pierce), after which the mixture is mixed briefly, kept at 37° for 2 h, and alkylated with iodoacetamide (30 mg), followed by a 30-min incubation at 37°.

Fat Aspirates

Subcutaneous fat underlying the periumbilical area is aspirated under local anesthesia with an 18-gauge needle affixed to a 10-ml syringe (Westermark *et al.*, 1989). The aspirate is smeared between two glass microscopic slides and the presence of amyloid documented by Congo red staining. The coverslips are removed by placing the slides for 2–4 h in a Coplin jar containing xylene (occasionally, this may require as long as 24–48 h). The material is rehydrated in a series of 5-min washes with 100%, 100%, 100%, 95%, and finally, 70% ethanol in water and air-dried. The aspirates are then scraped into a 1.8-ml Eppendorf tube, treated with 1 ml of the Tris-EDTA-Gd HCl solution, and reduced and alkylated as detailed previously.

Protein/Peptide Preparation

The protein components in the extracted specimen are separated by reverse-phase high-performance liquid chromatography (HPLC) on a new (to avoid possible contamination) Aquapore 300 C_8 (30 × 4.0 mm) column (Brownlee Columns, Perkin Elmer, Norwalk, CT) that has been washed over a 45-min period with a linear 0.1% trifluoroacetic acid (TFA) water–7% acetonitrile to a 0.1% TFA water–70% acetonitrile gradient at a flow rate of 1 ml/min using a Perkin-Elmer model 200 solvent delivery system (Applied Biosystems, Foster City, CA). The absorbance is read at 220 nm with a Perkin-Elmer model 785 UV/Vis programmable absorbance detector. This step is repeated using the Tris-EDTA-Gd HCl solution until the baseline is flat. To ensure reproducibility, the column is first injected with a protein standard, for example, horse myoglobin (Sigma, St. Louis, MO), that elutes between 50 and 60% acetonitrile.

For sample analysis, the reduced-alkylated protein is centrifuged at 16,000g for 10 min to obtain a particulate-free solution, and both the supernatant and pellet are saved. After filtration through a 0.22-μm centrifugal filtration device (Utrafree-MC, Millipore, Bedford, MA), the supernatant (or pellet) dissolved in 99% formic acid is subjected to reverse-phase HPLC using the conditions described previously. The fractions corresponding to UV-absorbing peaks are collected manually, and each is reduced in volume to ~60 μl in a Speed Vac sample concentrator (Savant Instruments, Farmingdale, NY). For those having an absorbance >0.02, 25% of the material is put aside for Edman sequence determination. Because amyloid-related proteins generally elute between acetonitrile concentrations of 35 and 45%, all material from this region is saved and partially dried. The samples are then transferred to a 50-μl autosampler (AS) vial insert (Thermofinnigan, Thousand Oaks, CA) and the drying continued to completion.

For tryptic digestion, the fraction in the AS vial insert is dissolved in 20 μl of a 0.02 μg/ml solution of trypsin (Promega, Madison, WI) in 50 mM sodium bicarbonate buffer, pH 8.0. The enzyme is forced to the bottom of the vial by quick centrifugation, after which the insert is loaded into a 1.8-ml AS vial kit (Thermofinnigan), incubated at 37° for 4 h, and, for subsequent mass spectrometric analyses, placed into the Surveyor Autosampler that is maintained at 4°.

Tandem Mass Spectrometry (MS/MS)

Our laboratory currently uses an ion-trap tandem mass spectrometer (MS/MS) (Thermofinnigan LCQ DECA XP) equipped with a nanospray ion source (NSI), together with a Surveyor HPLC and AS. The instrument is calibrated at least monthly with a mixture containing caffeine (Sigma), the tetramer peptide Met-Arg-Phe-Ala, and Ultramark 1621 (Heraus, Karlsruhe, Germany) using the automatic calibration feature of the LC-Tune program according to the specifications of the manufacturer. To tune the instrument and for analyses of tryptic peptides, the spectrometer is infused with 5 pmol/μl of human [Glu[1]]-fibrinopeptide B (Sigma) dissolved in 50% acetonitrile/0.1% formic acid at a flow rate of 2 μl/min, a capillary temperature of 135°, and a m/z setting of 786 amu, as detailed by the auto-tune program. To ensure that all elements of the instrument are properly functioning, a 5-μl volume that contains 500 fmol/μl of trypsin-digested bovine cytochrome C (Dionex, Sunnyvale, CA) is injected by the AS (using its "no waste" feature) into the LC stream. Samples are chromatographed first through a 1.0 × 0.3 mm C_{18}, 5 μm, 100 Å, Micro Guard column (Dionex) and subsequently, a 150 × 0.3 mm C_{18}, 5 μm, 300 Å Vydac column (Dionex) with a pump speed of 100 μl/min and a pressure of 300 psi. The stream is split ~50:1, so that the flow rate is maintained at 2 μl/min (as verified by collecting the output from the emitter in a 10-μl capillary tube for 2 min). After injection, the composition of the eluent is held at 10% acetonitrile/0.1% formic acid for 10 min, and then the concentration of acetonitrile is increased to 90% over 30 min and held for an additional 10 min. During this 50-min period, data are collected (~3000 scans) from the mass spectrometer in three successive segments: MS, MS zoom, and MS/MS. The most robust mass from the MS scan (that is not on either a dynamic- or operator-selected exclusion list) is chosen for the zoom scan to determine the charge state and then subjected to MS/MS for fragmentation analyses. The cytochrome C trypsin-derived peptide standards are run with each test group and identified using the BioWorks (Thermofinnigan) data analysis software. If at least four peptides have Xcorr values equal to or greater than 2.5, the instrument is deemed ready for sample injection.

The mass and fragmentation patterns of specimen-derived tryptic peptides are compared with those obtained hypothetically through "*in silico*" trypsin digestion of human proteins contained in the nonredundant NCBI.NLM.NIH.GOV database. The magnitude of the Xcorr values reflects the extent of similarity between unknown and theoretical tryptic peptides.

MS/MS Analyses of Amyloid Deposits

To illustrate that MS/MS technology can be used to identify the chemical nature of amyloid extracted from tissue biopsies or fat aspirates, analytic data are provided on specimens obtained from patients with different forms of systemic amyloidosis; namely, ALλ, ALκ, AA, ALys, AApo A-I, and ATTR that resulted from pathologic deposition of λ or κ light chain (LC)-, serum amyloid A protein (SAA)-, lysozyme (Lys)-, apolipoprotein A-I (apo A-I)-, and transthyretin (TTR)-related molecules, respectively (Table I).

In the first case, the reverse-phase-HPLC profile of the reduced-alkylated protein extracted from a formalin-fixed, paraffin-embedded tissue biopsy from a patient (Mcg) who had cutaneous amyloid deposits yielded three major peaks that eluted at acetonitrile concentrations of 28, 34, and 42% (Fig. 1). Each was digested with trypsin and analyzed by MS/ MS where, as indicated in Fig. 2, seven peptides were found to be identical in sequence to portions of the variable (V), joining (J), and constant (C) segments of the amyloidogenic precursor $\lambda 2$ LC (Bence Jones protein Mcg), for which the primary structure had been determined (Fett and Deutsch, 1974). As is evident, the peak 3–derived tryptic peptides contained more of the Mcg sequence than did those from earlier eluting peaks 1 and 2. The scarcity of V-region data reflected the limited number of trypsin-sensitive lysyl and arginyl residues, as well as peptide length. For example, two peptides (1–44 and 69–106) were too large to be identified by our M/MS instrument, and one (45–47) was too small to be retained by the HPLC column. In addition, the presence of two prolyl residues in the putative tryptic peptide 56–63 may have prevented its fragmentation.

The results of MS, MS zoom, and MS/MS scans of the V-region peptide VIIYEVNK (mass, 977.5) are provided in Fig. 3. The HPLC elution profile of this peptide (Fig. 3A) plotted in the range of 977–979 amu and the MS scan from 300–2000 m/z revealed this molecule to have two charged states (977.5 and 489.2) (Fig. 3B). The MS zoom scans centered around 978 amu showed m/zs of 977.5, 978.5, 979.4, and 980.4 (Fig. 3C), whereas those around 490 amu were 489.2, 489.6, and 490.1 (Fig. 3D) (in each case, the range of m/z peaks reflected different contents of naturally occurring isotopes of C, H, N, and O). The ~1-Da and 0.5-Da increments in the former

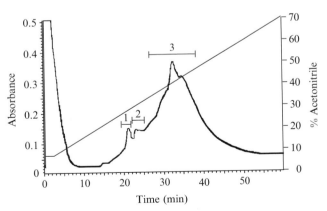

Fig. 1. Purification of ALλ amyloid by reverse-phase HPLC. Profile of reduced, alkylated protein extracted by 6 *M* guanidine HCl from sections of a formalin-fixed, paraffin-embedded cutaneous nodule biopsy from patient Mcg. The eluates were pooled as indicated in the figure (Peaks 1, 2, and 3).

HPLC Peak	Peptide sequence	MH[a]	Charge	Xcorr[b]	DelCn[c]	Ions[d]
#1 194–208	SYSCQVTHEGSTVEK	1712.77	2	4.49	0.51	21/28
▼ 107–114	VTVLGQPK	842.02	2	2.68	0.29	12/14
#2 107–114	VTVLGQPK	842.02	2	2.65	0.40	11/14
▼ 48–55	VIIYEVNK	978.17	2	2.65	0.18	12/14
#3 48–55	VIIYEVNK	978.17	2	2.46	0.31	12/14
107–114	VTVLGQPK	842.02	2	2.33	0.32	11/14
115–133	ANPTVTLFPPSSEELQANK	2044.25	2	4.13	0.55	21/36
134–153	ATLVCLISDFYPGAVTVAWK	2212.55	2	5.53	0.64	27/38
154–160	ADGSPVK	673.74	2	1.31	0.13	7/12
▼ 161–170	AGVETTKPSK	1018.15	2	1.75	0.20	12/18

[a] Mass plus 1 proton.
[b] X corr: value computed from cross correlation of experimental MS/MS data *vs.* the candidate peptide in the database (significant if >2).
[c] DelCn: delta correlation score between the top two candidate peptide matches (significant if >0.2).
[d] Number of peptide fragment ions matched/total number of expected fragment ions.

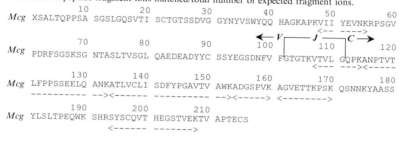

FIG. 2. Identification of ALλ amyloid by MS/MS. (Upper) Data generated by mass spectrometric analyses of tryptic peptides derived from Peaks 1, 2, and 3 (see Fig. 1). (Lower) The amino acid sequences of Peak 3–related tryptic peptides are compared with that of the amyloidogenic precursor λ LC Mcg. The variable (V), joining (J), and constant (C) region segments are as indicated.

and latter, respectively, indicated the presence of singly and doubly charged peptides. In the MS/MS scans of the two charged peptides shown in Fig. 3E and F, the m/zs ± 3 amu of each were isolated in the ion trap, and their kinetic energy increased to induce fragmentation. The resultant singly charged daughter ions, including the N-terminal valine (*b ions*) and the fragments containing the C-terminal lysine (*y ions*), are shown in Fig. 3G.

Representative data generated from MS/MS analyses of a κ LC-containing extract are illustrated in Fig. 4 and that from four additional types of amyloid-associated protein in Figs. 5–8. In the first case, the ALκ nature of the deposits was evidenced by the presence of tryptic peptides that encompassed V_κ and, predominately, C_κ region residues (Fig. 4). In two others, amyloid isolated from fine-needle kidney biopsies revealed, in one case, SAA- (Fig. 5) and in the second, lysozyme- (Fig. 6) related proteins, confirming the diagnosis of AA and ALys, respectively (in the latter, DNA

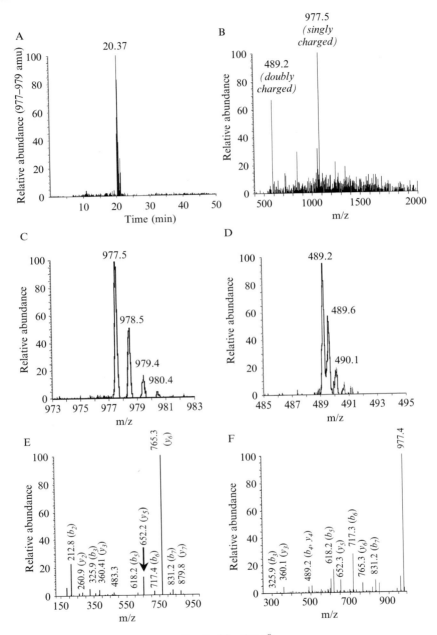

FIG. 3. (*Continued*)

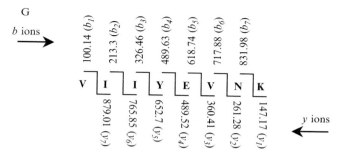

FIG. 3. Analyses of an eight-residue V_λ-related tryptic peptide encompassing amino acids from positions 48–55. (A) HPLC-MS electrospray ion trap mass spectrometer m/z plot ranging from 977–979 amu; (B) MS scan from 300–2000 amu of the peak eluting at 20.37 min; (C) and (D) MS zoom scans from 973–983 amu of the singly charged (977.5) and 485–495 amu of the doubly charged (489.2) peptide, respectively; (E) and (F) MS/MS fragmentation scans from 150–1000 amu of the singly (977.5 amu) and doubly (489.2) charged peptides, respectively; the observed b and y ions are as indicated. (G) Theoretical masses of the b and y ions calculated on the basis of the known peptide sequence.

Peptide	Peptide sequence	MH	Charge	Xcorr	DelCn	Ions
1–18	DIQMTQSPSTLSASVGDR	1894.05	2	3.22	0.32	15/34
25–42	ASQGIGNDLGWYQQKPGK	1948.13	2	3.04	0.38	13/34
127–142	SGTASVVCLLNNFYPR	1798.99	2	3.74	0.42	18/30
170–183	DSTYSLSSTLTLSK	1503.63	2	2.47	0.25	15/26
191–207	VYACEVTHQGLSSPVTK	1877.06	2	3.84	0.35	19/32

```
                 10          20          30          40          50          60
V_κl  DIQMTQSPSS LSASVGDRVT ITCRASQGIG NDLGWYQQKP GKAPKRLIYA ASSFQSGVPS
Sto   <--------- ------->               <----- ---------- ->

                                                    ← V ─── J ───┐ C →
                 70          80          90         100        110         120
V_κl  RFSGSGSGTE FTLTISGLQP EDFATYYCLQ HNSYPLTFGG GTRVEIKRTV AAPSVFIFPP

                130         140         150         160         170         180
V_κl  SDEQLKSGTA SVVCLLNNFY PREAKVQWKV DNALQSGNSQ ESVTEQDSKD STYSLSSTLT
Sto   <--- ---------- ->                                     < ----------

                190         200         210
V_κl  LSKADYEKHK VYACEVTHQG LSSPVTKSFN RGEC
Sto   -->                   <--------- ------>
```

FIG. 4. ALκ amyloidosis. Identification by MS/MS of a κ light chain (LC)–related molecule encompassing variable (V) and constant (C) region residues from protein extracted from a lymph node biopsy (patient *Sto*). The amino acid sequence of the five tryptic peptides was compared with that of a κ1 LC (gi:441 395) V_κ domain and that specified by the Cκ gene (Km3).

Peptide	Peptide sequence	MH	Charge	Xcorr	DelCn	Ions
2–15	SFFSFLGEAFDGAR	1551.68	2	3.27	0.53	19/26

```
                10            20            30            40            50            60
SAA RSFFSFLGEA FDGARDMWRA YSDMREANYI GSDKYFHARG NYDAAKRGPG GVWAAEAISD
Saw <-------- ---->
                70            80            90           100
SAA ARENIQRFFG HGAEDSLADQ AANEWGRSGK DPNHFRPAGL PEKY
```

Fig. 5. AA amyloidosis. Identification by MS/MS of serum amyloid A (SAA)–related protein extracted from a fine-needle kidney biopsy (patient *Saw*). Comparison of the amino acid sequence of SAA with that of a trypsin-derived 14-residue N-terminal peptide.

Peptide	Peptide sequence	MH	Charge	Xcorr	DelCn	Ions
15–21	LGMDGYR	811.93	1	1.54	0.13	7/12
51–62	STDYGIFQINSR	1401.51	2	1.84	0.21	12/22
108–113	AWVAWR	788.92	1	1.53	0.22	8/10
123–130	QYVQGCGV	910.97	1	2.19	0.34	11/14

```
                    10            20            30            40            50            60
Lysozyme KVFERCELAR TLKRLGMDGY RGISLANWMC LAKWESGYNT RATNYNAGDR STDYGIFQIN
Moo                 <----- >                                   <---------
                    70            80            90           100           110           120
Lysozyme SRYWCNDGKT PGAVNACHLS CSALLQDNIA DAVACAKRVV RDPQGIRAWV AWRNRCQNRD
Moo ->                                                          <-- -->
                   130
Lysozyme VRQYVQGCGV
Moo      <------>
```

Fig. 6. ALys amyloidosis. Identification by MS/MS of a lysozyme (Lys)-related protein extracted from a fine-needle kidney biopsy (patient *Moo*). Comparison of the amino acid sequence of lysozyme with four trypsin-derived peptides.

analyses showed a Trp→Arg substitution at position 64; however, the tryptic peptide encompassing this alteration was not found). Another patient with a familial history of amyloidosis was diagnosed with AApoA-I when an N-terminal fragment of apo A-I with a novel point mutation was detected in an extract prepared from liver biopsy sections (Coriu *et al.*, 2003) (Fig. 7). Although the substitution was not in the nonredundant NCBI.NLM.NIH.GOV database, the predicted mass and fragmentation pattern of tryptic peptides containing the mutation were evident by MS and MS/MS. As for material extracted from a subcutaneous fat aspirate from a patient with a polyneuropathy, eight TTR-related peptides were identified, including two spanning residues 49–70, one of which contained the T60A "Appalachian" mutation (Benson, 2001) (Fig. 8).

Peptide	Peptide sequence	MH	Charge	Xcorr	DelCn	Ions
11–23	VKDLATVYVDVLK	1463.74	2	3.76	0.94	21/24
24–40	DSGRDYVSQFEGSALGK	1816.91	3	1.34	0.19	14/32
28–40	DYVSQFEGSALGK	1401.5	2	4.04	0.96	20/24
46–59	LLDNWDSVTSTFSK	1613.75	2	3.98	1.00	21/26
62–77 (L64P)	EQPGPVTQEFWDNLEK	1918.07	2	1.60		15/30

```
              10          20          30          40          50          60
ApoAI DEPPQSPWDR VKDLATVYVD VLKDSGRDYV SQFEGSALGK QLNLKLLDNW DSVTSTFSKL
Riz              <--------- -->><------ --------->          <---- -------->
                                <-- --------->

              70          80          90          100         110         120
ApoAI REQLGPVTQE FWDNLEKETE GLRQEMSKDL EEVKAKVQPY LDDFQKKWQE EMELYRQKVE
Riz   <-P------ ------>

              130         140         150         160         170         180
ApoAI PLRAELQEGA RQKLHELQEK LSPLGEEMRD RARAHVDALR THLAPYSDEL RQRLAARLEA

              190         200         210         220         230         240
ApoAI LKENGGARLA EYHAKATEHL STLSEKAKPA LEDLRQGLLP VLESFKVSFL SALEEYTKKL

ApoAI NTQ
```

FIG. 7. AApoA-I amyloidosis. Identification by MS/MS of a mutated apolipoprotein A-I (apoA-I)–related protein in a liver biopsy–derived extract (patient *Riz*). Comparison of the amino acid sequence of wild-type ApoA-I with those of five tryptic-derived peptides, one of which encompassed residues 62–77 and contained a Leu→Pro substitution at position 64.

Peptide	Peptide sequence	MH	Charge	Xcorr	DelCn	Ions
10–15	CPLMVK	747.94	2	2.03	0.67	9/10
16–21	VLDAVR	672.80	1	1.54	0.14	7/10
22–34	GSPAINVAVHVFR	1367.58	2	4.34	0.97	19/24
36–48	AADDTWEPFASGK	1395.46	2	2.82	1.00	16/24
49–70	TSESGELHGLTTEEEFVEGIYK	2455.62	2	2.50	0.28	17/42
49–70 (T60A)	TSESGELHGLTAEEEFVEGIYK	2426.58	2	2.99	0.53	19/42
81–103	ALGISPFHEHAEVVFTANDSGPR	2452.67	2	2.40	0.15	18/44
104–126	RYTIAALLSPYSYSTTAVVTNPK	2517.86	2	4.86	0.61	22/44

```
              10          20          30          40          50          60
TTR GPTGTGESKC PLMVKVLDAV RGSPAINVAV HVFRKAADDT WEPFASGKTS ESGELHGLTT
Zum     < ----><---- ><-------- ---> <---- ------->><- ----------
                                                        <- --------A

              70          80          90          100         110         120
TTR EEEFVEGIYK VEIDTKSYWK ALGISPFHEH AEVVFTANDS GPRRYTIAAL LSPYSYSTTA
Zum --------->          <-------- ---------- --><------ ----------
    --------->

TTR VVTNPK
Zum ----->
```

FIG. 8. ATTR amyloidosis. Identification by MS/MS of a mutated transthyretin (TTR)–related protein extracted from a subcutaneous fat aspirate (patient *Zum*). Comparison of the amino acid sequence of wild-type TTR with those of eight trypsin-derived peptides (two of which encompassed residues 49–70 and one that contained the Thr→Ala substitution at position 60).

Comments

The ability to identify by MS/MS the chemical nature of the fibrillar protein present in amyloid extracts is dependent on quantitative, as well as qualitative, factors. First and foremost is the amount of tissue available for extraction, which may be quite limited in fine-needle biopsy specimens, which generally are only ~1 × 15 mm in size (on the basis of present technology, ~1–10 pmol of extracted protein is required for successful analysis). Given that most of such specimens are used for routine microscopic and immunohistochemical analyses (and additionally, in the case of the kidney, embedded for electron microscopy), it may be necessary to obtain another sample, which we request be furnished to us in a fresh or frozen state (i.e., is not placed in a tissue fixative) that is then homogenized and extracted with the Tris-EDTA-Gd HCl solution. We also have discovered that if HPLC of the water supernatant yields scant or nondiagnostic material, the requisite protein may be found in the formic acid–dissolved water pellet. With regard to MS/MS analyses, if the instrumentation lacks *de novo* sequencing capability, it is necessary that the database include pertinent algorithms and mass data; however, these may not be available for novel amyloidogenic proteins (Solomon *et al.*, 2003). In such instances, amino acid sequencing by Edman degradation of enzymatically or chemically derived peptides is required. Furthermore, it must be recognized that extracts, especially those derived from fresh (unfixed) tissue, often contain nonfibrillar intracellular or extracellular proteins that are codeposited with the amyloid but do not represent amyloidogenic precursor molecules (Table II).

TABLE II
NON-AMYLOIDOGENIC PROTEINS FOUND BY MS/MS IN AMYLOID EXTRACTS

Actin
Apolipoprotein E
Clusterin
Collagen
Defensin
Heat shock protein
Hemoglobin
Heparin sulfate proteoglycan
Histones
Keratin
P-component
Vitronectin

Conclusions

The development of micromethods to extract and analyze by MS/MS the amyloid present in tissue biopsy specimens or fat aspirates can provide a clinically important diagnostic tool. Furthermore, through use of even more sensitive proteomic technology (i.e., capillary nano-LC and highly sensitive and high-resolution mass spectrometers [LTQ-FT]), it should be possible to obtain relevant data on subpicomole quantities of material. Although immunochemistry has been used to identify the type of amyloid present in pathologic deposits, given the false-positive or negative results found with this method, we recommend that it be used only to confirm the nature of the amyloid established chemically.

Acknowledgments

This work was supported in part by USPHS Research Grant 10056 from the National Cancer Institute and the Aslan Foundation. A. S. is an American Cancer Society Clinical Research Professor.

References

Arbustini, E., Verga, L., Concardi, M., Palladini, G., Obici, L., and Merlini, G. (2002). Electron and immuno-electron microscopy of abdominal fat identifies and characterizes amyloid fibrils in suspected cardiac amyloidosis. *Amyloid.* **9**, 108–114.

Benson, M. D. (2001). Amyloidosis. *In* "The Metabolic and Molecular Bases of Inherited Disease" (C. R. Scriver, A. L. Beaudet, W. S. Sly, and D. Valle, eds.), 8th ed., pp. 5345–5378. McGraw-Hill, New York.

Coriu, D., Dispenzieri, A., Stevens, F. J., Murphy, C. L., Wang, S., Weiss, D. T., and Solomon, A. (2003). Hepatic amyloidosis resulting from deposition of the apolipoprotein A-I variant Leu75Pro. *Amyloid.* **10**, 215–223.

Fett, J. W., and Deutsch, H. F. (1974). Primary structure of the Mcg lambda chain. *Biochemistry.* **13**, 4102–4114.

Gallo, G. R., Feiner, H. D., Chuba, J. V., Beneck, D., Marion, P., and Cohen, D. H. (1986). Characterization of tissue amyloid by immunofluorescence microscopy. *Clin. Immunol. Immunopathol.* **39**, 479–490.

Ikeda, K., Monden, T., Kanoh, T., Tsujie, M., Izawa, H., Haba, A., Ohnishi, T., Sekimoto, M., Tomita, N., Shiozaki, H., and Monden, M. (1998). Extraction and analysis of diagnostically useful proteins from formalin-fixed, paraffin-embedded tissue sections. *J. Histochem. Cytochem.* **46**, 397–403.

Kaplan, B., Yakar, S., Kumar, A., Pras, M., and Gallo, G. (1997). Immunochemical characterization of amyloid in diagnostic biopsy tissues. *Amyloid.* **4**, 80–86.

Kaplan, B., Vidal, R., Kumar, A., Ghiso, J., and Gallo, G. (1999a). Immunochemical microanalysis of amyloid proteins in fine-needle aspirates of abdominal fat. *Am. J. Clin. Pathol.* **112**, 403–407.

Kaplan, B., Hrncic, R., Murphy, C. L., Gallo, G., Weiss, D. T., and Solomon, A. (1999b). Microextraction and purification techniques applicable to chemical characterization of amyloid proteins in minute amounts of tissue. *Methods Enzymol.* **309,** 67–81.

Kaplan, B., Murphy, C. L., Ratner, V., Pras, M., Weiss, D. T., and Solomon, A. (2001). Micromethod to isolate and purify amyloid proteins for chemical characterization. *Amyloid.* **8,** 22–29.

Kaplan, B., Martin, B. M., Livneh, A., Pras, M., and Gallo, G. R. (2004). Biochemical subtyping of amyloid in formalin-fixed tissue samples confirms and supplements immunohistologic data. *Am. J. Clin. Pathol.* **121,** 794–800.

Lachmann, H. J., Booth, D. R., Booth, S. E., Bybee, A., Gilbertson, J. A., Gillmore, J. D., Pepys, M. B., and Hawkins, P. N. (2002). Misdiagnosis of hereditary amyloidosis as AL (primary) amyloidosis. *N. Engl. J. Med.* **346,** 1786–1791.

Layfield, R., Bailey, K., Lowe, J., Allibone, R., Mayer, R. J., and Landon, M. (1996). Extraction and protein sequencing of immunoglobulin light chain from formalin-fixed cerebrovascular amyloid deposits. *J. Pathol.* **180,** 455–459.

Layfield, R., Bailey, K., Dineen, R., Mehrotra, P., Lowe, J., Allibone, R., Mayer, R. J., and Landon, M. (1997). Application of formalin fixation to the purification of amyloid proteins. *Anal. Biochem.* **253,** 142–144.

Linke, R. P. (1985). Immunochemical typing of amyloid deposits after microextraction from biopsies. *Appl. Pathol.* **3,** 18–28.

Murphy, C. L., Eulitz, M., Hrncic, R., Sletten, K., Westermark, P., Williams, T., Macy, S. D., Wooliver, C., Wall, J., Weiss, D. T., and Solomon, A. (2001). Chemical typing of amyloid protein contained in formalin-fixed paraffin-embedded biopsy specimens. *Am. J. Clin. Pathol.* **116,** 135–142.

Olsen, K. E., Sletten, K., and Westermark, P. (1999). The use of subcutaneous fat tissue for amyloid typing by enzyme linked immunosorbent assay. *Am. J. Clin. Pathol.* **111,** 355–362.

Röcken, C., Schwotzer, E. B., Linke, R. P., and Saeger, W. (1996). The classification of amyloid deposits in clinicopathological practice. *Histopathology.* **29,** 325–335.

Solomon, A., and Westermark, P. (2002). Hereditary amyloidosis. *N. Engl. J. Med.* **347,** 1206–1207.

Solomon, A., Murphy, C. L., Weaver, K., Weiss, D. T., Hrncic, R., Eulitz, M., Donnell, R. L., Sletten, K., Westermark, G., and Westermark, P. (2003). Calcifying epithelial odontogenic (Pindborg) tumor-associated amyloid consists of a novel human protein. *J. Lab. Clin. Med.* **142,** 348–355.

Westermark, P., Benson, L., Juul, J., and Sletten, K. (1989). Use of subcutaneous abdominal fat biopsy specimen for detailed typing of amyloid fibril protein-AL by amino acid sequence analysis. *J. Clin. Pathol.* **42,** 817–819.

Westermark, G. T., Johnson, K. H., and Westermark, P. (1999). Staining methods for identification of amyloid in tissue. *Methods Enzymol.* **309,** 3–25.

Westermark, P., Benson, M. D., Buxbaum, J. N., Cohen, A. S., Frangione, B., Ikeda, S., Masters, C. L., Merlini, G., Saraiva, M. J., and Sipe, J. D. (2005). Amyloid: Toward terminology clarification. *Amyloid.* **12,** 1–4.

[5] Proteomics of Polyglutamine Aggregates

By Kenichi Mitsui, Hiroshi Doi, and Nobuyuki Nukina

Abstract

In nine members of polyglutamine (polyQ) diseases, CAG repeat expansions of their responsible genes are observed. The disease is considered to be caused by the formation of polyQ aggregates that sequester proteins essential for cell viability. To understand the pathological process of polyQ diseases, a proteomic approach was used to identify aggregate interacting proteins (AIPs). Constructs were designed to express EGFP-fused, CAG-expanded (150Q) huntingtin exon1 under the control of an ecdysone-inducible promoter and either lacking or containing a nuclear localization signal (NLS). After induction of a stably transfected Neuro 2A cell line with ecdysone, aggregates form in either the cytoplasm or the nucleus. The aggregates in these two different compartments were isolated with different methods. Cytoplasmic aggregate particles were purified using a fluorescence-activated cell sorter (FACS) by monitoring EGFP fluorescence, whereas nuclear aggregates were purified by using the detergent insoluble nature of aggregates. The resulting highly pure aggregates were subjected to SDS-PAGE followed by Coomassie blue staining. Bands containing AIP candidates were excised, and, after in-gel digestion with trypsin, were analyzed by mass spectrometry to identify the proteins. Novel candidates were confirmed as AIPs by immunocytological analysis to observe colocalization with polyQ aggregates.

This chapter describes methods for the establishment of stable mutant cells, the purification of polyQ aggregates, and sample preparation for mass spectrometry analysis in detail.

Proteomics of Polyglutamine Aggregates

Intracellular inclusions are a common pathological feature in the group of neurodegenerative disorders known as CAG repeat diseases or polyglutamine (polyQ) diseases. This group contains at least nine members, including spinocerebellar ataxia (SCA) types-1, 2, 3, 6, 7, and 17; dentatorubropallindo-luysian atrophy (DRPLA); spinal and bulbar muscular atrophy (SBMA); and Huntington (HD) disease. It is hypothesized that the expanded polyQ tract causes a toxic gain of function through abnormal protein–protein interactions. Therefore, the identification of aggregate-interacting proteins (AIPs) will

METHODS IN ENZYMOLOGY, VOL. 412
0076-6879/06 $35.00
DOI: 10.1016/S0076-6879(06)12005-4

allow the process of aggregate formation to be investigated and will help to explain the role of AIPs in HD pathogenesis.

To study aggregate formation in cells, a mutant cell line was established that forms enhanced green fluorescence protein (EGFP)–fused aggregate proteins. This system allows aggregates to be easily visualized as green fluorescent particles by fluorescence microscopy. Furthermore, the process of aggregate formation in living cells can be analyzed. We and others have identified AIPs from cell culture systems expressing a huntingtin (htt)-polyQ-EGFP fusion protein (Mitsui *et al.*, 2002; Suhr *et al.*, 2001). The strategy for analyzing AIPs adopted in both cases followed three major steps. First, htt-polyQ-EGFP–expressing cell lines were made. Second, the fluorescent aggregate particles formed by htt-polyQ-EGFP proteins in the cells were purified. Third, purified aggregates were subjected to SDS-PAGE, followed by mass spectrometric sequence analysis.

Suhr *et al.* (2001) made a construct for transient expression of htt-96Q-EGFP and transfected HEK293 cells. They used CsCl gradient procedures to isolate aggregates formed in cells and analyzed the AIPs by SDS-PAGE followed by MALDI mass spectrometry analysis.

We made constructs to express truncated N-terminal htt (tNhtt) exon1 containing a 150-repeat polyQ chain and fused with EGFP either with or without a nuclear localization signal (NLS). We named them as tNhtt-150Q-EGFP and tNhtt-150Q-EGFP-NLS, respectively. Then we made the stably transfected Neuro 2A cell line that expresses tNhtt-150Q-EGFP or tNhtt-150Q-EGFP-NLS after induction by ecdysone. We purified the fluorescent aggregate particles formed by tNhtt-150Q-EGFP in cytosol using a fluorescence activated cell sorter, and from tNhtt-150Q-EGFP-NLS in the nucleus by sequential removal of nuclear soluble proteins with different detergents. Then we subjected the purified aggregates to SDS-PAGE and analyzed the prominent bands by the high-performance liquid chromatography (HPLC)–tandem mass spectrometry (MS/MS) method (Mitsui *et al.*, 2002).

Preparing Neuro 2A Cells Carrying Ecdysone-Inducible tNhtt-150Q-EGFP or tNhtt-150Q-EGFP-NLS Gene

PolyQ aggregates are formed in the brains of patients or model mice throughout their lifetime. Although transient expression can lead to the rapid formation of aggregates that might sequester molecules nonspecifically rather than through specific protein-protein interactions, stable expression of polyQ results in a comparatively slower formation of aggregates. Therefore, to monitor aggregate formation over time, we prepared the stably transfected Neuro 2A cell line that expresses tNhtt-150Q-EGFP or tNhtt-150Q-EGFP-NLS and form aggregates after induction with ecdysone, as described in the following (Doi et al., 2004; Wang et al., 1999).

Using phage genomic clones containing the entire open reading frame of exon1 from HD patients, we introduced a nucleotide fragment encoding MATLEKLMKAFESLKSFQQQQQQQQQQQQQQQQQQQQQPPPPPPPP PPPPPPQLPQPPPQAQPLLPQPQPPPPPPPPPPPGPAVAEEPLHRP into pBluescript vector. The fragment was excised from pBluescript by digestion with Hind III and Bam HI and subcloned into pEGFP-N1 vector to make EGFP fusion protein (tNhtt-polyQ-EGFP). A highly expanded 150-CAG repeat or a normal range of 16-CAG repeat in the polyQ region of htt exon1 was obtained by PCR, using the unstable nature of CAG repeats during replication. The fragments encoding tNhtt-150Q-EGFP and tNhtt-16Q-EGFP were then cut with Hind III and Xba I from these constructs and subcloned into pIND vector.

For preparing cells forming nuclear aggregates, three repeats of the nuclear localization signal, GATCCAAAAAAGAAGAGAAAGGTA, were inserted into the Not I site at the 3' end of EGFP gene in the tNhtt-16Q-EGFP and tNhtt-150Q-EGFP constructs to express tNhtt-16Q-EGFP-NLS and tNhtt-150Q-EGFP-NLS.

For convenience, all four of these constructs are collectively named pIND-HD exon1-EGFPs in the following paragraphs.

The ecdysone-inducible mammalian expression system (Invitrogen Japan, K.K., Tokyo, Japan) was used for tNhtt expression. The procedure was performed according to the manufacturer's manual. Neuro 2A cells were first stably transfected with pVgRXR (Zeocin resistance) using Lipofectamine (Invitrogen Japan). Cells were transiently transfected with pIND-EGFP and induced with ecdysone derivatives such as muristerone A or ponasterone A to select cells that were tightly regulated. These selected cells (Neuro 2A-pVgRXR) were then transfected with pIND-HD exon1-EGFPs (neomycin resistance) to generate double stable cell lines expressing the desired HD exon1-EGFP proteins. Individual clones were selected using 0.4 mg/ml Zeocin and 0.4 mg/ml G418 (Neomycin). Zeocin and G418 resistant clones were then screened for transgene expression induced with ecdysone by microscopic observation of EGFP fluorescence and immunoblot analysis using monoclonal anti-EGFP antibody (BD Biosciences Clontech, Mountain View, CA). Clones with the highest expression and tightest regulation were picked for further study. Neuro 2A-pVgRXR cells without pIND-HD exon1-EGFP transfection were established as a control.

The stable transfectant cells were cultured in Dulbecco's modified Eagle's medium (DMEM; Invitrogen Japan) with 10% fetal bovine serum (FBS), 0.4 mg/ml Zeocin, and 0.4 mg/ml G418 at 37° with 5% CO_2.

To differentiate the cells, 5 mM $N^6,2'$-O-dibutyryl cyclic AMP (dbcAMP; Nacalai Tesque, Inc., Kyoto, Japan) was added, and the expression of HD exon1-EGFPs was induced with 1 μM muristerone A or ponasterone A.

Purification of PolyQ Aggregates

EGFP-Fused Cytoplasmic PolyQ Aggregates. EGFP-fused cytoplasmic polyQ aggregates were purified as described by Mitsui *et al.* (2002). tNhtt-150Q-EGFP–expressing Neuro 2A cells (5 × 10^7) were plated in 100 mm tissue culture dishes (5 × 10^6 cells/dish) and cultured overnight. After 2 days of treatment with 5 mM dbcAMP for differentiation and 1 μM muristerone A or ponasterone A for induction, 70% of cells contained bright green fluorescent aggregate particles in the cytosol. Cells were collected in an ice-cooled homogenization glass pot in 3 ml of PBS containing 10 mM MgCl$_2$, 1500 U of DNase I (Nacalai Tesque, Inc.), and 3 U of RNase A (Nacalai Tesque, Inc.). These nucleases were essential for removing contaminating nucleic acids entangled around the aggregates. Homogenization was performed with a Potter-Elvehjem–type homogenizer with 30 up-and-down strokes of the glass pot under the spinning Teflon pestle (at 3000 rpm) jointed to a Digital Homogenizer (As One Corp, Osaka, Japan). To avoid loss of AIPs by severe physical and chemical treatment of the aggregates, we chose to use a glass homogenizer for milder homogenization of the cells rather than sonication. The temperature of the glass pot was kept below 4° by occasional dipping in an ice bath. The homogenate was applied directly to a fluorescence-activated cell sorter containing an argon laser (Epics Elite ESP, Beckman Coulter, Fullerton, CA) with an outlet nozzle of 100 μm in diameter. The flow rate was adjusted to ~500 events/min, and EGFP fluorescence was monitored for sorting. The sorted aggregates were collected in 16 × 100-mm culture tubes (code 9834–1610, Iwaki Glass, Chiba, Japan) and spun down by centrifugation at 1500g for 30 min. For aggregate collection, glass tubes with an electrically uncharged surface were used to avoid loss of aggregate particles by attaching to the wall of the tubes. The green fluorescence of the precipitated aggregates in the bottom of the tube was observed by fluorescence microscopy (Fig. 1). For further mild purification of the precipitated aggregates, the precipitant was dissolved in PBS containing 4% sarcosyl and then transferred to a 1.5-ml Microfuge tube. Different concentrations of several different kinds of detergents such as SDS, Triton X-100, Tween 20, NP-40, and sarcosyl were tested for purification of the aggregates. Sarcosyl (4%) is often used for nuclear preparation, and we found this was suitable on the basis of the numbers and clarity of

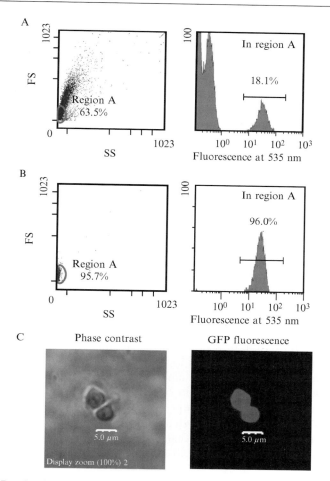

FIG. 1. Result of purification of tNhtt-150Q-EGFP aggregates with a cell sorter. (A) Sorting profiles of aggregates from lysates of tNhtt-150Q-EGFP-expressing neuro 2A are shown in a scattergram (*left*; forward scattering [FS] vs side scattering [SS]) and in a histogram for the fluorescence at 535 nm (*right*). Aggregates with bright fluorescence are observed mostly in the particles within region A of the *left panel*. Numbers in the *left* and *right panels* represent the relative frequency of particles in region A against the total particles and the frequency of bright particles in region A, respectively. (B) Results of the reanalysis of the sorted aggregates. Note that sorted particles are homogeneous with respect to the forward scattering, side scattering, and the fluorescence at 535 nm. (C) Microscopic observation of purified aggregates presented by phase contrast image (*left*) and the corresponding fluorescence image (*right*). Scale bars, 5 μm. (See color insert.)

silver-stained SDS-PAGE bands after each treatment. After washing twice more with PBS containing 4% sarcosyl to remove nonspecifically bound contaminants, the Microfuge tubes were spun down to pellet the purified aggregates, which were stored at $-80°$.

The protein content of the aggregate could not be estimated using normal protein assay methods because of its insolubility in buffers. In our case, the diameter of most aggregate particles was 5 μm on average, so the number of aggregate particles served as a good alternative indicator for estimating the amount of aggregated protein.

The frozen stock of the fluorescent aggregate particles was resuspended in PBS in the 1.5-ml Microfuge tube; 10 μl of the suspension was laid on a hemocytometer and the numbers counted under a fluorescent microscope. This gave the concentration of the particles per μl of the suspension.

Nuclear Aggregates. Nuclear aggregates were purified as demonstrated by Doi *et al.* (2004); 1.5 $\times$ 10^6 tNhtt-150Q-EGFP-NLS cells were plated in 150-mm dishes and cultured overnight. The medium was supplemented with 5 mM dbcAMP and 1 μM ponasterone A. After 3 days, 70% of whole cells developed green fluorescent nuclear aggregates. Cells were washed twice with ice-cold PBS and collected by centrifugation at 1000 rpm for 5 min. Nuclei were isolated as described previously (Emig *et al.*, 1995; Martelli *et al.*, 1992) with a slight modification; 1 $\times$ 10^8 cells were resuspended in 5 ml of 10 mM Tris-HCl, pH 7.4, containing 2 mM MgCl$_2$, 1% NP-40, 2 mM dithiothreitol (DTT), 1 mM phenylmethylsulfonyl fluoride (PMSF), and Complete protease inhibitor EDTA-free (Roche Diagnostics, K.K., Tokyo, Japan). After 5 min on ice, the same volume of deionized distilled water was added and kept on ice for an additional 5 min. Cells were lysed by five passages through a 22-μm-gauge needle, and nuclei were recovered by centrifugation at 300g for 6 min. at 4°. Collected nuclei were resuspended in Buffer A (10 mM Tris-HCl, pH 7.4, 5 mM MgCl$_2$, 0.25 M sucrose, 2 mM DTT, 1 mM PMSF, and Complete protease inhibitor EDTA-free) containing 1% NP-40, 500 U/ml DNase I (Sigma Aldrich Japan, Tokyo, Japan), and 15 U/ml RNase A (Nacalai Tesque, Inc.), mixed for 60 min at 37°, and centrifuged at 1500g for 5 min. The pellets were resuspended in Buffer A containing 2 M NaCl, mixed for 60 min at 4°, and centrifuged at 1500g for 5 min. The pellets were resuspended in Buffer A containing 4% sarcosyl (Nacalai Tesque, Inc.), mixed for 60 min at 4°, and centrifuged for 30 min at 186,000g. The pellets were further washed with Buffer A containing 4% sarcosyl, centrifuged for 30 min at 186,000g, and collected as the final insoluble nuclear fraction. The isolated aggregate particles were observed under a fluorescence microscope (Fig. 2).

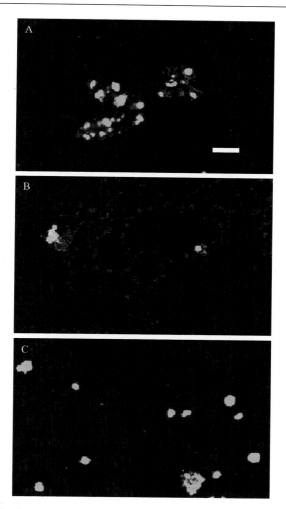

FIG. 2. Result of purification of tNhtt-150Q-EGFP-NLS aggregates. (A) Immunofluorescence of aggregates treated with nucleases. Red fluorescence indicates the localization of lamin B as a marker of the nuclear matrix. (B) Nuclear aggregates after 2 M NaCl treatment. (C) Nuclear aggregates after 4% sarcosyl treatment. Note that aggregate particles are clearly removed from nuclear structures. Scale bars, 20 μm. (See color insert.)

SDS-PAGE OF THE PURIFIED AGGREGATES. Purified aggregates were subjected to SDS-PAGE. We obtained 7000 particles for one lane of 8 cm × 8 cm 5–20% polyacrylamide gradient gels, which gave reasonable result in staining bands by Coomassie brilliant blue (CBB) or Sypro Ruby. The numbers of the particles were adjusted from the suspension of which concentration was measured as described previously. Because polyQ aggregates were highly ubiquitinated, the stained gel showed a remarkable smear on the edge of each band. At the top of the gel, intense staining of aggregated proteins that were retained in the wells was observed.

The condition of the SDS-PAGE was as follows:

Sample buffer component: 0.05 M Tris-HCl (pH 6.8), 2% SDS, 6% β-mercaptoethanol, 10% glycerol, 0.05% bromophenol blue

We generally make five times concentrated stock solution.

Running buffer component: 25 mM Tris base, 192 mM glycine, 0.1% SDS.

We generally make 10 times concentrated stock solution.

Gel: PAGEL model AE-6000 SPG-520L* (ATTO Corp., Tokyo, Japan) *5–20% gradient premade gel (8 × 8 cm)

Charged volume per lane: 15 μ

Charged current: 30 mA constant current

Running time: 1.5 h

Electrophoresis apparatus: ATTO PAGERUN AE-6531

Proteins corresponding to gel bands were annotated (Fig. 3) by mass spectrometry analysis as described in the following.

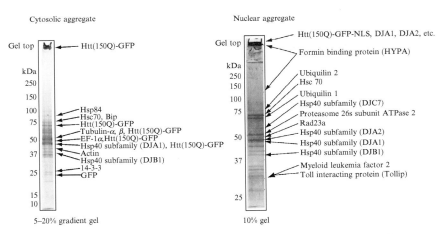

FIG. 3. SDS-PAGE of purified cytosolic and nuclear aggregates and protein assignment to the CBB-stained bands. Cytosolic aggregates were run on 5–20% gradient gel, and nuclear aggregates were run on a 10% continuous gel.

HPLC-MS/MS Analysis

After intensive washing of the gel with sterile water, gel slices of prominent bands were excised. The eluate from the gel piece containing enzyme-digested peptides was transferred to a 96-well plate, and the plate was placed in an auto sampler (HTC-PAL, CTC Analytics, Zwingen, Switzerland) linked to a HPLC (Magic 2002 system, AMR Inc., Tokyo, Japan)-MS (LCQ, Thermo Finnigan, San Jose, CA) system. The ionization of the enzyme-digested peptides was induced by the nano-electrospray method immediately after passing through a reverse-phase HPLC capillary column. The MS/MS spectrum data were analyzed by searching mouse proteins in the NCBI nr database (National Center for Biotechnology Information, Bethesda, MD) using the mass spectrometry analysis software Mascot (Matrix Sciences Inc., Boston, MA).

The step-by-step protocol for sample preparation, including in-gel digestion, is shown in the following. The key for successful analysis is to minimize contamination and sample loss.

Sample Preparation. Some tips for gel cutting and in-gel digestion:

1. Clean your bench thoroughly with ethanol or methanol.
2. Avoid possible gel contamination by exhaling and talking over the gel.
3. Cut gels on glass plates or petri dishes only; no plastic.
4. Use only powder-free nitrile gloves; no Latex gloves.

PROCEDURE

1. Excise the protein spot or band.
2. Dice each gel slice into small pieces (1 mm^2) and place into each 1.5-ml-TPX (polymethyl pentene copolymer) tube (Hi-tech Inc., Tokyo, Japan)
3. Washing:
   Add 100–200 μl of MilliQ water, incubate for 10 min at 37°. Remove supernatant. (Repeat three times.)
4. Destaining:
   For Coomassie Blue and SYPRO Ruby fluorescent staining:
   Add 50–100 μl of 50 mM NH$_4$HCO$_3$/50% CH$_3$CN and incubate for 10 min at 37°.
   Remove supernatant.
   For silver staining:
   a. Add 100 μl of 1:1 mixture of the stock solution A (30 mM potassium ferricyanide) and B (100 mM sodium thiosulfate). Vortex until no bands or spots are visible.
   Remove supernatant.

     b. Add 200 μl of MilliQ water, vortex, and incubate for 10 min at 37°. Remove supernatant. (Repeat three times.)

5. Dehydration:
   Add 50–100 μl (or enough to cover) of CH_3CN, vortex, and remove supernatant.

6. Incubate for 10–30 min at 50°.

7. Reduction:
   Add 50 μl (or enough to cover) of 0.01 M DTT/100 mM NH_4HCO_3 and incubate for 15 min at 50°.

8. Alkylation:
   Cool to room temp, add 2 μl of 0.25 M iodoacetoamide/100 mM NH_4HCO_3 and incubate for 15 min in the dark at room temp. Remove supernatant.

9. Washing:
   Add 50 μl (or enough to cover) of 100 mM NH_4HCO_3 and vortex. Remove supernatant.

10. Washing:
    Add 50 μl (or enough to cover) of 50 mM NH_4HCO_3/50% CH_3CN and vortex. Remove supernatant.

11. Dehydration:
    Add 50–100 μl (or enough to cover) CH_3CN, vortex, and remove supernatant. Incubate for 10–30 min at 50°.

12. Digestion:
    Re-swell gel pieces at 4° for 15 min in 20 μl of 12.5 ng/μl trypsin solution. (Add 100 mM NH_4HCO_3 as needed to cover the gel pieces.) Spin each tube briefly at 5000 rpm with MX-300 centrifuge (Tomy Seiko Co, Ltd., Tokyo, Japan) to fully immerse the gels, then incubate overnight at 37°.

13. Extraction:
    Add 50 μl (or enough to cover) 50% CH_3CN/5% trifluoroacetic acid and incubate for 10 min at 37°. Pool supernatants in new TPX tubes. (Repeat three times.)

14. Dry the pooled supernatants in a CVE-3100 centrifugal evaporator (EYELA, Tokyo, Japan).

15. These samples were resuspended in 10 μl of 2% acetonitrile with 0.1% trifluoroacetic acid solution and then applied to the HPLC-MS system.

Condition of HPLC. Ten microliters of sample in each well was taken by HTC-PAL autosampler and injected to the line of Magic 2002 HPLC system with a reverse-phase capillary column. Peptides retained in the column were eluted by gradient of % concentration of organic solvent.

Those separated peptides were ionized by nanoelectrospray from the on-line pico tip and then were introduced to inlet of LCQ-deca. The column feature, flow rate, sample volume, and gradient condition were as follows;

Column: 0.2-mm diameter × 50-mm long Magic C18 column (Microm Bioresources, Inc., Auburn, CA) packed with silanol group residues of 3 μ and 200 Å of particle size and pore size, respectively.
Column temperature: 25°
Solvent A: 2% Acetonitrile/0.1% formic acid
Solvent B: 90% Acetonitrile/0.1% formic acid
Gradient: % Solvent B increased from 5–65% during first 30 min and then reached 90% in next 10 min.
Flow rate: 1 μl/min
Sample volume: 10 μl

Condition of MS/MS. Peptide ions introduced into LCQ-deca are trapped in ion trap space that is constructed with a central ring electrode and a set of end cap electrodes. When the ring electrode is charged with a low voltage of alternate current, ions in certain range of m/z can stably stay in the electric field. With increasing voltage, ions start to get out through a hole in one end of the end cap electrode sequentially in proportion to the size of m/z, from smaller to larger, thus forming mass spectrum.

MS/MS spectrum is obtained using the same ion trap space by repeating the following cycle; selection of parent ion for MS/MS analysis, collision of the parent ion with inactive gas to generate fragment peptide ions, and detection of the spectrum of the fragment peptide ions.

Parent ions in each mass spectrum are scanned, and MS/MS spectra data corresponding to those parent ions are sequentially obtained. Those spectra are captured by PC in about 1-sec interval, and eventually more than 1800 spectra data are stored during 30 min of running time.

The main parameters of LCQ-deca for the current experiment were as follows:

Spray voltage: 2000 V
Capillary temperature: 150°
N_2 gas flow rate: 40 (arbitrary unit at LCQ)
Capillary voltage: 10 V
Collision energy: 35
Tube lens offset voltage: −5 V

Data Analysis. Each MS/MS spectrum data are automatically converted to peak list file (DTA file) in LCQ-daca system. Mascot software can take DTA file for protein identification analysis.

There are two ways of using Mascot. One is to access the Internet home page at http://www.matrixscience.com/search_form_select.html to submit your query for sequence analysis, and the other is to use the commercial version of the program that is to be installed in the local PC server.

In either case, the following parameters should be input to the Mascot query form.

1. Database to be searched: Select the sequence database to be searched. (In this particular experiment, we used NCBInr.)
2. Taxonomy: Select taxonomy of the protein source to limit the species to be searched so that the search speeds up. (Although Neuro 2A cells were originated from mouse, we selected "all entries" that covered all proteins in the database regardless of taxonomy.)
3. Digestion enzyme: Indicate what digestion enzyme has been used. (We used trypsin in this experiment.)
4. Numbers of miss cleavage: Indicate how many miss cleavage sites in a peptide can be allowed. (We allowed up to two miss cleavage sites.)
5. Modifications: If it is known that posttranslational modifications are included, indicate what the modification is. (In this experiment, it was not necessary to consider any modifications.)
6. Peptide and MS/MS tolerance: Indicate the allowance of error between measured m/z and calculated m/z for precursor ion and fragment ion, respectively. (Because DTA format contains such information, it was not necessary to consider.)
7. Indicate which mass value should be used, monoisotopic value or average value.

In the main page of Mascot data, significant hits to the proteins in the database are lined up in order of the probability-based Mowse score. Graphical representation of the probability score indicates the significant score to distinguish statistically significant identifications from random hits. Each hit links to peptide summary report that contains information such as accession number of the protein, molecular weight, total score, numbers of peptides that hit, measured m/z of precursor ion, peptide mass, error between calculated and measured m/z, number of miss cleavage, score of each peptide, and sequence of peptide. Each set of the information of peptide is headed by a query number that links to the peptide view window. The peptide view window shows the MS/MS spectrum data and deduced amino acid sequence of the fragment peptide with its reliability information. Each accession number in the peptide summary report links

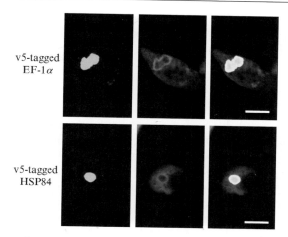

FIG. 4. Immunochemical analysis of AIP candidates in tNhtt-150Q-EGFP–expressing Neuro 2A cells. Cells were transfected with v5-tagged EF-1α or v5-tagged HSP84 in pcDNA3.1 expression vector, then differentiation and tNhtt-150Q-EGFP expression were induced with dbcAMP and ecdysone, respectively. Two days later, the cells were fixed and incubated with antibody against the v5-epitope. *Left column,* fluorescence of tNhtt-150Q-EGFP aggregates. *Middle column,* v5-tagged proteins detected with anti-v5 antibody and Alexa Fluor 546-labeled secondary antibody. *Right column,* merged images of the two signals. Scale bars 20 μm. (See color insert.)

the protein view window that shows the sequence coverage of the measured peptide in the entire sequence of the hit protein.

From the preceding information, promising candidates were carefully selected. We generally check the following points.

1. The higher score of each peptide gives better reliability.
2. Number of miss cleavage to be allowed should be 1 or 2.
3. If the total score is close to significance score, carefully check the peptide view window for each peptide to confirm whether the MS/MS spectrum is clear (low S/N) for deducing amino acid sequence or not.

We repeated the experiment with the same condition three times, and the candidates observed with the similar score in all three experiments were selected.

After this procedure, aggregates that were purified from tNhtt-150Q-EGFP–expressing Neuro 2A cells were revealed to contain at least two new AIP candidates, heat shock protein-84 and translational elongation factor-1α, together with several already known AIPs. From nuclear aggregates in tNhtt-150Q-EGFP-NLS–expressing Neuro 2A cells, tollip, ubiquitin 1, and 2, etc. were identified as new AIPs.

Confirmation of AIP Candidates

Once AIP candidates are picked up by MS/MS analysis, they should be confirmed by several other methods. If antibodies for the AIP candidate molecules are available, Western blotting and immunocytochemistry are commonly used. The colocalization of each AIP with aggregates should be examined. If antibodies are not available, candidate cDNAs fused with a tag such as V5 protein can be expressed in cells and an anti-tag antibody used for confirming the colocalization of the AIP candidate with aggregates. Fig. 4 is an example of immunocytochemical confirmation that heat shock protein-84 and translational elongation factor-1α are AIPs of cytosolic polyQ aggregates.

References

Doi, H., Mitsui, K., Kurosawa, M., Machida, Y., Kuroiwa, Y., and Nukina, N. (2004). Identification of ubiquitin-interacting proteins in purified polyglutamine aggregates. *FEBS Lett.* **571,** 171–176.

Emig, S., Schmalz, D., Shakibaei, M., and Buchner, K. (1995). The nuclear pore complex protein p62 is one of several sialic acid-containing proteins of the nuclear envelope. *J. Biol. Chem.* **270,** 13787–13793.

Martelli, A. M., Gilmour, R. S., Bertagnolo, V., Neri, L. M., Manzoli, L., and Cocco, L. (1992). Nuclear localization and signalling activity of phosphoinositidase C beta in Swiss 3T3 cells. *Nature* **358,** 242–245.

Mitsui, K., Nakayama, H., Akagi, T., Nekooki, M., Ohtawa, K., Takio, K., Hashikawa, T., and Nukina, N. (2002). Purification of polyglutamine aggregates and identification of elongation factor-1α and heat shock protein 84 as aggregate-interacting proteins. *J. Neurosci.* **22**(21), 9267–9277.

Suhr, S. T., Senut, M.-C., Whitelegge, J. P., Faull, K. F., Cuizon, D. B., and Gage, F. H. (2001). Identities of sequestered proteins in aggregates from cells with induced polyglutamine expression. *J. Cell Biol.* **153**(2), 283–294.

Wang, G. H., Mitsui, K., Kotliarova, S., Yamashita, A., Nagao, Y., Tokuhiro, S., Iwatsubo, T., Kanazawa, I., and Nukina, N. (1999). Caspase activation during apoptotic cell death induced by expanded polyglutamine in Neuro 2a cells. *NeuroReport* **10,** 2435–2438.

[6]   Merger of Laser Capture Microdissection and
Mass Spectrometry: A Window into the
Amyloid Plaque Proteome

By Yair M. Gozal, Dongmei Cheng, Duc M. Duong,
James J. Lah, Allan I. Levey, and Junmin Peng

Abstract

The occurrence of protein accumulation and aggregation in the brain is one of the pathological hallmarks of neurodegenerative diseases such as Alzheimer's disease (AD). Although it is instructive to analyze the aggregated proteins in the brain, biochemical purification and identification of these proteins have been challenging. Recent developments in laser capture microdissection (LCM) and mass spectrometry (MS) enable large-scale protein profiling of captured tissue samples. We present here the method of analyzing senile plaques from postmortem AD brains by coupling LCM and highly sensitive liquid chromatography–tandem mass spectrometry (LC-MS/MS). First, the senile plaques were stained with thioflavin-S and precisely isolated by adjusted laser beams under a microscope. Total proteins in the isolated tissues were extracted and resolved on an SDS gel. To identify all proteins in the samples, the gel was excised into multiple pieces followed by trypsin digestion. The resulting peptides were further separated by reverse-phase chromatography and analyzed by tandem mass spectrometry. A database search of acquired MS/MS spectra allowed the identification of hundreds to thousands of peptides/proteins in the original samples. Moreover, quantitative comparison of protein composites in different LCM samples could be achieved by MS strategies. For instance, the comparison between plaques and surrounding nonplaque tissues from the same specimen revealed tens of proteins specifically enriched in the plaques. Finally, the data were corroborated by independent experiments using the approach of immunohistochemistry. Taken together, the merger of LCM and MS is a powerful tool to probe the proteome of any given pathological lesions.

Introduction

Alzheimer's disease (AD) is a progressive neurodegenerative disorder that accounts for most cases of senile dementia in the United States. Pathologically, AD is characterized by neuronal loss, gliosis, and the accumulation

METHODS IN ENZYMOLOGY, VOL. 412
0076-6879/06 $35.00
DOI: 10.1016/S0076-6879(06)12006-6

of amyloid plaques and intraneuronal neurofibrillary tangles. The discovery that the major component of amyloid plaques is amyloid β (Aβ), a 39–43 amino acid peptide produced by proteolysis of amyloid precursor protein (APP) (Glenner and Wong, 1984) ushered in two decades of research focused on discerning the details of plaque formation, composition, and neurotoxicity. This result, the product of brute-force biochemical purification techniques, fueled a field of research and generated the amyloid cascade hypothesis. This hypothesis asserts that neuronal dysfunction is directly related to alter Aβ metabolism and deposition in plaques and has emerged as a leading theory of AD pathogenesis (Verdile *et al.*, 2004). However, this broad framework fails to account for several important observations including apparent incongruities between the plaque density and degree of cognitive impairment in AD patients (Hardy and Selkoe, 2002). In addition, the presence of significant oxidative damage (Nunomura *et al.*, 2001), inflammation (Tuppo and Arias, 2005), and plaque histological diversity (Atwood *et al.*, 2002) implies further complexity of disease pathogenesis with concurrent involvement of diverse, currently unknown, cellular pathways. A systematic analysis of plaque proteins is, therefore, necessary to explain the mechanisms underlying plaque formation.

Mass spectrometry–based proteomic profiling is a particularly revealing approach to characterizing Aβ plaques. This technique is complementary to previous Aβ plaque analyses using genetic, immunochemical, and biochemical methods. Genetic studies, in particular those based on familial cases of AD where analysis of specific mutations is possible, have helped identify several molecules critical to amyloidogenesis and plaque formation (Tsuji *et al.*, 2002). However, in contrast with proteomics, methods based on mRNA expression cannot provide direct information regarding functionally active proteins. Conversely, although immunochemical methods focus specifically on protein expression and colocalization in plaques, they only allow the characterization of a limited number of predetermined targets, and the reliability of results are highly dependent on the specificity of antibodies used. Finally, biochemical isolation and purification of plaque components has been critical to the analysis of the composition of amyloid plaques (Atwood *et al.*, 2002). The necessity for stringent extraction conditions, however, may remove Aβ-associated proteins resulting in inadequate material for subsequent analysis and a failure to detect many nonabundant proteins despite their potential functional importance. The shortcomings in these various techniques support the need for a global protein analysis strategy, a niche that can be filled with the seamless combination of laser capture microdissection (LCM) technology and highly sensitive mass spectrometry.

Laser capture microdissection has proven successful in applications requiring the extraction of specific cells, or small biologically relevant areas, from complex tissues (Emmert-Buck et al., 1996). LCM is ideal for use in the purification of amyloid plaques, because it circumvents the shortcomings of traditional biochemical isolation methods. First, the high precision of the LCM technique allows for isolation of regions as small as 3–5 μm in diameter, thereby reducing the capture of nonplaque debris (Simone et al., 1998). Second, contamination is further restricted as a result of limited handling and processing of the captured material (Bonner et al., 1997). Finally, material isolated by LCM not only maintains its morphology for subsequent visual verification (Bonner et al., 1997) but also maintains the quality of the captured tissues to minimize the impact on downstream analyses (Ornstein et al., 2000). One of the major limitations of LCM is that only a small amount of proteins can be recovered from captured tissues. However, the limitation is largely alleviated by extremely high sensitivity provided by the proteomics platform such as nanoscale capillary liquid chromatography–tandem mass spectrometry (LC-MS/MS).

The technique of LC-MS/MS has emerged as the method of choice for large-scale proteomic analysis (Aebersold and Mann, 2003) of amyloid plaques. The development of quantitative strategies, data processing software, and instrumentation has boosted the field by markedly increasing detection sensitivity, throughput, and robustness of the technology. Its ability to analyze hundreds to thousands of proteins present at the low femtomole or even attomole level has been exploited with great success in a variety of samples (Aebersold and Mann, 2003). In addition, LC-MS/MS has routinely been used to identify and quantify peptides in complex mixtures (Peng and Gygi, 2001), which minimizes purification steps of plaque proteins extracted from LCM-procured samples. In fact, large-scale analyses to identify key proteome changes between AD and control patients were conducted using traditional proteomic approach of 2D gel separation followed by mass spectrometry (Cottrell et al., 2005; Schonberger et al., 2001; Tsuji et al., 2002). Although this technique resulted in the identification of several differentially expressed proteins between AD and controls, the human error inherent in selecting 2D gel spots for sequencing, the narrow sensitivity of this technique, and its limitations in identifying proteins of extreme pI, molecular weight, and hydrophobicity, suggested the need for a different approach. Recently, we performed comprehensive proteomic analyses of senile plaques from post-mortem AD brain tissues by coupling LCM and the LC-MS/MS approach (Liao et al., 2004). A total of 488 proteins were identified to coisolate with the plaques, including more than 80% plaque proteins previously documented. More significantly, quantitative comparison of plaques and nonplaque tissues revealed at least twofold enrichment of 26 proteins in the plaque regions, suggesting the complexity and diversity of cellular processes involved in the plaque formation.

This chapter will highlight the use of LCM in the large-scale proteomic analysis of AD amyloid plaques. Particular focus will be directed to detailing the methods required to isolate, identify, and quantify plaque proteins. Finally, the merits of similar or complementary methods will be discussed.

Methods

Preparation of Tissue for Laser Capture Microdissection

Postmortem blocks of fresh frontal and temporal cortex obtained at autopsy from neuropathologically diagnosed AD patients were embedded in cryostat mounting medium (Tissue-Tek O. C. T.; Jed Pella Inc., Redding, CA), frozen on dry ice, and stored at $-80°$ in preparation for sectioning. Although laser capture microscopy on Pixcell II Laser Capture apparatus (Arcturus, Mountain View, CA) may be conducted on tissue sections ranging from 5–10 μm (Cao *et al.*, 2001; Elenitoba-Johnson *et al.*, 2000; Landis *et al.*, 2003; Simone *et al.*, 1998) in thickness, too thin a section may prevent complete removal of an Aβ plaque, whereas too thick a section may allow the acquisition of contaminating excess material. As such, the tissues were sectioned at 10 μm in a cryostat and mounted on uncoated and uncharged glass slides. After thawing the frozen sections at room temperature, the tissue was briefly fixed for 1 min with 75% ethanol to histologically preserve the section while maintaining the quality of macromolecules for subsequent proteomic profiling. In contrast, despite being considered the "gold standard" for histopathological preservation, the use of aldehyde-based fixatives (e.g., formalin) was avoided, because the high level of covalently cross-linked proteins typically observed with such fixation is incompatible with current proteomic analysis protocols (Ahram *et al.*, 2003). After fixation of the tissue, Aβ plaques were visualized by 1% thioflavin-S staining for 1 min. The sections were consequently differentiated in 75% ethanol for 1 min, subjected to dehydration in graded alcohols, cleared for 5 min in fresh xylene, and air-dried for 5 min. Finally, sections were desiccated before LCM (Emmert-Buck *et al.*, 1996).

Technical Comment

LCM allows for efficient transfer of a variety of tissue samples including frozen tissues, formalin- and ethanol-fixed tissues, and even archival paraffin-embedded tissues (Emmert-Buck *et al.*, 1996; Simone *et al.*,

1998). Preparation of sections for LCM from paraffin-embedded tissues requires dewaxing in consecutive baths of fresh xylenes (5 min each), rehydration in decreasing concentrations of ethanol (100%, 95%, 70%) for 10 sec each, staining with the method of choice, dehydration in series of graded ethanol (70%, 95%, 100%) for 10 sec each, and clearing in xylene for 1 min (Ahram et al., 2003). It is worthwhile to note that although it allows for effective preservation of proteins, paraffin embedding may affect the condition of the macromolecules and may lead to inconsistencies in the quantity of material recovered (Bonner et al., 1997). More recently, analysis of proteins from formalin-fixed paraffin-embedded samples has also been reported (Hood et al., 2005). More details of thioflavin staining and dehydration of tissue sections are described in companion chapters in this volume.

Laser Capture Microdissection

LCM was performed under a fluorescent microscope attached to a Pixcell II laser capture facility. $A\beta$ plaques were visualized using thioflavin-S (excitation wavelength 495 nm) (Dickson and Vickers, 2001) and were captured from cortex with a short-duration pulse (1 msec) of an infrared laser (laser spot size: 7.5 μm) using laser power setting of 60–80 milliwatts, depending on the individual plaque diameter. Because senile plaques are heterogeneous in size and typically ~60 μm (Davis and Robertson, 1997), 500 individual plaques were captured on a single Cap-Sure Macro LCM cap (Arcturus). A typical capture is shown in Fig. 1. Four cortical sections were processed from each patient to capture 2000 plaques per AD case. Similarly, an equivalent amount of nonplaque material was procured on separate LCM caps to serve as a control. The tissue was derived from areas surrounding the plaques.

Protein Extraction from LCM Caps

Before processing of the captured tissue, each cap was visually inspected for the capture of any contaminating material removed during the lifting of the transfer film of the cap from the slide. $A\beta$ plaques were clearly visualized on the cap, because the morphology of captured material was unaltered during the transfer from the slide to the cap (Fig. 1). The plaques on the caps were then extracted with 20 μl of lysis buffer composed of 2% sodium dodecyl sulfate (SDS), 10% glycerol, 10 mM dithiothreitol (DTT), 1 mM ethylenediaminetetraacetic acid (EDTA), and protease inhibitor cocktail (Roche Applied Science) in phosphate-buffered saline, pH 7.2. After a 15-min extraction at 65°, the buffer was collected and replaced with a fresh 20-μl aliquot of lysis buffer for a second incubation

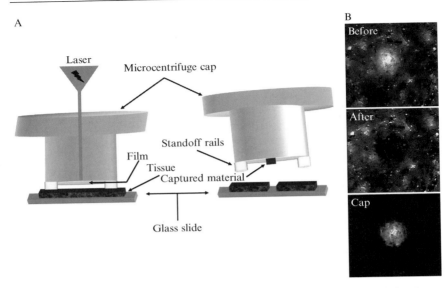

Fig. 1. Isolation of amyloid plaques by laser capture microdissection. (A) When a polymer-coated cap is placed on the top of fixed tissue, standoff rails prevent the direct contact of the coat with tissue to reduce contamination. Under a microscope, a laser pulse generates enough energy to melt the coated film and to glue the area of interest to the cap. Lifting the cap leads to the removal of targeted area of tissue. (B) The "before" and "after" images indicate the removal of a plaque region from a thioflavin-S–stained AD brain section. The isolated plaque was attached to the cap as shown. Moreover, the surrounding nonplaque regions were also captured as control on a different cap (not shown). This figure is adopted from our previously published sources (Liao *et al.*, 2004). (See color insert.)

at 65° for 15 min. To minimize the volume of extracted protein samples, the lysis buffer was reused in the sequential extractions of up to four caps, thus representing the pooled captured material (Aβ plaques or control tissue) from the four slides for each case. Because of evaporation during each of the incubations at 65°, lysis buffer volume was replenished with H_2O (warmed to 65°) during the transfer of buffer from one cap to the next so as to maintain appropriate lysis buffer solute concentrations for the after extraction. Finally, to increase recovery of cysteine-containing peptides and promote the identification of thiol-containing peptides, the samples were alkylated in the dark at room temperature with 50 mM iodoacetamide for 30 min after the extractions.

To normalize total proteins submitted for proteomic analysis, a small fraction of the total sample (1–5%) was resolved on a 6–12% SDS gel and analyzed by a modified silver staining (Shevchenko *et al.*, 1996). Proteins were quantified by densitometric analysis using Scion Image software (Scion Corporation, Frederick, MD) in comparison with standard protein markers previously quantified using known concentrations of BSA. The protein yield per Aβ plaque was found to be ~2 ng and ~4 μg of proteins were harvested from a total of 2000 captured plaques.

Technical Comment

The composition of the lysis buffer used to extract captured materials from LCM caps is highly dependent on the nature of the tissue and proteins to be solubilized. LCM offers the advantage of accessibility to microdissected tissue embedded in the thermoplastic film (Bonner et al., 1997). Thus, any aqueous buffer can be applied directly to the cap in small volumes to facilitate various downstream molecular analyses (Craven and Banks, 2001). Therefore, ensuring proper solubility of proteins in the extraction buffer is the limiting factor for consequent proteomic investigation. We chose 2% SDS among several tested extraction conditions including formic acid, because this buffer led to the best yield of total proteins despite the fact that SDS dissolves Aβ fibrils poorly (Masters et al., 1985). Nevertheless, the small amount of solubilized Aβ warranted strong signals in subsequent MS analysis, strongly suggesting that other major protein aggregates in the plaques can hardly elude the detection. Moreover, only 20 μl of lysis buffer was used to extract proteins each time. If a large volume of lysis buffer is used, the extracted proteins would be highly diluted and have to be concentrated, in which extra sample loss may occur.

Preparation of Aβ Plaque Proteins for Gel LC-MS/MS

For mass spectrometry analysis, the proteins in each sample were first separated by their molecular weight by SDS gel electrophoresis. An equal amount of protein from each sample was loaded, along with molecular weight markers, on a 6–12% SDS gel (~12 cm in length, 0.75 mm in thickness) and run slowly at 250 mV for approximately 4 h. When a small amount of protein is loaded (<1 μg), the gel may be run for only 1 h, allowing protein separation over a smaller distance (3 cm). The gel was then stained with Coomassie Blue G-250 to visualize the proteins in each lane and was subsequently soaked in water at room temperature to reduce the residual SDS content. The water was changed at least three times during a minimum 3-h rinse. Sample lanes were then excised from the gel and cut into 6–40 pieces according to the flanking markers to achieve sufficient molecular weight resolution (Fig. 2A). The number of pieces was dependent on the total amount of proteins loaded on the gel, and in

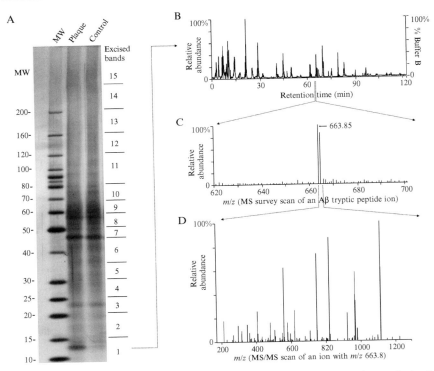

FIG. 2. Identification of LCM-isolated proteins by gel LC-MS/MS. (A) Protein in the captured plaques and nonplaque control were extracted and a small fraction (~5%) was analyzed on a SDS gel followed by silver staining. Most samples (~95%) were resolved on another SDS gel and stained with Coomassie blue G250 (data not shown). Each sample lane was cut into 15 pieces that were subjected to trypsin digestion and LC-MS/MS analysis. The gel excision pattern is shown on the right according to the marker. (B) The elution profile of the #1 gel piece from plaques on reverse-phase liquid chromatography, indicating the base peak intensity on *y*-axis and retention time on *x*-axis. (C) The MS survey scan at 65 min. Only a small *m/z* range (620.0–700.0) was shown for simplicity. (D) The MS/MS spectrum of the ion with *m/z* 663.8, which was identified to be an Aβ tryptic peptide after database search. This figure is adopted from our previously published sources (Liao *et al.*, 2004).

general at least 0.2 μg proteins per piece were recommended. Gel pieces were further fragmented into 1-mm^3 cubics in preparation for in-gel trypsin digestion, which was performed as previously reported (Shevchenko *et al.*, 1996). The fragments were first destained during 10-min incubation in a buffer containing half acetonitrile and half 50 mM sodium bicarbonate (NH$_4$HCO$_3$) by volume. Next, acetonitrile (100%) was applied to shrink the gel pieces by dehydration. At the end of the incubation, the acetonitrile was removed, and the gel fragments were vacuum-dried. Trypsin was easily absorbed by the dry gel pieces by incubating with digestion buffer containing 12.5 ng/μl sequencing grade trypsin (Promega Corporation, Madison, WI) in 50 mM NH$_4$HCO$_3$ on ice for 45 min. An additional 50 mM NH$_4$HCO$_3$ ($\sim$10 μl) was added to each digestion reaction to keep the gel pieces wet during overnight proteolysis at 37°. Peptides were extracted from the gel pieces at room temperature by centrifugation for 1 min at 15,000g in a tabletop microcentrifuge (Eppendorf Centrifuge 5417C, Eppendorf, Westbury, NY) and subsequent collection of the supernatant. The extraction was repeated three times per sample using an extraction buffer containing 5% formic acid and 50% acetonitrile. Supernatants of each extraction were pooled, vacuum dried, and stored at −20° until LC-MS/MS analysis.

Reverse-Phase LC-MS/MS

Sample peptides corresponding to the original gel pieces cut from each lane were dissolved in 5 μl buffer A containing 0.4% acetic acid, 0.005% heptafluorobutyric acid (HFBA), and 5% acetonitrile. Typically we analyzed 3 μl (60%) of each sample and used the remainder as backup. Each sample was individually analyzed by reverse phase LC-MS/MS using 50-μm or 75-μm inner diameter × 12-cm self-packed fused silica C$_{18}$ (5-μm beads) capillary columns at a flow rate of $\sim$300 nl/min. Samples were loaded onto the column off-line by a pressure cell to maximize the sensitivity for analysis (Peng and Gygi, 2001). The column was initially washed with buffer A, and then eluted in a 120 min gradient using an increasing concentration (10%–30%) of buffer B (0.4% acetic acid, 0.005% HFBA, and 95% acetonitrile) (Fig. 2B). High voltage (1.8–2 kV) was applied to ionize the peptides eluted in the run. The ionized peptides were detected in MS survey scans (400–1700 m/z, 2 microscans) performed on an LCQ-DECA XP-Plus ion trap mass spectrometer (Thermo Finnigan, San Jose, CA), followed by data-dependent MS/MS scans of the three most abundant ions (three microscans each, isolation width 3 m/z, 35% normalized collision energy, dynamic exclusion for 3 min) in completely automated fashion (Fig. 2C, D). That is, MS/MS spectra were only acquired for ions, the abundance of which was among the top three during the MS survey scan. It should be noted that tandem mass spectrometry itself is a high-resolution separation tool, because it can physically isolate a peptide ion according to its mass-to-charge ratio (m/z) value despite the presence of many other co-eluting peptides.

Aβ Plaque Protein Identification

Tandem mass spectrometry (MS/MS) allows for the accurate identification of proteins in a complex mixture on the basis of fragmentation patterns (MS/MS spectra) of peptides. Bioinformatics software is used to match the experimental MS/MS spectra with theoretical (computer-produced) spectra of predicted peptides from a database. In this case, all MS/MS spectra were searched using SEQUEST algorithm (Eng *et al.*, 1994) against the human reference database (ftp.ncbi.nih.gov/genbank). The SEQUEST program matches an MS/MS spectrum with a peptide sequence in the database by several consecutive steps. (1) The program extracts all peptides from a protein database with their masses matching the parent ion selected for MS/MS analysis. A mass difference can be tolerated according to the user's setting (e.g., 3 Da if using a LCQ ion trap mass spectrometer). If protein samples are digested by a protease with certain specificity, the same restriction can be applied on the theoretical cleavage of proteins in the database. We did not use this protease option, because it reduces the number of theoretical peptides and forces a spectrum to match fully digested peptides, although this option is useful to save time in computation. With regard to posttranslational modifications, the corresponding mass shifts are dynamically assigned to the specified amino acid residues, such as oxidized methionine (+16 Da); carboxymethylated cysteine (+57 Da); and phosphorylated serine, threonine, and tyrosine (+80 Da). (2) After initial matching, each peptide is then assigned a preliminary score by comparing its predicted fragment ion pattern to the MS/MS spectrum. (3) The 500 best-matching peptides are subjected to a more rigorous ion-matching algorithm that generates a cross-correlation score (X_{corr}). A list of peptides with good X_{corr} scores is generated with the top peptide generally considered as the best candidate. The difference of X_{corr} scores between the first and second peptide matches is represented by a score termed ΔCn. The SEQUEST algorithm always assigns a peptide sequence to a MS/MS spectrum, but not all assignments are statistically significant. For example, MS/MS spectra with many noisy peaks, poor fragmentation patterns, and/or unexpected modifications can cause false assignments, and some other false positives are simply generated by random matches. Therefore, it is of importance to remove these false positives to improve the data set quality.

To achieve a list of peptide matches with a minimum number of false positives, we used the target/decoy database method (Peng *et al.*, 2003a) to evaluate the false-positive (FP) rate associated with filtering criteria. A composite database was made by attaching a decoy database to the end of the target human database. Two types of decoy databases were used with similar results: a reverse database made by reversing each protein sequence in the target database and a random database made by randomly shuffling each protein sequence in the target database. The computer-generated peptide matches from the decoy sequences were defined as false positives that were used to derive FP rate. Only the peptides with at least seven amino acid residues were considered in the analysis, because a shorter peptide was possibly shared in both target and decoy databases. In general, the assigned peptides were classified into nine groups according to the combination of tryptic state (fully, partially, and non-tryptic) and parent ion charge state (singly, doubly, and triply charged) (Peng et al., 2003a). In each group, we filtered the peptide assignments by ΔCn and X_{corr} values to reduce the FP rate to 1%.

Several additional modifications to the previously established criteria (Peng et al., 2003a) were introduced to maximize reliability of identification. First, only fully tryptic peptides were considered. Because trypsin is a highly specific protease, the exclusion of partially- or non-tryptic peptides (even with high X_{corr}) limited sources of potential false positives (Olsen et al., 2004). Moreover, manual verification of all proteins matched by less than three peptides was conducted as a final sweep for false positives, because recent work has suggested that confirmation of proteins identified by one or two peptides may be sufficient in eliminating false-positive matches from the data set (Peng et al., 2003b). Furthermore, redundancy resulting from the sharing of peptides within several proteins, which may occur during the proteolytic cleavage of protein variants expressed from the same gene locus, was corrected (Rappsilber and Mann, 2002). Finally, after removal of obvious contaminants, including trypsin and keratins, the data sets were merged to create the final "Aβ plaque" and "nonplaque" protein lists.

Aβ Plaque Protein Quantification

The accurate, comprehensive profiling of protein expression in complex digests requires not only the correct identification of peptides in the mixture but also subsequent quantification of identified proteins. The quantification was based on comparison of paired peptides from the plaque and nonplaque samples. The extracted ion current of each pair of peptides was recorded from their corresponding MS survey scans, and a ratio of the peak intensities for the peptide precursor ion was calculated (Andersen et al., 2003; Bondarenko et al., 2002; Chelius et al., 2003). The ratio is a measure of the relative abundance of the peptide in the two separate samples (Wang et al., 2003). It is important to note that although peptide identification relies on the peptide fragmentation pattern, which results in a unique MS/MS spectrum, the peptide quantification is primarily derived from the MS survey scan. Because MS/MS analysis was data-dependent, only a fraction of ionized peptides were selected for sequencing. In fact, only the three predominant ion peaks in each MS survey scan were subsequently sequenced. As such, the MS survey scans contained many nonsequenced

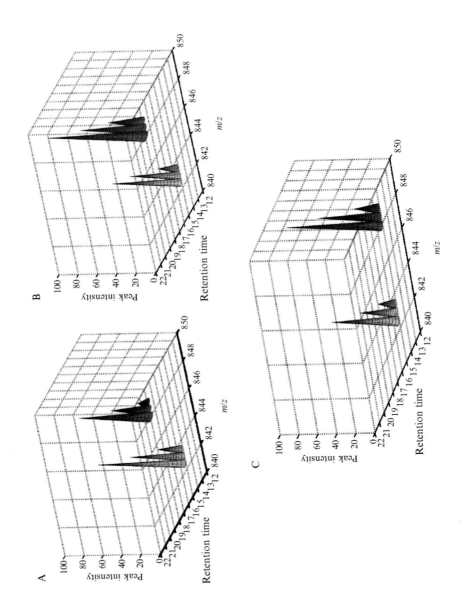

ion peaks that, after localization using the predicted m/z value and adjusted retention time, were used to maximize the number of quantified peptides.

To minimize the inevitable quantitative variation in a comparison between a pair of LC-MS/MS runs (Fig. 3A, B), the trypsin autocleavage peptide VATVSLPR (m/z 842.5 for singly charged monoisotopic ion) was used as an internal control. Paired LC-MS/MS runs corresponded to the consecutive analysis of the peptide mixtures extracted from either plaque or nonplaque sample gel pieces containing proteins of identical molecular weight ranges. By performing consecutive analyses on the same column for sample pairs, variation was limited to less than twofold, as determined by the quantification of the aforementioned trypsin autocleavage peptide. Consequently, a minimum twofold change was required in both AD cases to consider a protein enriched in the plaques. If a peptide was identified in the plaque sample but its peak was not reliably located in the control sample, a twofold enrichment was assumed. In addition to the threshold for protein enrichment, the trypsin autocleavage peptide was also used to normalize the retention time of peptides to allow peptide peak matching across separate LC-MS/MS runs.

Technical Comment

The trypsin auto-cleavage peptide was chosen as an internal control for the relative quantification of proteins, because an equivalent amount of trypsin was used for in-gel digests during the preparation of plaque and non-plaque samples for mass spectrometry. This peptide, therefore, should have been found in identical concentrations in both plaque and control samples. Thus, normalization of the peak intensities of this trypsin auto-cleavage peptide was expected to correct for pipetting errors, incomplete digestions, and other experiment-dependent parameters (Chelius and Bondarenko, 2002). This principle extends to the use of any proteins

---

FIG. 3. Illustration of the quantitative strategies in mass spectrometry. The data acquired in LC-MS/MS are three-dimensional (peptide ion current indicated by peak intensity, retention time of LC, and the m/z measurement every $\sim$0.5–2 sec during the entire LC). Each peptide ion was indicated by three cone-shaped peaks caused by naturally occurring isotope distribution. (A) and (B) The peptide ion with monoisotopic m/z 842.5 (filled in light gray) was eluted between 15 and 16 min, whereas the other ion with monoisotopic m/z 848.0 was between 18.5 and 19.5 min. The peak intensities in run A and B provide the measurement of the peptide abundance. If one peptide is known to be constant in two runs, it can be used to normalize the run-to-run variation. (C) Elution of isotopically labeled peptide pairs. They typically have the same retention time (15–16 min) but differ in mass (monoisotopic m/z 842.5 and 848.0).

known to exist at a constant concentration in two experimental samples (e.g., housekeeping proteins). In the case of complex mixtures, however, it may be incorrect to speculate which, if any, specific peptide concentration is sufficiently steady to be used as a reference for data analysis. Therefore, if one can assume that there is a consistency in abundance of a majority of peptides in the samples, then normalization of peptide peak intensities can be based on all identified proteins in the mixtures (Bondarenko *et al.*, 2002; Chelius *et al.*, 2003). When such an assumption cannot be made, other relative quantification methods based primarily on stable isotope labeling may be used (Fig. 3C). One of the most common labeling techniques is the isotope-coded affinity tag (ICAT) in which cysteine-containing peptides are labeled with either an isotopically light or heavy reagent, mixed, and analyzed by mass spectrometry (Gygi *et al.*, 1999). Both light and heavy peptides can be readily compared in the resultant MS survey scan. However, the expensive labeling reagents, multistep isolation of labeled peptides, and consequent identification of only cysteine-containing peptides are significant disadvantages of this technique. As a result, many variants of isotope labeling approaches have been developed that extend the range of the technique (Aebersold and Mann, 2003).

Validation by Immunohistochemistry

To independently verify the accuracy of the LC-MS/MS generated data, several of the most promising candidate proteins were analyzed by immunohistochemistry (IHC). The IHC technique was ideal, because it allowed visual confirmation of localization of protein candidates within amyloid plaques with specific antibodies. However, the IHC is dependent on the availability of reliable antibodies. Nevertheless, the technique allows for the characterization of candidates in a variety of tissues, including both in autopsy-confirmed human AD cases and in mouse models that mimic plaque deposition and amyloidosis. Immunostaining was, therefore, performed in both human cases and in double transgenic C3/B6 mice expressing the APP695 isoform with the "Swedish" double mutation (APPswe) and the PS1ΔEx9 mutation coding for a functional PS1 mutant with a deletion of amino acids 290–319 in exon 9 (Borchelt *et al.*, 1997).

Conclusion

The marriage of laser capture microdissection, a method that facilitates the enrichment of specific anatomical regions from complex tissues, and mass spectrometry may have a profound impact on the study of neurodegenerative diseases. The methods described herein offer unique opportunities to study neurodegenerative diseases that are often characterized by defined neuro-pathological lesions that are difficult to isolate biochemically. These methods can serve as the primary tools to reveal the proteomes of microanatomical structures characteristic of these pathoses. In addition, these techniques may also be applied in the temporal proteomic profiling of yet undefined pathoses including analysis of the subtle changes that occur in amyloid plaque composition during the progression of AD. Further improvements in both LCM and MS technologies may allow for more extensive identification and quantification of previously physically inaccessible or scarce proteins. For example, improved optics and specificity of capture in LCM may subsequently produce purer samples for analysis. In parallel, recent introduction of LTQ-FT and LTQ-Orbitrap hybrid mass spectrometers has dramatically increased the sensitivity, mass resolution, and throughput of mass spectrometry that requires less quantity of samples. Application of this technology may lead to significant advances in the investigation of the pathophysiological pathways underlying various neurodegenerative diseases, providing potential novel target proteins for therapeutic intervention.

Acknowledgments

We thank Dr. Marla Gearing and Dr. Howard D. Rees for sample preparation and Dr. Michael Iuvone and James Wessel for their help in LCM. We also thank Elizabeth A Gieseker for her generous help in the formatting of figures. This work was supported by the Emory Alzheimer's Disease Research Center (P50 AG025688).

References

Aebersold, R., and Mann, M. (2003). Mass spectrometry-based proteomics. *Nature* **422**, 198–207.

Ahram, M., Flaig, M. J., Gillespie, J. W., Duray, P. H., Linehan, W. M., Ornstein, D. K., Niu, S., Zhao, Y., Petricoin, E. F., 3rd, and Emmert-Buck, M. R. (2003). Evaluation of ethanol-fixed, paraffin-embedded tissues for proteomic applications. *Proteomics* **3**, 413–421.

Andersen, J. S., Wilkinson, C. J., Mayor, T., Mortensen, P., Nigg, E. A., and Mann, M. (2003). Proteomic characterization of the human centrosome by protein correlation profiling. *Nature* **426**, 570–574.

Atwood, C. S., Martins, R. N., Smith, M. A., and Perry, G. (2002). Senile plaque composition and posttranslational modification of amyloid-beta peptide and associated proteins. *Peptides* **23**, 1343–1350.

Bondarenko, P. V., Chelius, D., and Shaler, T. A. (2002). Identification and relative quantitation of protein mixtures by enzymatic digestion followed by capillary reversed-phase liquid chromatography-tandem mass spectrometry. *Anal. Chem.* **74**, 4741–4749.

Bonner, R. F., Emmert-Buck, M., Cole, K., Pohida, T., Chuaqui, R., Goldstein, S., and Liotta, L. A. (1997). Laser capture microdissection: Molecular analysis of tissue. *Science* **278**, 1481–1483.

Borchelt, D. R., Ratovitski, T., van Lare, J., Lee, M. K., Gonzales, V., Jenkins, N. A., Copeland, N. G., Price, D. L., and Sisodia, S. S. (1997). Accelerated amyloid deposition in the brains of transgenic mice coexpressing mutant presenilin 1 and amyloid precursor proteins. *Neuron* **19**, 939–945.

Cao, Z., Wanagat, J., McKiernan, S. H., and Aiken, J. M. (2001). Mitochondrial DNA deletion mutations are concomitant with ragged red regions of individual, aged muscle fibers: Analysis by laser-capture microdissection. *Nucleic Acids Res.* **29**, 4502–4508.

Chelius, D., and Bondarenko, P. V. (2002). Quantitative profiling of proteins in complex mixtures using liquid chromatography and mass spectrometry. *J. Proteome Res.* **1**, 317–323.

Chelius, D., Zhang, T., Wang, G., and Shen, R. F. (2003). Global protein identification and quantification technology using two-dimensional liquid chromatography nanospray mass spectrometry. *Anal. Chem.* **75**, 6658–6665.

Cottrell, B. A., Galvan, V., Banwait, S., Gorostiza, O., Lombardo, C. R., Williams, T., Schilling, B., Peel, A., Gibson, B., Koo, E. H., Link, C. D., and Bredesen, D. E. (2005). A pilot proteomic study of amyloid precursor interactors in Alzheimer's disease. *Ann. Neurol.* **58**, 277–289.

Craven, R. A., and Banks, R. E. (2001). Laser capture microdissection and proteomics: Possibilities and limitation. *Proteomics* **1**, 1200–1204.

Davis, R. L., and Robertson, D. M. (1997). "A Textbook of Neuropathology." Lippincott Williams & Wilkins, San Francisco.

Dickson, T. C., and Vickers, J. C. (2001). The morphological phenotype of beta-amyloid plaques and associated neuritic changes in Alzheimer's disease. *Neuroscience* **105**, 99–107.

Elenitoba-Johnson, K. S., Bohling, S. D., Mitchell, R. S., Brown, M. S., and Robetorye, R. S. (2000). PCR analysis of the immunoglobulin heavy chain gene in polyclonal processes can yield pseudoclonal bands as an artifact of low B cell number. *J. Mol. Diagn.* **2**, 92–96.

Emmert-Buck, M. R., Bonner, R. F., Smith, P. D., Chuaqui, R. F., Zhuang, Z., Goldstein, S. R., Weiss, R. A., and Liotta, L. A. (1996). Laser capture microdissection. *Science* **274**, 998–1001.

Eng, J., McCormack, A. L., and Yates, J. R., 3rd. (1994). An approach to correlate tandem mass spectral data of peptides with amino acid sequences in a protein database. *J. Am. Soc. Mass Spectrom.* **5**, 976–989.

Glenner, G. G., and Wong, C. W. (1984). Alzheimer's disease: Initial report of the purification and characterization of a novel cerebrovascular amyloid protein. *Biochem. Biophys. Res. Commun.* **120**, 885–890.

Gygi, S. P., Rist, B., Gerber, S. A., Turecek, F., Gelb, M. H., and Aebersold, R. (1999). Quantitative analysis of complex protein mixtures using isotope-coded affinity tags. *Nat. Biotechnol.* **17**, 994–999.

Hardy, J., and Selkoe, D. J. (2002). The amyloid hypothesis of Alzheimer's disease: Progress and problems on the road to therapeutics. *Science* **297**, 353–356.

Hood, B. L., Darfler, M. M., Guiel, T. G., Furusato, B., Lucas, D. A., Ringeisen, B. R., Sesterhenn, I. A., Conrads, T. P., Veenstra, T. D., and Krizman, D. B. (2005). Proteomic analysis of formalin fixed prostate cancer tissue. *Mol. Cell. Proteomics.* **4**, 1741–1753.

Landis, W. J., Jacquet, R., Hillyer, J., and Zhang, J. (2003). Analysis of osteopontin in mouse growth plate cartilage by application of laser capture microdissection and RT-PCR. *Connect. Tissue Res.* **44**(Suppl. 1), 28–32.

Liao, L., Cheng, D., Wang, J., Duong, D. M., Losik, T. G., Gearing, M., Rees, H. D., Lah, J. J., Levey, A. I., and Peng, J. (2004). Proteomic characterization of postmortem amyloid plaques isolated by laser capture microdissection. *J. Biol. Chem.* **279**, 37061–37068.

Masters, C. L., Simms, G., Weinman, N. A., Multhaup, G., McDonald, B. L., and Beyreuther, K. (1985). Amyloid plaque core protein in Alzheimer disease and Down syndrome. *Proc. Natl. Acad. Sci. USA* **82,** 4245–4249.

Nunomura, A., Perry, G., Aliev, G., Hirai, K., Takeda, A., Balraj, E. K., Jones, P. K., Ghanbari, H., Wataya, T., Shimohama, S., Chiba, S., Atwood, C. S., Petersen, R. B., and Smith, M. A. (2001). Oxidative damage is the earliest event in Alzheimer disease. *J. Neuropathol. Exp. Neurol.* **60,** 759–767.

Olsen, J. V., Ong, S. E., and Mann, M. (2004). Trypsin cleaves exclusively C-terminal to arginine and lysine residues. *Mol. Cell Proteomics* **3,** 608–614.

Ornstein, D. K., Gillespie, J. W., Paweletz, C. P., Duray, P. H., Herring, J., Vocke, C. D., Topalian, S. L., Bostwick, D. G., Linehan, W. M., Petricoin, E. F., 3rd, and Emmert-Buck, M. R. (2000). Proteomic analysis of laser capture microdissected human prostate cancer and *in vitro* prostate cell lines. *Electrophoresis* **21,** 2235–2242.

Peng, J., Elias, J. E., Thoreen, C. C., Licklider, L. J., and Gygi, S. P. (2003a). Evaluation of multidimensional chromatography coupled with tandem mass spectrometry (LC/LC-MS/MS) for large-scale protein analysis: The yeast proteome. *J. Proteome Res.* **2,** 43–50.

Peng, J., and Gygi, S. P. (2001). Proteomics: The move to mixtures. *J. Mass Spectrom.* **36,** 1083–1091.

Peng, J., Schwartz, D., Elias, J. E., Thoreen, C. C., Cheng, D., Marsischky, G., Roelofs, J., Finley, D., and Gygi, S. P. (2003b). A proteomics approach to understanding protein ubiquitination. *Nat. Biotechnol.* **21,** 921–926.

Rappsilber, J., and Mann, M. (2002). What does it mean to identify a protein in proteomics? *Trends Biochem. Sci.* **27,** 74–78.

Schonberger, S. J., Edgar, P. F., Kydd, R., Faull, R. L., and Cooper, G. J. (2001). Proteomic analysis of the brain in Alzheimer's disease: Molecular phenotype of a complex disease process. *Proteomics* **1,** 1519–1528.

Shevchenko, A., Wilm, M., Vorm, O., and Mann, M. (1996). Mass spectrometric sequencing of proteins silver-stained polyacrylamide gels. *Anal. Chem.* **68,** 850–858.

Simone, N. L., Bonner, R. F., Gillespie, J. W., Emmert-Buck, M. R., and Liotta, L. A. (1998). Laser-capture microdissection: Opening the microscopic frontier to molecular analysis. *Trends Genet.* **14,** 272–276.

Tsuji, T., Shiozaki, A., Kohno, R., Yoshizato, K., and Shimohama, S. (2002). Proteomic profiling and neurodegeneration in Alzheimer's disease. *Neurochem. Res.* **27,** 1245–1253.

Tuppo, E. E., and Arias, H. R. (2005). The role of inflammation in Alzheimer's disease. *Int. J. Biochem. Cell Biol.* **37,** 289–305.

Verdile, G., Fuller, S., Atwood, C. S., Laws, S. M., Gandy, S. E., and Martins, R. N. (2004). The role of beta amyloid in Alzheimer's disease: Still a cause of everything or the only one who got caught? *Pharmacol. Res.* **50,** 397–409.

Wang, W., Zhou, H., Lin, H., Roy, S., Shaler, T. A., Hill, L. R., Norton, S., Kumar, P., Anderle, M., and Becker, C. H. (2003). Quantification of proteins and metabolites by mass spectrometry without isotopic labeling or spiked standards. *Anal. Chem.* **75,** 4818–4826.

[7]   MALDI MS Imaging of Amyloid

By Markus Stoeckli, Richard Knochenmuss,
Gregor McCombie, Dieter Mueller, Tatiana Rohner,
Dieter Staab, and Karl-Heinz Wiederhold

Abstract

Label-free molecular imaging by mass spectrometry allows simultaneous mapping of multiple analytes in biological tissue sections. In this chapter, the application of this new technology to the detection $A\beta$ peptides in mouse brain sections is discussed.

Introduction

Spatially resolved analysis of biological systems often delivers key information for the understanding of a disease. Considerable effort, therefore, continues to be directed toward improvement of modern molecular imaging techniques. Key performance characteristics are higher sensitivity, better spatial resolution, and greater specificity. In Alzheimer's disease (AD), the ability to localize and characterize amyloid β ($A\beta$) peptides in biological systems is critical for the study of the disease. A number of different techniques are now available and routinely applied on tissue sections and *in vivo*. On tissue sections, staining techniques and immunohistochemistry are widely used for the analysis of plaques and single $A\beta$ peptide species (Dixon, 1997). Detection of $A\beta$ plaques by magnetic resonance imaging (MRI) (Vanhoutte *et al.*, 2005), PET (Nordberg, 2004), or, more recently, near-infrared fluorescence (NIRF) (Hintersteiner *et al.*, 2005) enables *in vivo* study of disease progression (Sair *et al.*, 2004).

A common feature of the preceding molecular imaging techniques is that they require labeling. The information content of the image, therefore, strongly depends on the specificity of the labels, which may range from unspecific for stains to highly specific if using antibodies. A major advantage of the mass spectrometric technology described here is that it does not require labels. Instead, molecular weight can be used to distinguish between analytes such as different $A\beta$ peptides. Matrix-assisted laser desorption/ionization mass spectrometric imaging (Caprioli *et al.*, 1997) (MALDI MSI) has the ability to detect analytes over a nearly unlimited mass range at a point defined by a laser focused on the matrix-coated sample. By rastering the laser, spatial distributions are obtained. An initial

METHODS IN ENZYMOLOGY, VOL. 412
0076-6879/06 $35.00
DOI: 10.1016/S0076-6879(06)12007-8

application of MALDI MSI to the detection of Aβ peptides in AD has been described (Stoeckli *et al.*, 2002). The methods discussed here represent further developments improving robustness and sensitivity, as well as new techniques for data analysis and molecular identification.

Protocol for Processing Tissue Specimen

The sample preparation process for MSI needs to preserve the original state of the tissue and avoid contamination. This is achieved by decapitation of the animals and fast freezing of the excised brains by immersion in isopentane/dry ice. Alternately, animals can be sacrificed by microwave exposure (Pierson *et al.*, 2004), which has been reported to preserve neuropeptide levels and distributions because of protease denaturation. It is important to note that the tissues should not be embedded in a polymeric medium (e.g., Tissue-Tek O.C.T.), because this will lead to serious artifacts in the mass spectra: The polyethylene glycol present in O.C.T. results in numerous strong MALDI signals, which can obscure an extended mass region. Once frozen, the brains are stored at $-80°$ before further processing.

Cutting of the tissue sections was performed on a HM 560 (Microm International, Walldorf, Germany) cryostatic microtome. Special care must also be taken in mounting the tissues with Tissue-Tek O.C.T compound on the back side of the tissue. Because conductive sample support is a requirement for MALDI, stainless steel or gold-coated plates, which are commonly used with commercial MALDI sources, are preferred to conventional glass slides. However, if the sections have to be compared with staining methods, coated glass slides may be used (Chaurand *et al.*, 2004). Sagittal sections are cut to a thickness of 10 μm at $-17°$ and thawed on the metal plates. Alternating sections are placed on glass slides for staining. Sections are air dried at room temperature and then stored at $-80°$.

MALDI Matrix Deposition

For MSI, the section is fixed by immersion into 96% ethanol for 1 min followed by rinsing with an equal solution. After fixation, the sections are dried at room temperature in a small desiccator, which takes not longer than 1 h. During this process, the brain tissue adheres to the stainless steel plate, and no detaching was observed.

A matrix layer is required to absorb the laser energy and induce desorption and ionization of analytes. Homogeneous co-crystallization of the analyte with matrix is crucial for high sensitivity and artifact-free imaging. Several coating methods for biological tissues have been reported, including electrospray deposition (Caprioli *et al.*, 1997), picoliter deposition (Schwartz *et al.*, 2003), sublimation, pneumatic spray (Reyzer *et al.*, 2003), and pipetting (Chaurand and Caprioli, 2002; Stoeckli *et al.*, 2001). Despite their simplicity, the last two methods achieve the best sensitivities, possibly because of the relatively large crystals formed, but this in turn limits the lateral resolution. Sinapinic acid (SA) and alpha-cyano-4-hydroxy-cinnamic (CHCA) acid have been found to be the most suitable matrices for $A\beta$ peptides. The matrices were obtained at the highest available quality from Fluka (85429) and Sigma (C8982), respectively, and used without any further purification in our experiments. Modifications caused by aging of the matrix powder were observed in MALDI mass spectra for matrices that were kept for a long period on the shelf. Matrix supplies were only used for 1 year after purchase. Matrix solutions deteriorate fast, so they were freshly prepared before each matrix coating.

For this study, the matrix solution [15 mg/ml in 50:50:0.1 (v/v/v) acetonitrile, water, trifluoroacetic acid] was pneumatically sprayed on the tissue section. Reproducible results were achieved using a thin-layer chromatography sprayer (Typ A, 6 ml, 552–6002, VWR) operated at 0.5 bar air pressure. The pressure was adjusted to the lowest value that would result in fine droplets without spitting. For matrix deposition, the sprayer is moved across the sample at a distance of 15 cm, with a velocity of 10 cm/sec. Before sample coating, the sprayer was tested and adjusted by coating a blank metal plate. It is extremely crucial to deposit a fine layer of matrix without forming a continuous solvent film over the sample to prevent lateral diffusion of the analytes. Multiple layers are sprayed, the new one being applied when the previous is nearly dry, typically every 30 sec. After 40–50 layers, a continuous layer of matrix covers the tissue and the sample support. The entire process is readily observed by illuminating the sample plate with a bright light. When the matrix crystals grow to a usable size, color diffraction of the crystals can be observed. The homogeneity of the coating is then checked by optical microscopic inspection using polarized light.

Alternately to the spray coating, the matrix may be deposited by crystallization at 4° from bulk matrix solution (70 μl/cm^2).

Once the section is coated with matrix, it is stored in an amber dry chamber (no vacuum) to avoid contamination and degeneration of the light-sensitive matrix. Storage up to several days under these conditions did not lead to a significant decrease in signal quality of subsequent MSI analyses.

Fixation, Staining

Most generic staining procedures are poorly or not compatible with MALDI (Chaurand *et al.*, 2004). Therefore, MSI analysis control can be achieved either by staining MSI sections after analysis or with stained adjacent sections.

Sections were fixed at room temperature for about 5 min using 4% paraformaldehyde in 0.1 M phosphate-buffered saline (PBS). After rinsing in PBS for 3 min, sections were rehydrated in 50% ethanol solution and stained with 0.0125% Thioflavin S in 50% ethanol for 5 min. Excess staining solution was removed from the sections by washing in 50% ethanol for 2 min. Sections were rehydrated in distilled water, and a cover slip was added with Vectashield solution for fluorescence microscopy.

Staining of amyloid plaques after the MSI procedure required removal of matrix by extensive washing with ethanol. The staining procedure described previously was then applied, with an extended final washing step with ethanol of 2–30 min.

Fluorescence micrographs were then taken at 420–490 nm excitation and 510 nm emission wavelengths. The example of a stain shown in Fig. 1 is a fluorescence image of a brain section from a 20-month-old APP23 transgenic mouse showing AD-like plaque pathology. The Aβ deposits in APP23 mice are extracellular and mostly of the compact type. The section was imaged with MALDI MSI before the staining procedure previously described was applied. One can clearly see a robust and prominent plaque deposition throughout the entire neocortex (Cx) and hippocampus (Hi). No plaques are visible in caudate putamen (CPu), globus pallidus (GP),

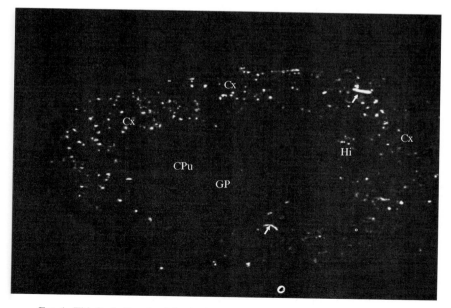

Fig. 1. Thioflavin S–stained forebrain section of a 20-month-old APP23 mouse.

and cerebellum. The resolution and quality of this image is limited because of remaining matrix material on the section surface. Some auto fluorescence of dust particles and parts of the cerebellum are also visible (arrows).

MALDI Mass Spectrometric Imaging

MALDI combination with a time-of flight (TOF) mass analyzer has typically been developed for proteomic studies, where high sensitivity over a wide mass range is required. For a classic application, the sample is co-spotted with the matrix on a sample plate, which is introduced in the MS after solvent evaporation. A single mass spectrum is then acquired from this spot. In a single mass spectrum, MALDI-TOF achieves extremely high sensitivity, down to attomols of analyte. MSI is just a logical extension of that concept, where instead of a single mass spectrum, multiple spectra are acquired from different positions of a flat surface covered with MALDI matrix. By moving the sample under the laser beam, the region to be analyzed is scanned following a user-defined raster. From every position in the raster, or in terms of imaging from every pixel, a mass spectrum is acquired. These mass spectra are then stored together in a data format compatible with the image analysis software.

Although many commercial MALDI-TOF instruments exist, currently only a few are suitable for imaging. These systems have a motorized sample stage, fast lasers, and can be controlled through image acquisition software. At present, the following commercial MSI systems are available: 4700 Proteomics Analyzer (TOF/TOF geometry, Applied Biosystems, Framingham, MA), AXIMA QUIT (ion trap TOF, Shimadzu, Kyoto, Japan), Pulsar XL (quadrupole TOF, MDS Sciex, Concord, ON), ultraflex II, (TOF-TOF geometry, Bruker Daltonics, Bremen, Germany) and sSTR (TOF, Applied Biosystems).

The data shown in this chapter was acquired on an sSTR and a 4700 instrument from Applied Biosystems. Both use the same sample plates with the dimensions of $44 \times 44 \times 0.75$ mm made from stainless steel. These "Opti-TOF" plates are commercially available with a blank polished surface, suitable for placement of up to four mouse brain sections on a single plate. Custom-made water jet-cut plates made from magnetic stainless steel offer an economic alternative for high sample numbers. No performance difference was observed between the commercial and custom-made plates.

The sSTR instrument was operated in the linear mode to obtain best sensitivity, whereas the 4700 was operated in reflector mode. The digitizer gain was set to maximum sensitivity, and the number of data points was adjusted to be less than 32,000 per spectrum by selecting the matching time bin. The laser intensity necessary to obtain a high signal for the Aβ

peptides is typically about 20% higher than dried drop conventional MALDI measurements. It was individually adjusted for every image to optimize signals from $A\beta$ peptide ions while keeping the background signal low. With a high laser intensity one can clearly observe a rise in the baseline at the low mass range.

MSI data were acquired using in-house software (available from http://maldi-msi.org), written for the sSTR and 4700 instruments. Analyze 7.5 (Mayo Foundation, Rochester, MN), has been selected as image data format. The acquisition software allows us to define the area to be imaged by visually inspecting the sample outline. A key characteristic for MSI is the maximal spatial resolution, which depends on sample preparation, laser spot size, and the accuracy of the sample stage. Lateral resolutions of 40–300 μm are currently achievable with commercial instruments, where the laser spot size is the limiting factor. The matrix deposition step also introduces a limitation to the resolution. For the images shown here a resolution of 100 μm was chosen. This matches with the matrix crystal dimensions and results in image files with a manageable size below 1 GByte, as the mouse brain image of 120 × 90 points shown in Fig. 2.

MSI Data Analysis

During a MALDI MSI data acquisition, thousands of complete mass spectra may be acquired, resulting in data sets easily exceeding one gigabyte in size. To make full use of this wealth of information, powerful image analysis software is required. The in-house program BioMap (available from http://maldi.ms) has been adapted to meet the needs of MSI data post processing as previously described (Stoeckli et al., 2002).

Data analysis using BioMap starts by loading the MSI file (File-Import-MSI). Because the entire file is copied from the disk to the memory, this process may take several seconds. Once the file is loaded into BioMap, the menu "Analysis-Plot-Point" is selected, and a mass spectrum window is then opened. By left-clicking with the mouse on the MS image, the corresponding mass spectrum is now displayed for this particular location. On the other hand, right-clicking on the mass spectrum displays the MS image corresponding to this particular mass. One can now go back and forth between the two modes to browse through the data and to get a quick evaluation. By adjusting the minimum and maximum values of the color scale, the distribution can be represented in an optimal way. To get high-quality MS images as shown in Fig. 2, multiple MS data points related to a single analyte have to be combined. This is achieved by dragging the cursor over the peak of interest while holding down the middle mouse button. In the pop-up window "Mean with Baseline Correction" is selected as analysis method. This leads to a

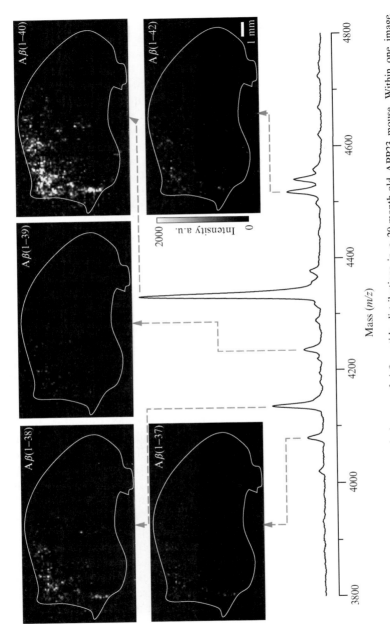

FIG. 2. MALDI mass spectrometric images of Aβ peptide distributions in a 20-month-old APP23 mouse. Within one image acquisition, multiple analytes are measured simultaneously. Here individual mass peaks are used to generate separate images.

molecular image representing the intensity integral of the selected mass range with baseline subtraction. This procedure is repeated for every peak of interest.

Automated Data Analysis

Multiple single peak images can be overlaid to find the spatial correlation of various signals such as different Aβ peptides co-localizing in plaques. However, this method is partly subjective and becomes increasingly impractical as more signals are present in a data set. A recently developed approach applies clustering and multivariate pattern recognition to MALDI-MSI data (McCombie et al., 2005) to increase, identify, and visualize such correlations.

The workflow for clustering and multivariate analysis of MALDI-MSI data is as follows; more details can be found in McCombie et al. (2005). The data set is a sagittal mouse brain section of an AD model mouse.

Alignment. Apparent mass shifts because of uneven surfaces can lead to artifacts in the images. The offset between spectra can be corrected using an internal calibrant or by cross correlation with a reference spectrum. For Fig. 3, insulin was added to the matrix solution as reference.

Clustering. Clustering seeks to identify unique regions of a sample by grouping pixels that have similar mass spectra. If the regions of interest are already known, it can be omitted.

Clustering algorithms performed directly on mass spectra of the images have a limited efficiency if the data present a high noise level. Therefore, MALDI MSI data were filtered before clustering. Principal component analysis (PCA) is well suited for this purpose. Reducing the information of the spectra into a few principal components not only discards noise but reduces the dimensionality of the data. Clustering on the PCA eigenvalues then often clearly shows areas on a tissue with similar mass spectra that can be used to define regions of interest (ROIs). However, PCA itself is computationally intensive, so data compression before PCA may include:

1. Selecting a limited mass range known to be of interest. This may discard important information.
2. Combining adjacent points in the spectrum (binning). This trades resolution for computability if the bins are wider than the peaks.

In Fig. 3, a limited mass range of 4000–8500 Da (7200 data points from a total of 23,000, without binning) was used. After PCA, the significant components are selected (200 for Fig. 3), and the coefficients of each component for every pixel were used for the following clustering. Total data compression at this stage was, therefore, 200/23,000.

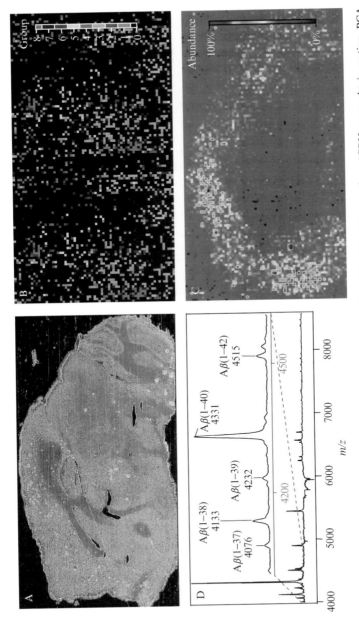

Fig. 3. Clustering and multivariate processing of MS imaging data. (A) Optical image of an APP23 mouse brain section. PCA followed by ISODATA clustering groups pixels with similar mass spectra (B). PCA-DA was performed on clusters four (gray) and eight (pink), the resulting image is shown (C). Contrast is based on the complex peak pattern shown in the spectrum of (D), in which various $A\beta$ peptides play a significant role. (See color insert.)

Many clustering algorithms are known (Jain *et al.*, 1999). For MALDI-MSI, k-means and a variant known as ISODATA offer optimal results. A lot more information on clustering algorithms is readily available in books and the web. The algorithms used here are based on Euclidian distances between the pixels in multidimensional space. Fig. 3B shows the result of ISODATA clustering on the component coefficients found by PCA. Clustering identifies pixels with similar spectra, allowing the definition of meaningful ROIs. In the example, clusters four and eight were selected. The former is outside the tissue and represents background, whereas the latter has a high content of Aβ fragments. The other clusters are not analyzed in more detail here.

PCA-DA on ROIs. Linear discriminant analysis (DA) is a method for analyzing data sets with known class membership. DA is used here to identify those ion signals that differentiate two or more ROIs. Because the spectra contain more data points than there are pixels, the system is over-determined and dimensionality reduction is again needed, or the DA result is not unique. The spectra of the pixels belonging to the ROIs under investigation are taken out of the data set and PCA is applied to them before DA. N-1 relevant DA spectra result from the analysis of N groups. The spectra are aligned (calibrated) for maximal overlap with adjacent spectra. This processing will allow for the spectra to be interpretable as regional differences. The DA spectrum showing the spectral differences between clusters four and eight is shown in Fig. 3D.

DA Image. The DA spectra are differentiating patterns. These patterns may be used to interpret data outside the ROIs, or even from other images. Projection of each pixel's spectrum onto the DA spectra gives coefficients that can then be plotted as an image.

Figure 3C shows an image based on the DA spectrum in Fig. 3D. Bright pixels of Fig. 3C have mass spectra similar to that of 3D. As well as identifying the correct Aβ peptide fragment ratios, the spectrum shows that a variety of other species have the same spatial correlation and may, therefore, be biochemically related. This entire pattern is the basis for the image, greatly enhancing its value for further biological interpretation. However, the identification of unknown signals from a MALDI-MSI data set is still a demanding task and has only been done unambiguously for the Aβ peptides.

Identification

On the basis of the molecular weight of the detected species, the Aβ peptides can be tentatively assigned. This finding can be confirmed by analyzing the fragmentation pattern on excitation of the ions in the mass spectrometer (MS/MS mode). Because this fragmentation follows predictable

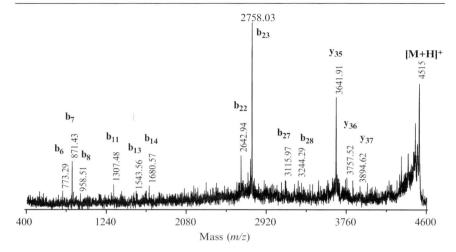

FIG. 4. Fragment spectrum of Aβ(1–42) measured directly from a tissue section. The strongest signals are a result of fragmentation adjacent to aspartic acid.

patterns, it can be used for identification. Fragmentation is induced by applying CHCA as a "hot" matrix (i.e., enhancing fragmentation) (Macht *et al.*, 2004). Raising the laser energy during the MALDI process generates metastable ions or dissociation induced by collision in the flight tube. Figure 4 shows a MS/MS spectrum acquired with the TOF/TOF instrument. The spectrum was acquired from a tissue section after MS image acquisition. Regions with high Aβ concentration were localized during the imaging experiment. The laser was carried back to these locations, and MS/MS data acquisition was performed by accumulating 20,000 mass spectra over a region of 1 mm². Kinetic energy was set to 1 keV, and products of metastable fragmentation were recorded. The MS/MS spectrum of Aβ(1–42) shows the typical peaks associated with fragmentation adjacent to aspartic acid (Bailey *et al.*, 2003). The high number of laser shots is required to obtain a spectrum with an acceptable signal-to-noise ratio. This MS/MS method may be applied to any peak in the spectrum whose sequence information is desired. The spectrum shown in Fig. 4 demonstrates the power of this technique by unambiguous identification of the Aβ(1–42) peptide.

Conclusions

This chapter has described methods for the spatial analysis of Aβ peptides in tissue sections using MALDI MSI. The procedures were optimized for detection of Aβ peptides, but are applicable to other tissues and targets. Clustering and multivariate analysis helped to identify and visu-alize spatial and spectral correlations in MALDI images. Applied to an AD mouse model, it was possible to automatically detect co-localized $A\beta$ peptides and related species, resulting in images with high contrast and information content. It was also shown that the sensitivity of modern TOF and TOF-TOF equipment in combination with optimized preparation methods allows direct and unambiguous identification of $A\beta$ peptides from tissue sections.

References

Bailey, T. H., Laskin, J., and Futrell, J. H. (2003). Energetics of selective cleavage at acidic residues studied by time- and energy-resolved surface-induced dissociation in FT-ICR MS. *Int. J. Mass Spectrom.* **222,** 313–327.

Caprioli, R. M., Farmer, T. B., and Gile, J. (1997). Molecular Imaging of Biological Samples: Localization of Peptides and Proteins Using MALDI-TOF MS. *Anal. Chem.* **69,** 4751–4760.

Chaurand, P., and Caprioli, R. M. (2002). Direct profiling and imaging of peptides and proteins from mammalian cells and tissue sections by mass spectrometry. *Electrophoresis* **23,** 3125–3135.

Chaurand, P., Schwartz, S. A., Billheimer, D., Xu, B. J., Crecelius, A., and Caprioli, R. M. (2004). Integrating histology and imaging mass spectrometry. *Anal. Chem.* **76,** 1145–1155.

Dixon, D. W. (1997). The pathogenesis of senile plaques. *J. Neuropath. Exp. Neur.* **56,** 321–339.

Hintersteiner, M., Enz, A., Frey, P., Jaton, A.-L., Kinzy, W., Kneuer, R., Neumann, U., Rudin, M., Staufenbiel, M., Stoeckli, M., Wiederhold, K.-H., and Gremlich, H.-U. (2005). *In vivo* detection of amyloid-b deposits by near-infrared imaging using an oxazine-derivative probe. *Nat. Biotechnol.* **23,** 577–583.

Jain, K. A., Murty, M. N., and Flynn, P. J. (1999). Data clustering: A review. *ACM Comput. Surv.* **31,** 264–323.

Macht, M., Asperger, A., and Deininger, S. O. (2004). Comparison of laser-induced dissociation and high-energy collision-induced dissociation using matrix-assisted laser desorption/ionization tandem time-of-flight (MALDI-TOF/TOF) for peptide and protein identification. *Rapid Commun. Mass Spectrom.* **18,** 2093–2105.

McCombie, G., Staab, D., Stoeckli, M., and Knochenmuss, R. (2005). Spatial and spectral correlations in MALDI mass spectral images by clustering and multivariate analysis. *Anal. Chem.* **77,** 6118–6124.

Nordberg, A. (2004). PET imaging of amyloid in Alzheimer's disease. *Lancet Neurol.* **3,** 519–527.

Pierson, J., Norris, J. L., Aerni, H. R., Svenningsson, P., Caprioli, R. M., and Andren, P. E. (2004). Molecular profiling of experimental Parkinson's disease: Direct analysis of peptides and proteins on brain tissue sections by MALDI mass spectrometry. *J. Proteome Res.* **3,** 289–295.

Reyzer, M. L., Hsieh, Y., Ng, K., Korfmacher, W. A., and Caprioli, R. M. (2003). Direct analysis of drug candidates in tissue by matrix-assisted laser desorption/ionization mass spectrometry. *J. Mass Spectrom.* **38,** 1081–1092.

Sair, H. I., Doraiswamy, P. M., and Petrella, J. R. (2004). *In vivo* amyloid imaging in Alzheimer's disease. *Neuroradiology* **46,** 93–104.

Schwartz, S. A., Reyzer, M. L., and Caprioli, R. M. (2003). Direct tissue analysis using matrix-assisted laser desorption/ionization mass spectrometry: Practical aspects of sample preparation. *J. Mass Spectrom.* **38**, 699–708.

Stoeckli, M., Chaurand, P., Hallahan, D. E., and Caprioli, R. M. (2001). Imaging mass spectrometry: A new technology for the analysis of protein expression in mammalian tissues. *Nat. Med.* **7**, 493–496.

Stoeckli, M., Staab, D., Staufenbiel, M., Wiederhold, K. H., and Signor, L. (2002). Molecular imaging of amyloid .beta. peptides in mouse brain sections using mass spectrometry. *Anal. Biochem.* **311**, 33–39.

Vanhoutte, G., Dewachter, I., Borghgraef, P., Van Leuven, F., and Van der Linden, A. (2005). Noninvasive *in vivo* MRI detection of neuritic plaques associated with iron in APP transgenic mice, a model for Alzheimer's disease. *Magn. Res. Med.* **53**, 607–613.

[8] Imaging Polyglutamine Deposits in Brain Tissue

By Alexander P. Osmand,
Valerie Berthelier, and Ronald Wetzel

Abstract

The formation of polyglutamine aggregates occupies a central role in the pathophysiology of neurodegenerative diseases caused by expanded trinucleotide repeats encoding the amino acid glutamine. This chapter describes sensitive histological methods for detection of tissue sites that are capable of further recruitment of polyglutamine and for sites rich in polyglutamine defined immunohistochemically. These methods have been found to be applicable in a number of diseases and animal models of disease. Recruitment, which is a property of highly ordered, amyloid-like aggregates, is most commonly found in punctate sites, termed aggregation foci (AF), in the neuronal perikaryonal cytoplasm. As expected, these AF correspond to sites containing polyglutamine aggregates detected using the antibody 1C2. Interestingly, however, many of the latter sites, including most neuropil aggregates and neuronal intranuclear inclusions, exhibit a limited ability to support polyglutamine recruitment. Thus there is limited correlation between the distribution of polyglutamine aggregates and recruitment activity, suggesting functional heterogeneity among polyglutamine aggregates. These methods should prove useful in explaining the relationship between aggregation reactions, aggregate formation, and the development of symptomatic disease and should be adaptable to the study of other protein aggregation disorders.

METHODS IN ENZYMOLOGY, VOL. 412   0076-6879/06 $35.00
          DOI: 10.1016/S0076-6879(06)12008-X

Introduction

Trinucleotide-repeat Diseases

A number of hereditary degenerative diseases are caused by mutations that involve expanded trinucleotide repeats in the affected gene. The largest number of these entails expansion of the trinucleotide, CAG, in a frame that encodes the amino acid glutamine and includes nine neurodegenerative disorders of which Huntington's disease (HD), dentatorubral and pallidoluysian atrophy (DRPLA), and several spinocerebellar atrophies (SCA1, 2, 3, 6, 7, and 17) are conventional autosomal dominant diseases, whereas spinal bulbar muscular atrophy (SBMA) is, in effect, a sex-limited recessive disorder (Wells *et al.*, 1998).

Intranuclear and Neuropil Inclusions

A common histopathological feature of each of the CAG-repeat diseases is the detection at autopsy of inclusion bodies. Initially detected in SCA1 (Cummings *et al.*, 1998), SCA3 (Paulson *et al.*, 1997), and HD (DiFiglia *et al.*, 1997) as ubiquitinylated intraneuronal intranuclear inclusions, these structures have invariably been detected in glutamine-encoding CAG repeat disorders and in transgenic or knock-in animal models of these diseases and have been found to include the polyglutamine-containing segment of the affected gene product, most commonly as a proteolytic fragment of the protein (Bates *et al.*, 2002). Although only intranuclear inclusions are seen in most of these diseases, in HD numerous cytoplasmic inclusions are the dominant inclusion body in cortex being found in neuronal axons and dendrites; these inclusion bodies have been termed neuropil aggregates, because they are only rarely seen in the neuronal perikaryon (Gutekunst *et al.*, 1999).

Polyglutamine Aggregate Formation

The physical properties of polyglutamine aggregates have been considered to reflect an amyloid-like structure (i.e., the appearance of fibrillar or ribbon-like structures in electron micrographs) (Scherzinger *et al.*, 1999) and a composition rich in beta-pleated sheets (Perutz *et al.*, 1994). Because the inclusion bodies in CAG-repeat diseases display the tinctorial properties of amyloid (Huang *et al.*, 1998), that is, the appearance of birefringence when stained with Congo red, these inclusions can be considered to meet the formal histopathological criteria for amyloid.

The invariable presence of polyglutamine-containing inclusion bodies taken together with the discovery that polyglutamine readily aggregates into essentially insoluble amyloid-like structures gave rise to the idea that aggregate formation is the common cause of the expanded CAG repeat diseases. Subsequently, various studies have been interpreted to suggest that polyglutamine aggregates can be toxic, inert, or even protective (Arrasate *et al.*, 2004; Scherzinger *et al.*, 1997). The frequent presence of intranuclear inclusions in large numbers of seemingly unaffected neurons in these diseases indicates that these structures are not invariably neurotoxic. Nevertheless, there remain no *a priori* reasons to reject the hypothesis that the propensity of critical lengths of repetitive polyglutamine sequences to readily form aggregates is, indeed, the underlying cause of these diseases. Considerations of aggregate etiology, coupled with mounting experimental evidence, suggest that polyglutamine proteins can form a variety of aggregates of different sizes, morphologies, and functional characteristics (Wetzel, 2006). Thus, although large inclusions may prove to not have a toxic role, other aggregated states, and/or the aggregation process itself, may be implicated in the disease mechanism. Such arguments draw attention to the critical need for improved methods for detection of aggregates, and in particular requirements for enhanced sensitivity and the ability to discriminate among different aggregate types.

Detection of Polyglutamine Aggregates

Inclusion bodies were initially detected by immunoreactivity with antibodies to ubiquitin or with antibodies to the respective mutant protein (e.g., ataxin-3 in SCA3, ataxin-1 in SCA1, and huntingtin in HD). Subsequently, in HD, antibodies with selective reactivity for the aggregated fragments of the protein were used to demonstrate that anti-ubiquitin antibodies, although reactive with all intranuclear inclusions, only detected a subset of neuropil aggregates (Gutekunst *et al.*, 1999); furthermore, the use of antibodies specific to various regions of the huntingtin molecule showed that only fragments of the N-terminal region of the molecule was contained within the inclusion bodies (Lunkes *et al.*, 1999). The presence of the polyglutamine-containing segment in inclusion bodies was confirmed by the use of polyglutamine-specific monoclonal antibodies and one of these, 1C2, has become the *de facto* standard reagent for the immunochemical detection of polyglutamine (Trottier *et al.*, 1995).

Polyglutamine Aggregation Reactions In Vitro *and* In Situ

Because of solubility problems, initial observations on the formation of amyloid-like fibrils were based on studies with short lengths of polyglutamine peptides (Perutz *et al.*, 1994). Development in this laboratory of novel peptides together with procedures to solubilize and maintain pathological lengths of polyglutamine in solution (Chen and Wetzel, 2001) permitted the study of both solution-phase and solid-phase aggregation reactions of such peptides *in vitro*, demonstrating for the first time the kinetics of recruitment and elongation reactions of various lengths of polyglutamine (Berthelier *et al.*, 2001; Chen *et al.*, 2002). The logical application of these methods to the detection of polyglutamine recruitment in HD brain tissue and in animal models of HD, demonstrating sites at which active aggregate formation was ongoing, was successfully undertaken. Discrete intraneuronal structures were readily demonstrated, and these have been termed *aggregation foci*. The selective and sensitive methods that have been developed to detect these sites are the subject of this chapter, and the methods should be applicable to all glutamine-encoding CAG repeat diseases, although these have only been thoroughly investigated in HD and in a number of rodent models of HD (Menalled *et al.*, 2003; Slow *et al.*, 2003; von Horsten *et al.*, 2003) and recently found in a mouse model of Machado-Joseph disease (spinocerebellar ataxia-3) (Goti *et al.*, 2004).

Methods

Introduction

A previous volume in this series included articles presenting protocols for detecting amyloid proteins in tissues using conventional immunohisto-chemical methods (Westermark *et al.*, 1999) and for detecting inclusion bodies in CAG-repeat diseases (Davies *et al.*, 1999). The deposition of radiolabeled amyloid-β onto tissue sites and detection by autoradiography was also described (Esler *et al.*, 1999), an approach that is analogous in principle to the recruitment method described in the following. The methods described here have been developed specifically for use with free-floating sections of formalin-fixed tissue and depart sufficiently from conventional methods to warrant presentation in detail.

Glutamine is unreactive to aldehyde fixation, and polyglutamine-containing fragments of proteins would be preserved *in situ* by cross-linkage of adjacent reactive amino acids. The polyglutamine segment could thus be preserved in a native configuration, and aggregation-competent sites present in cells and tissues would be potentially reactive with added polyglutamine-containing peptides. The use of "antigen retrieval" techniques, such as thermal, denaturing, or enzymatic treatments, might be expected to expose additional epitopes that would serve as sites for poly-glutamine recruitment, but these sites would not necessarily represent biologically relevant sites of active polyglutamine-mediated aggregation or aggregate formation. In addition, the use of animal sera as conventional blocking agents for immunohistochemical methods should be avoided, because these may contain polyglutamine-binding proteins.

Human and Animal Tissues and Tissue Preparation

Formalin-fixed HD brain and control tissues are available from numerous archival collections; in the United States, these include the Harvard Brain Tissue Resource Center (http://www.brainbank.mclean.org), the National Brain and Spinal Fluid Resource Center (http://www.loni.ucla.edu/~nnrsb/NNRSB), and the New York Brain Bank (http://www.nybb.hs.columbia.edu). More than two dozen mouse models of HD have been created (Levine *et al.*, 2004). Some of these are available from the Jackson Laboratories; others should be available as breeding stock from the laboratories of origin.

Fixed human tissue can be cut into sections suitable for polyglutamine recruitment reactions either with a vibrating microtome or by freeze-cutting with a sliding microtome. Vibrating microtomes (e.g., Vibratome, Technical Products International; OTS-5000, Electron Microscopy Sciences) should be used according to the manufacturer's instructions. We routinely cut 40-μm sections of 5-mm-thick blocks of tissue up to 1 to 2 cm square with sapphire knives under cold phosphate-buffered saline (PBS; Fluka, BioChemika Ultra grade); sections are transferred into and stored under refrigeration in PBS containing 0.02% sodium azide.

Procedures for cryoprotection and freeze-cutting fixed brain tissue have been described in detail by others in an article that includes a detailed discussion of the factors involved in cryoprotection (Rosene *et al.*, 1986); these methods permit the preparation of much larger sections than can be obtained with a vibrating microtome. Blocks of tissue up to the size of coronal hemispheric sections of human brain can be sectioned using the following adaptation of the Rosene and Rhodes procedure:

1. Tissue blocks are equilibrated for 24–48 h in at least 10 volumes of a solution of 10% glycerol and 2% DMSO in neutral phosphate-buffered formalin (PBF) with gentle constant agitation at room temperature.
2. Blocks are then equilibrated in a solution of 20% glycerol and 2% DMSO in PBF for 24–72 h, depending on the thickness of the block, also at room temperature with agitation. After this time, the blocks may be stored refrigerated for a few days before sectioning.
3. The specimen stage of a sliding microtome is cooled with powdered dry ice and a layer of specimen mounting medium (e.g., Cryomatrix, Shandon; Tissue Freezing Medium, Triangle Biomedical Sciences) placed on the stage. As the layer freezes, blocks are placed in a

suitable orientation onto the freezing surface and frozen slowly by cooling with powdered dry ice.

4. The base of the block should be maintained below $-30°$ by constant addition of dry ice to the surface of the stage surrounding the specimen; a J-type or K-type thermistor probe embedded in the mounting medium below the block and a thermometer are used to monitor the temperature of the stage.

5. Thirty-five or 40-μm sections are cut with a knife maintained at room temperature, and thawed sections are removed from the surface of the knife with a paintbrush moistened with buffer.

6. Sections are transferred to PBS containing 0.02% sodium azide and stored at $4°$ until processed.

Sections that have been exposed to cryoprotection and freezing should be stored in buffer for 3–4 wks before testing for the presence of polyglutamine recruitment sites to recover full activity.

Both vibrating microtome and freeze-cut sectioning can be used with rodent tissues; however, for studies on polyglutamine recruitment and aggregation sites in animal models of HD, we have used Multibrain embedding, a proprietary service of Neuroscience Associates (http://www.neuroscienceassociates.com/multibrain.htm) that involves embedding brains in a matrix and collecting freeze-cut sections containing multiple brains, up to 16 rat or 25 mouse brains in a single block, usually into a series in which every 24th section is collected into a separate container. Each series can then be processed as a batch of free-floating sections containing representative sections of most anatomical subdivisions of the brains of numerous individual animals. Rodent brains for polyglutamine recruitment studies should be fixed *in situ* by perfusion with buffered paraformaldehyde and stored until embedding in buffered saline; limited evidence suggests that both fixation in the presence of glutaraldehyde and prolonged storage in 30% sucrose considerably reduce the reactivity of polyglutamine recruitment sites.

Peptide Design, Synthesis, and Purification

Previous studies from this laboratory have demonstrated that synthetic polypeptides containing long segments of polyglutamine require the presence of flanking residues of basic amino acids to maintain solubility. Peptides containing approximately 30 glutamine residues have been found to remain in monomeric form under the conditions used, while being readily recruited into elongating fibrillar assemblies. The addition of biotin moieties to these peptides permits the use of sensitive avidin-based reagent systems for the detection and quantification of recruited peptide. Biotinylated

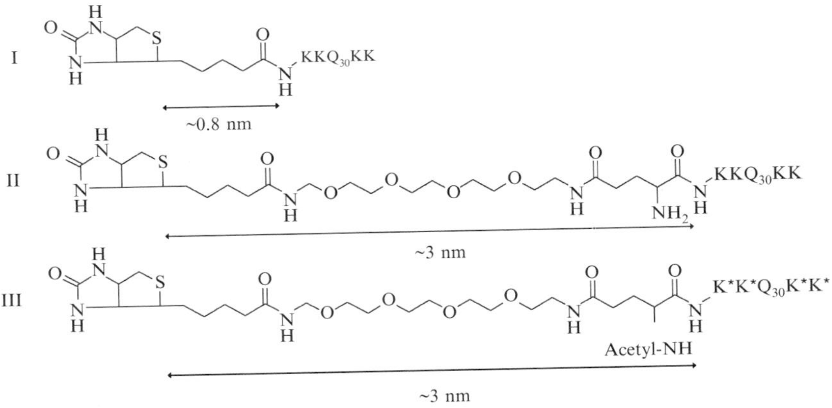

Fig. 1. Biotinylated synthetic peptides. Based on a general formula of lysyl-lysyl-polyglutaminyl-lysyl-lysine, biotin is either added directly, amide-linked to the α-amino acid of the N-terminal lysine (I) referred to as bQ30. Addition of an N-terminal glutaminyl residue with the γ-amide group substituted with a biotinyl-polyethylene glycol (II), referred to as bPEGQ30, provides reduced steric hindrance to polyglutamine aggregation and enhanced accessibility to avidin-based reagents; to avoid potential reactivity of amino groups with residual aldehyde or other reactive groups in fixed tissue. In (III) the α-amino group of the N-terminal substituted glutaminyl residue is acetylated, and the lysyl residues are added as ε-N-monomethyl lysine, indicated by K*. This peptide derivative is referred to as modified biotinylated PEG Q30, abbreviated to mbPEGQ30. Interatomic distances are approximate maxima for extended configurations.

polypeptides are prepared for this laboratory by custom solid-phase peptide synthesis at the Keck Biotechnology Center at Yale University (http://info.med.yale.edu/wmkeck); the structures of peptides used for *in situ* recruitment and elongation are represented in Fig. 1. For optimization of reactivity, the accessibility of the biotin moiety for streptavidin conjugates has been enhanced by inserting a polyethylene glycol spacer between biotin and the peptide (bPEGQ30 and mbPEGQ30). As a further refinement, potential reactivity of peptide amino groups with residual aldehyde derivatives, Schiff bases, or Amadori rearrangement products in fixed tissue sections has been eliminated by acetylation of the N-terminal group and the use of ε-N-monomethyl lysyl residues flanking the polyglutamine sequence (mbPEGQ30).

Peptides are purified by HPLC and disaggregated, as described in Chapter 3 of Volume 413 (O'Nuallain *et al.*, 2006), and stored at $-80°$ in 1-ml aliquots of 500 nM stock solutions in 5% DMSO in PBS.

Polyglutamine Recruitment Methods

Sensitive and specific methods to detect sites of recruitment in tissues have been developed that incorporate concepts from several immuno-histochemical protocols, using in particular the approach of Hoffman and colleagues (Berghorn *et al.*, 1994) in which the use of proteinaceous blocking agents has been avoided by destroying aldehyde-derived sites with borohydride treatment and preventing nonspecific binding of peptides and protein reagents by the inclusion of relatively high concentrations of detergent. The use of diluted reagents, together with extensive washings between steps, further reduces the potential for nonspecific or background staining reactions. Horseradish peroxidase–based enzymatic methods using avidin-biotinylated peroxidase complexes (ABC) enable the ready detection of biotinylated polyglutamine peptides, and sensitive and permanent detection is provided by the use of nickel-enhanced diaminobenzidine as the enzymatic substrate. Adams (1992) introduced a peroxidase-based procedure to enhance the sensitivity by depositing additional biotin through the action of the enzyme on biotinylated tyramine. Although numerous posttreatment methods for the intensification of the density and for the enhancement of the sensitivity of peroxidase-based procedures have been developed, the use of tyramide amplification provides both high sensitivity and ease of use under controlled conditions.

Solutions

1. PBS: Phosphate-buffered saline (Fluka, BioChemika grade) is supplied as a $10\times$ concentrate.
2. Triton X-100 (Sigma): Triton X-100 is stored as a 10% solution and is freshly filtered (0.2 μm) before use.
3. 0.4% Triton X-100 in PBS (PBSTx).
4. Tris-imidazole buffer (TI): 0.05 M Tris, 0.05 M imidazole, adjusted to pH 7.4 with 1 N HCl.
5. Substrate buffer (SB): 0.6% nickelous ammonium sulfate in TI, pH 7.4.
6. Diaminobenzidine (DAB): prepared as a 1% aqueous solution and stored as small (ca. 1 ml) aliquots at $-80°$.
7. Biotinylated tyramine (BT) is prepared as follows as described by Adams (1992): 100 mg sulfosuccinimidyl-6-(biotinamide)hexanoate (Pierce) is added to 32 mg tyramine hydrochloride dissolved in 50 mM borate buffer (pH 8.0) and the mixture stirred overnight. The solution is filtered (0.2 μm) and dispensed into small aliquots and stored at $-80°$.

Protocol for Polyglutamine Recruitment and Elongation. Sections are processed free-floating with gentle continuous agitation at room temperature. Sections are transferred at each step either by gently lifting with a bent glass rod or with a basket fabricated from inert plastic screen (e.g., Spectra/Mesh Fluorocarbon macroporous filter, Spectrum Laboratories Inc.) attached with silicone adhesive to the base of a glass cylinder; multiple sections can then be readily processed simultaneously.

1. Sections are washed twice in PBS for 10 min each.
2. Sections are treated with 1% sodium borohydride in PBS for 30 min.
3. Borohydride is removed by washing three times in PBS.
4. Sections are permeabilized and blocked by washing three times in PBSTx for 10 min, 30 min, and 10 min, respectively.
5. Sections are incubated overnight in a solution of biotinylated peptide in PBSTx at concentrations ranging from 10–25 nM (for bPEGQ30 or mbPEGQ30) to 100 nM (for bQ30).
6. Sections are washed several times over 1 h with PBSTx.
7. Sections are incubated for 1 h in avidin-biotinylated peroxidase, ABC "Elite" reagent (Vector Laboratories), prepared at least 30 min before use as a 1:200 dilution of reagents A and B, and diluted twofold immediately before use.
8. Unbound ABC reagent is removed by washing three times for 10 min each with PBSTx and twice with TI.
   a. When BT amplification is required, sections are washed once again for 10 min with TI.
   b. Sections are incubated for 10–15 min in a solution of BT diluted 1:100 in TI from stock.
   c. Sections are washed twice for 10 min in TI and three times for 10 min each in PBSTx.
   d. Sections are incubated for 1 h in avidin-biotinylated peroxidase reagent prepared as in (7).
   e. Sections are washed three times in PBSTx and twice in TI for 10 min each.
9. Sections are washed once for 10 min in SB.
10. Sections are incubated in nickel-DAB-H_2O_2 prepared by diluting stock DAB to a concentration of 0.01% in SB and adding H_2O_2 to a final concentration of 0.0006–0.001%. Incubation times range from a few minutes, for amplified procedures using rodent tissues and mbPEGQ30 or bPEGQ30, to 30–60 min for HD tissue without amplification.
11. Sections are washed once in TI, washed once for 30 min in mounting medium consisting of 20 mM ammonium acetate containing

0.01% Triton X-100 or 20% ethanol as wetting agents, transferred to fresh medium, mounted on gelatin-subbed glass microscope slides, and air-dried.

12. When required, sections may be counterstained for Nissl substance as described later. Mounted sections are processed through graded alcohols, xylenes, and coverslipped with Entellan (Merck).

Preparation of Slides and Staining for Nissl Substance. Glass microscope slides are washed in detergent, rinsed in deionized water, treated briefly with acid alcohol (10% concentrated HCl in 95% ethanol), rinsed in deionized water, dipped in a warm (50°) solution of 0.5% gelatin (Sigma) in water, drained, and dried at 60°. Several of the classical "Nissl substance" stains give suitable contrast for counterstaining the blue-black nickel-DAB peroxidase product, including cresyl violet, neutral red, and thionine. A light counterstain with the latter dye gives good contrast, although the conditions for optimum staining vary considerably. For this thionine stain, slide-mounted sections are washed in 70% ethanol and rinsed in water; the sections are stained with thionine (0.01–0.05%) in 0.04 M acetic acid/ 0.04 M sodium acetate pH 4.6, for 2–10 min, then rinsed extensively with water and inspected before processing for coverslipping. If inspection indicates insufficient staining, the sections can be returned to the thionine solution; if staining is excessive, sections can be differentiated in 0.1% glacial acetic acid in 95% ethanol for 2–3 min, rinsed in 95% ethanol, processed through ethanol and xylenes, and coverslipped as previously.

Examples of polyglutamine recruitment in HD tissue and in three animal models of HD, and in an animal model of SCA3 (Machado-Joseph disease), are shown in Figs. 2 and 3.

Immunohistochemical Detection of Polyglutamine

The sensitive detection of polyglutamine within inclusion bodies in HD tissue and in other CAG-repeat diseases requires prior unfolding of the amyloid-like structure to enhance access of antibodies to the peptide; this necessitates the use of aggressive antigen retrieval methods using conditions that have been shown to dissociate polyglutamine aggregates. Although numerous procedures have been used for this purpose, it has been generally found that treatment with concentrated formic acid provides maximal solubilization of amyloid fibrils and should thus give optimal exposure of buried epitopes. Although it is expected that aldehyde fixation should provide extensive covalent cross-linking of unfolded proteins within tissue sections, there is evidence that small amounts of amyloid proteins can be dissociated from formalin-fixed tissue (Murphy *et al.*, 2001). Care should be taken to exclude the possibility that solubilized proteins bind back to tissue on dilution of the formic acid.

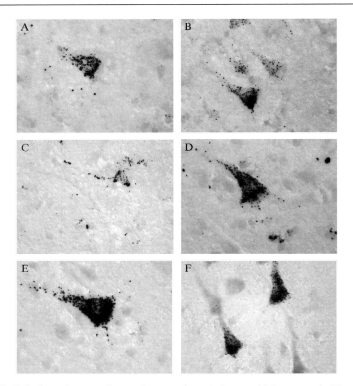

Fig. 2. Polyglutamine recruitment in several cortical pyramidal neurons in Huntington's disease (HD) brain (A–E) and in presymptomatic HD (F). Aggregation foci are seen widely distributed in the cytoplasm of affected cells, with the absence of nuclear involvement (A, C, E) and occasionally appearing ordered along presumably cytoskeletal elements (C). There is a noticeable accumulation of foci toward the axonal pole (B, E, F), although foci are frequently observed in proximal dendrites (D, E); bPEGQ30, 10 nM (A–E) and 20 nM (F) without tyramide enhancement. (See color insert.)

The density of polyglutamine epitopes in inclusion bodies in HD tissues is sufficiently high that biotinylated tyramide amplification is not required; however, in all animal models of CAG-repeat diseases thus far studied, under the conditions described here, amplification has been necessary to detect the modest levels of polyglutamine in aggregation foci, while dramatically enhancing reactivity with other inclusion bodies.

The monoclonal antibody, 1C2, was generated against the immunogen TATA box-binding protein (TBP) (Trottier *et al.*, 1995), a protein that includes a relatively long polyglutamine tract. By using this antibody with human tissues, we have seen a weak reactivity with cell nuclei, most

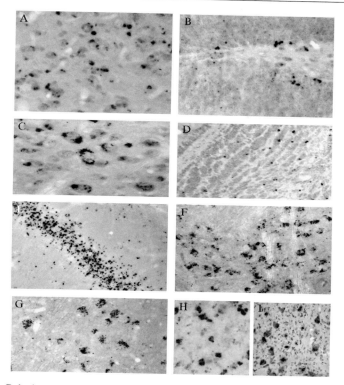

Fɪɢ. 3. Polyglutamine recruitment in animal models of CAG–repeat diseases; recruitment sites in these models are only revealed after biotin tyramide amplification. In the YAC 128 mouse (A–D), recruitment is widely distributed, shown here in thalamus (A), in the neurogenic layer of the hippocampal dentate gyrus (B), in magnocellular neurons of the red nucleus (C), and in the granule cell layer of the olfactory bulb (D). In the CAG140 mouse, recruitment foci were also widely distributed and are shown in the CA3 region of hippocampus (E), in neurons in the ventral tegmental area (F), and in the motor trigeminal nucleus (G). Recruitment sites were less widely distributed in the transgenic rat being observed in the basal ganglia and thalamus (not shown) and cortex (H). In the SCA3 mouse, nuclear and cytoplasmic recruitment were seen most strikingly in the vestibular nucleus (I); bPEGQ30 20 nM with tyramide recruitment (A–H) and mbPEGQ30 10 nM with tyramide recruitment (I). (See color insert.)

markedly in glial cells, presumably caused by the presence of TBP; this reactivity is essentially eliminated by the formic acid treatment.

Solutions. Reagents identical to (1) to (7) in the preceding section are used. In addition, the monoclonal antibody to polyglutamine, 1C2, is obtained from Chemicon (Temecula, CA) and used at a concentration of 25 ng/ml; biotinylated antibody to mouse IgG is obtained from Vector Laboratories and used at a concentration of 0.5–1 μg/ml.

Protocol for Detection of Polyglutamine. Sections are processed as previously, free-floating with gentle continuous agitation at room temperature.

1. Sections are washed twice in PBS and once in water for 10 min each.
2. Sections are treated three times with 88% or 98% formic acid for 10 min each.
3. Formic acid is removed by two or three quick rinses in water, followed by two washes in PBS for 10 min each.
4. Sections are treated with 1% sodium borohydride in PBS for 30 min.
5. Borohydride is removed by washing three times in PBS.
6. Sections are permeabilized and blocked by washing three times in PBSTx for 10 min, 30 min, and 10 min, respectively.
7. Sections are incubated overnight in a solution of the monoclonal antibody, 1C2, (Chemicon) in PBSTx at a concentration of 25 ng/ml.
8. Sections are washed several times over 1 h with PBSTx.
9. Sections are incubated for 2 h in biotinylated anti-mouse IgG (Vector Laboratories, absorbed with rat IgG), 0.5–1.0 μg/ml in PBSTx.
10. Sections are washed several times over 1 h with PBSTx.
11. Sections are incubated for 1 h in avidin-biotinylated peroxidase, ABC "Elite" reagent (Vector Laboratories), prepared at least 30 min before use as a 1:200 dilution of reagents A and B, and diluted twofold immediately before use.
12. Unbound ABC reagent is removed by washing three times for 10 min each with PBSTx and twice with TI.
    a. When BT amplification is required, sections are washed once again for 10 min with TI.
    b. Sections are incubated for 10–15 min in a solution of BT diluted 1:100 in TI from stock.
    c. Sections are washed twice for 10 min in TI and three times for 10 min each in PBSTx.
    d. Sections are incubated for 1 h in avidin-biotinylated peroxidase prepared as in (11).
    e. Sections are washed three times in PBSTx and twice in TI for 10 min each.
13. Sections are washed once for 10 min in SB.
14. Sections are incubated in nickel-DAB-H_2O_2 prepared by diluting stock DAB 1:100–0.01% in SB and adding H_2O_2 to a final concentration of 0.0006–0.001%. Incubation times range from a few minutes to up to 30 min.
15. Sections are washed once in TI, washed once for 30 min in mounting medium consisting of 20 mM ammonium acetate containing 0.01% Triton X-100 or 20% ethanol as wetting agents, transferred to fresh medium, mounted on gelatin-subbed glass microscope slides, and air-dried.

16. Sections may be counterstained as described previously before processing through graded alcohols, xylenes, and coverslipping.

Examples of polyglutamine distribution in HD tissue and in a rat model of HD are shown in Fig. 4. Striking qualitative differences have been found between the immunoreactivity for polyglutamine using the 1C2 antibody and the ability of these structures to recruit synthetic polyglutamine

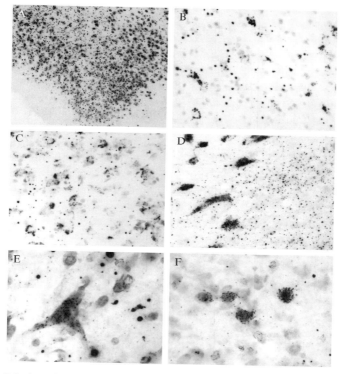

FIG. 4. Polyglutamine detected with 1C2 in 24-month-old heterozygous transgenic rat (A–D). Intranuclear accumulation and intranuclear inclusions, as well as neuropil aggregates are present densely in ventral striatum, shown here, in the olfactory tubercle (A). In cortex both cytoplasmic punctae, presumably corresponding to aggregation foci, and neuropil aggregates, occasionally appearing as "chains," are seen (B). Both weakly staining cytoplasmic sites and neuropil aggregates are shown in thalamus (C). In substantia nigra (D), neurons in the pars compacta show numerous minute cytoplasmic polyglutamine aggregates, whereas many neuropil aggregates are seen in the pars reticulata. In HD cortex (E, F), 1C2 staining demonstrates both punctate cytoplasmic staining in pyramidal neurons, as well as the presence of neuropil aggregates; weak peripheral staining of glial nuclei is also seen. 1C2 1:30,000 after formic acid treatment, with (A–D) and without (E, F) tyramide amplification. (See color insert.)

peptides; this is seen both for HD tissue and in the transgenic rat model. In HD tissue, recruitment activity is essentially confined to cytoplasmic aggregation foci (Fig. 2), whereas 1C2-reactivity is present weakly in cytoplasm but markedly in neuropil aggregates (Fig. 4E). Similarly, in cortex in the rat model, recruitment activity is predominantly cytoplasmic (Fig. 3H), whereas only modest reactivity is seen with 1C2 in the neuronal cytoplasm. There is additional 1C2 staining of scattered spherical aggregates in the neuropil, often appearing as linear arrays, which presumably indicate that these neuropil aggregates are present in neuronal processes (Fig. 4B).

Microscopy and Photomicrography

The procedures described here have the advantage of providing permanently stained sections, the optical quality of which improves over time as the mounting medium dries. However, imaging of small, frequently submicron, inclusions in thick sections can be challenging, because the particles are close to the limit of resolution of the light microscope, and only a small fraction of the section is in focus at the magnifications required. A high-definition digital camera coupled with Z-stage microscope control permits the rapid collection of a sequence of images through the full thickness of the section; image analysis software can then be used to generate a composite in-focus image of all features within that area of the section.

In this laboratory, we have used a Leica DMRB microscope (Leica Microsystems), equipped with a MAC-5000 Z-stage controller (Ludl Electronic Systems). Sequences of images were collected using a Spot RT Color camera (Diagnostic Instruments) as a through focus series taken at a Z-stage step interval determined by the numerical aperture of the objective and the refractive index of the medium used, according to the following formula: ΔZ (in microns) $= \lambda/4\eta\sin^2(0.5\sin^{-1}[NA/\eta])$, where λ is the wavelength (conventionally: 0.5 μm), η is the refractive index of the medium, and NA is the numerical aperture of the objective lens. For the objectives routinely used on the Leica DMRB microscope, the Z-stage spacing ranges from 1.87 μm (20×, NA:0.5, air), 0.75 μm (40×, NA:0.75, air), 0.34 μm (100×, NA:1.3, oil), to 0.27 μm (63×, NA:1.4, oil). A through-focus stack of images is flattened into a single in-focus image, a local contrast composite, using a maximum local contrast algorithm in the extended depth of field feature of Image-Pro Plus (Media Cybernetics, Version 5.0).

Acknowledgments

This work was supported by grants and contracts from the Hereditary Diseases Foundation to R. W. and A. O. The methods were established and validated using human tissues obtained from the Harvard Brain Tissue Resource Center, funded by R24-MH068855, from the Human Brain and Spinal Fluid Resource Center, and from the New York Brain Bank, and, using brain tissue from rodent models of polyglutamine diseases, from Michael Hayden and Elizabeth Slow, University of British Columbia, from Marie-Françoise Chesselet and Miriam Hickey, University of California at Los Angeles, from Stephan von Hörsten, Hannover Medical School, from Henry Paulson and Aislinn Williams, University of Iowa, and from Veronica Colomer, the Johns Hopkins University.

References

Adams, J. C. (1992). Biotin amplification of biotin and horseradish peroxidase signals in histochemical stains. *J. Histochem. Cytochem.* **40,** 1457–63.

Arrasate, M., Mitra, S., Schweitzer, E. S., Segal, M. R., and Finkbeiner, S. (2004). Inclusion body formation reduces levels of mutant huntingtin and the risk of neuronal death. *Nature* **431,** 805–810.

Bates, G., Harper, P. S., and Jones, L. (2002). "Huntington's Disease." Oxford University Press, Oxford, New York.

Berghorn, K. A., Bonnett, J. H., and Hoffman, G. E. (1994). cFos immunoreactivity is enhanced with biotin amplification. *J. Histochem. Cytochem.* **42,** 1635–42.

Berthelier, V., Hamilton, J. B., Chen, S., and Wetzel, R. (2001). A microtiter plate assay for polyglutamine aggregate extension. *Anal. Biochem.* **295,** 227–236.

Chen, S., Berthelier, V., Hamilton, J. B., O'Nuallain, B., and Wetzel, R. (2002). Amyloid-like features of polyglutamine aggregates and their assembly kinetics. *Biochemistry* **41,** 7391–7399.

Chen, S., and Wetzel, R. (2001). Solubilization and disaggregation of polyglutamine peptides. *Protein Sci.* **10,** 887–891.

Cummings, C. J., Mancini, M. A., Antalffy, B., DeFranco, D. B., Orr, H. T., and Zoghbi, H. Y. (1998). Chaperone suppression of aggregation and altered subcellular proteasome localization imply protein misfolding in SCA1. *Nat. Genet.* **19,** 148–154.

Davies, S. W., Sathasivam, K., Hobbs, C., Doherty, P., Mangiarini, L., Scherzinger, E., Wanker, E. E., and Bates, G. P. (1999). Detection of polyglutamine aggregation in mouse models. *In* "Methods in Enzymology," (R. Wetzel, ed.), Vol. 309, pp. 687–701. Academic Press, San Diego, CA.

DiFiglia, M., Sapp, E., Chase, K. O., Davies, S. W., Bates, G. P., Vonsattel, J. P., and Aronin, N. (1997). Aggregation of huntingtin in neuronal intranuclear inclusions and dystrophic neurites in brain. *Science* **277,** 1990–1993.

Esler, W. P., Mantyh, P. W., and Maggio, J. E. (1999). Deposition of soluble amyloid-β into amyloid templates: With application to the identification of amyloid fibril extension inhibitors. *In* "Methods in Enzymology" (R. Wetzel, ed.), Vol. 309, pp. 350–374. Academic Press, San Diego.

Goti, D., Katzen, S. M., Mez, J., Kurtis, N., Kiluk, J., Ben-Haiem, L., Jenkins, N. A., Copeland, N. G., Kakizuka, A., Sharp, A. H., Ross, C. A., Mouton, P. R., and Colomer, V. (2004). A mutant ataxin-3 putative-cleavage fragment in brains of Machado-Joseph disease patients and transgenic mice is cytotoxic above a critical concentration. *J. Neurosci.* **24,** 10266–10279.

Gutekunst, C. A., Li, S. H., Yi, H., Mulroy, J. S., Kuemmerle, S., Jones, R., Rye, D., Ferrante, R. J., Hersch, S. M., and Li, X. J. (1999). Nuclear and neuropil aggregates in Huntington's disease: Relationship to neuropathology. *J. Neurosci.* **19,** 2322–2334.

Huang, C. C., Faber, P. W., Persichetti, F., Mittal, V., Vonsattel, J. P., MacDonald, M. E., and Gusella, J. F. (1998). Amyloid formation by mutant huntingtin: Threshold, progressivity and recruitment of normal polyglutamine proteins. *Somat. Cell Mol. Genet.* **24,** 217–233.

Levine, M. S., Cepeda, C., Hickey, M. A., Fleming, S. M., and Chesselet, M. F. (2004). Genetic mouse models of Huntington's and Parkinson's diseases: Illuminating but imperfect. *Trends Neurosci.* **27,** 691–697.

Lunkes, A., Trottier, Y., Fagart, J., Schultz, P., Zeder-Lutz, G., Moras, D., and Mandel, J. L. (1999). Properties of polyglutamine expansion *in vitro* and in a cellular model for Huntington's disease. *Philos. Trans. R. Soc. Lond. B Biol. Sci.* **354**, 1013–1019.

Menalled, L. B., Sison, J. D., Dragatsis, I., Zeitlin, S., and Chesselet, M. F. (2003). Time course of early motor and neuropathological anomalies in a knock-in mouse model of Huntington's disease with 140 CAG repeats. *J. Comp. Neurol.* **465**, 11–26.

Murphy, C. L., Eulitz, M., Hrncic, R., Sletten, K., Westermark, P., Williams, T., Macy, S. D., Wooliver, C., Wall, J., Weiss, D. T., and Solomon, A. (2001). Chemical typing of amyloid protein contained in formalin-fixed paraffin-embedded biopsy specimens. *Am. J. Clin. Pathol.* **116**, 135–42.

O'Nuallain, B., Thakur, A., Williams, A., Bhattacharyya, A., Chen, S., Thiagarajan, G., and Wetzel, R. (2006). Kinetics and thermodynamics of amyloid assembly using an HPLC-based sedimentation assay. **413** (in press).

Paulson, H. L., Perez, M. K., Trottier, Y., Trojanowski, J. Q., Subramony, S. H., Das, S. S., Vig, P., Mandel, J. L., Fischbeck, K. H., and Pittman, R. N. (1997). Intranuclear inclusions of expanded polyglutamine protein in spinocerebellar ataxia type 3. *Neuron* **19**, 333–344.

Perutz, M. F., Johnson, T., Suzuki, M., and Finch, J. T. (1994). Glutamine repeats as polar zippers: Their possible role in inherited neurodegenerative diseases. *Proc. Natl. Acad. Sci. USA* **91**, 5355–5358.

Rosene, D. L., Roy, N. J., and Davis, B. J. (1986). A cryoprotection method that facilitates cutting frozen sections of whole monkey brains for histological and histochemical processing without freezing artifact. *J. Histochem. Cytochem.* **34**, 1301–1315.

Scherzinger, E., Lurz, R., Turmaine, M., Mangiarini, L., Hollenbach, B., Hasenbank, R., Bates, G. P., Davies, S. W., Lehrach, H., and Wanker, E. E. (1997). Huntingtin-encoded polyglutamine expansions form amyloid-like protein aggregates *in vitro* and *in vivo*. *Cell* **90**, 549–558.

Scherzinger, E., Sittler, A., Schweiger, K., Heiser, V., Lurz, R., Hasenbank, R., Bates, G. P., Lehrach, H., and Wanker, E. E. (1999). Self-assembly of polyglutamine-containing huntingtin fragments into amyloid-like fibrils: Implications for Huntington's disease pathology. *Proc. Natl. Acad. Sci. USA* **96**, 4604–4609.

Slow, E. J., van Raamsdonk, J., Rogers, D., Coleman, S. H., Graham, R. K., Deng, Y., Oh, R., Bissada, N., Hossain, S. M., Yang, Y. Z., Li, X. J., Simpson, E. M., Gutekunst, C. A., Leavitt, B. R., and Hayden, M. R. (2003). Selective striatal neuronal loss in a YAC128 mouse model of Huntington disease. *Hum. Mol. Genet.* **12**, 1555–1567.

Trottier, Y., Lutz, Y., Stevanin, G., Imbert, G., Devys, D., Cancel, G., Saudou, F., Weber, C., David, G., Tora, L., Agid, Y., Brice, A., and Mandel, J.-M. (1995). Polyglutamine expansion as a pathological epitope in Huntington's disease and four dominant cerebellar ataxias. *Nature* **378**, 403–406.

von Horsten, S., Schmitt, I., Nguyen, H. P., Holzmann, C., Schmidt, T., Walther, T., Bader, M., Pabst, R., Kobbe, P., Krotova, J., Stiller, D., Kask, A., Vaarmann, A., Rathke-Hartlieb, S., Schulz, J. B., Grasshoff, U., Bauer, I., Vieira-Saecker, A. M., Paul, M., Jones, L., Lindenberg, K. S., Landwehrmeyer, B., Bauer, A., Li, X. J., and Riess, O. (2003). Transgenic rat model of Huntington's disease. *Hum. Mol. Genet.* **12**, 617–624.

Wells, R. D., Warren, S. T., and Sarmiento, M. (1998). "Genetic Instabilities and Hereditary Neurological Diseases." Academic Press, San Diego, CA.

Westermark, G. T., Johnson, K. H., and Westermark, P. (1999). Staining methods for identification of amyloid in tissue. *In* "Methods in Enzymology," (R. Wetzel, ed.) Vol. 309, pp. 3–25. Academic Press, San Diego, CA.

Wetzel, R. (2006). Chemical and physical properties of polyglutamine repeat sequences. *In* "Genetic Instabilities and Hereditary Neurological Diseases," (T. Ashizawa and R. D. Wells, eds.), 2nd Ed. Elsevier-Academic Press, Amsterdam.

[9]   X-34 Labeling of Abnormal Protein Aggregates During the Progression of Alzheimer's Disease

By Milos D. Ikonomovic, Eric E. Abrahamson,
Barbara A. Isanski, Manik L. Debnath, Chester A. Mathis,
Steven T. DeKosky, and William E. Klunk

Abstract

Postmortem pathological diagnosis and basic research investigations of neurodegenerative disorders rely on histochemical staining procedures developed specifically to visualize abnormal protein conformation. In Alzheimer's disease (AD), two major pathological hallmarks are required to confirm the clinical diagnosis. Both consist of abnormally aggregated proteins that share the structural and histological properties common to all amyloid deposits. Amyloid-beta peptide (Aβ) of extracellular senile plaques (SP) and hyperphosphorylated tau of intracellular neurofibrillary tangles (NFT) are assembled in the abnormal β-pleated sheet (amyloid-like) structural conformation that can be visualized with histological staining procedures using Congo red or its derivatives. These histochemical dyes bind amyloid with high affinity and allow easy detection of amyloid structure in postmortem brain samples. This chapter focuses on the development and application of a histological protocol using the compound X-34, a highly fluorescent derivative of Congo red, for sensitive detection of pathological amyloid structures in histopathological investigations of postmortem brain tissue. This procedure provides a simple and effective method for detailed fluorescent visualization of the localization and distribution of the majority of currently known major histopathological structures in AD, including compact cored, neuritic, and diffuse-appearing SP, NFT, dystrophic neurites, neuropil threads, and cerebrovascular amyloidosis.

Introduction

Alzheimer's disease (AD) is the most prevalent of several etiologically distinct neurodegenerative disorders characterized neuropathologically by misfolding of specific peptides and proteins (amyloidosis) in the brain. In these disorders, constitutively produced, normally soluble protein monomers polymerize into insoluble fibrillar aggregates by arranging in the β-pleated sheet conformation (Selkoe, 2003). Neuropathological evidence of sufficient numbers of extracellular aggregates of amyloid-β (Aβ)

METHODS IN ENZYMOLOGY, VOL. 412
0076-6879/06 $35.00
DOI: 10.1016/S0076-6879(06)12009-1

peptide in neuritic senile plaques (SP) and intracellular aggregates of hyperphosphorylated tau in neurofibrillary tangles (NFT) defines the diagnosis of AD (National Institute on Aging and Reagan Institute Working Group on Diagnosis Criteria for Alzheimer's Disease, 1997). Postmortem histopathological analysis (Dickson, 2005) and, more recently, *in vivo* brain imaging (Mathis *et al.*, 2005) of these hallmark lesions of AD have been possible because of the unique dye-binding properties of the β-sheet structure assumed by Aβ peptide in SP and walls of blood vessels, and by hyperphosphorylated tau in NFT and neuropil threads (NT). The most widely used amyloid-binding dye, Congo red (Bennhold, 1922; Puchtler *et al.*, 1962), has been superseded by analogs that have several advantages, such as increased sensitivity and better optical quality. These new compounds are proving more applicable as potential *in vivo* neuroimaging agents (Scheme 1; Mathis *et al.*, 2004). One of these compounds is X-34 (1,4-*bis*(3-carboxy-4-hydroxyphenylethenyl)-benzene), which was derived from Congo red by replacing the naphthalenesulfonic acids with salicylic acids and the azo linkage with an ethenyl link (Scheme 1), and which emerged as a particularly sensitive marker of AD pathology (Styren *et al.*, 2000). Because it can reveal the full spectrum of pathological structures, including SP, NFT, dystrophic neurites (DN), and NT, X-34 can be used to examine multiple pathological structures in single tissue sections and distinguish different types of SP and NFT, and can be combined with other histological or immunohistochemical markers of normal or diseased brain structures. The simplicity of the procedure, its wide applicability for commonly used tissue processing techniques, and its high sensitivity for amyloid structure make X-34 an extremely versatile histological compound for use in neuropathological examinations and studies of AD pathology across the entire clinical spectrum of the disease progression (e.g., no cognitive impairment—mild cognitive impairment—early AD—late AD).

X-34 Histofluorescent Staining of Postmortem Brain: Basic Protocol

Synthesis of X-34

There is no commercial source of X-34 at present, but it can be readily synthesized without specialized equipment and with only basic organic chemistry skills. Details of the synthesis and physicochemical properties of X-34 have been previously reported (Styren *et al.*, 2000). The following is a simplified protocol that requires no isolation or purification procedures and results in 500 ml of ~100 μM stock solution of X-34. This solution is sufficiently pure for most histological applications.

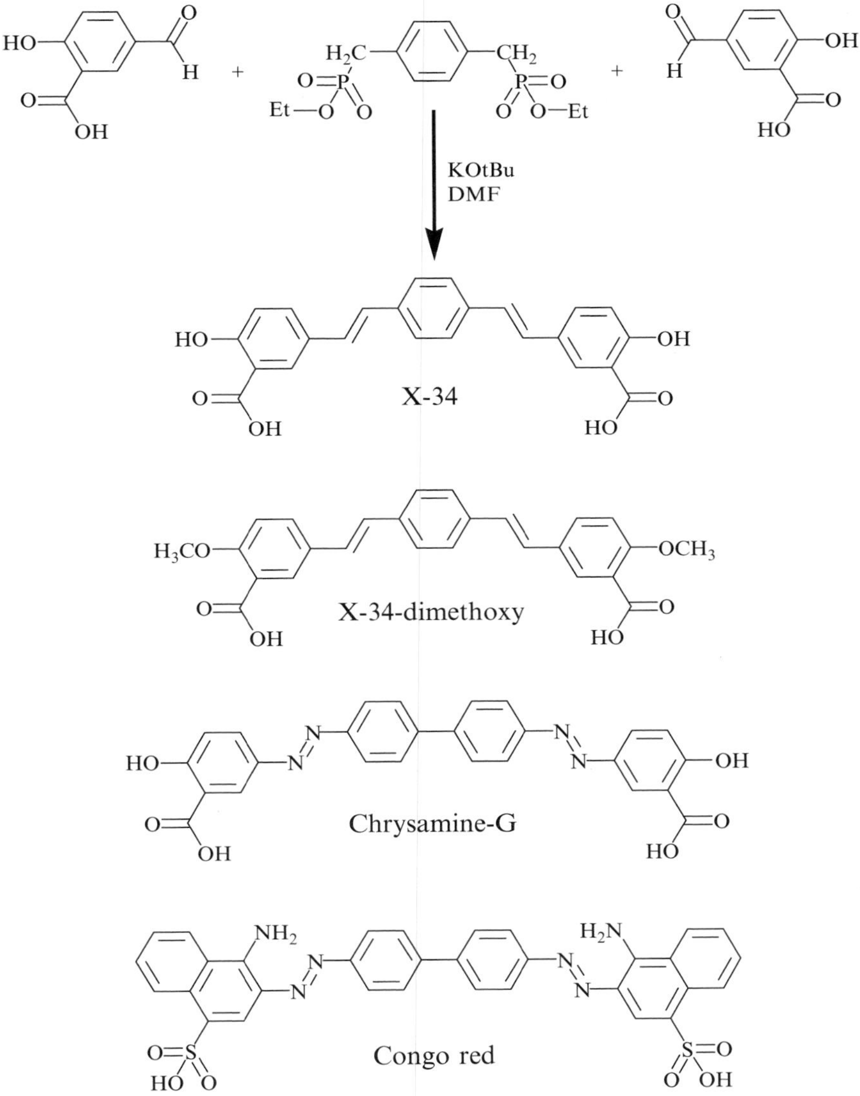

SCHEME 1. Synthetic scheme for X-34 and chemical structures of X-34-dimethoxy, chrysamine-G, and Congo red.

1. Wearing appropriate protective equipment (i.e., gloves, goggles, lab coat), place a Teflon-coated small stir bar in a 4-ml borosilicate glass vial with a PTFE-lined screw cap (such as the KIMAX, Kimble No. 60940A-4).

2. Weigh 32 mg (0.085 mmol) of tetraethyl *p*-xylylenediphosphonate (TCI America, Portland, OR, cat # T1582) into this vial, marked "vial-X", and add 2.0 ml of anhydrous dimethylformamide (DMF) from a freshly opened bottle (Aldrich, cat # 22,705–6). Cap and begin stirring.

3. Into a separate vial, marked "vial-5F", weigh 23 mg (0.17 mmol) 5-formylsalicylic acid (Aldrich, cat # F1,760–1) and add 1.0 ml anhydrous DMF to dissolve. Cap and set aside.

4. After the tetraethyl *p*-xylylenediphosphonate has dissolved, add 95 mg dry potassium *tert*-butoxide (0.85 mmol) (Aldrich, cat # 15,667–1) into vial-X. The mixture in vial-X will become dark red and opaque.

Note: A freshly opened container is best, and, once opened, the potassium *tert*-butoxide remains usable for no more than 1 mo. Potassium *tert*-butoxide is a *flammable solid* and should *not be brought directly into contact with water*. Therefore, care should be taken in the proper disposal of unused portions.

5. Add the contents of vial-5F to vial-X in four 0.25-ml aliquots. The mixture becomes thick. Cap and vortex occasionally over the next 2 h to ensure that the mixture remains liquid and the stir bar does not become jammed.

6. The mixture should turn a burnt-orange color over 30–60 min. Stir overnight at room temperature.

7. Prepare 500 ml of 40% ethanol (v/v) in distilled water.

8. Add 2.0 ml of the reaction mixture to the ethanol/water and mix thoroughly. The solution should become bright greenish yellow.

9. Adjust the pH to between 9.8 and 10.2 with 1 N NaOH or 1 N HCl as needed.

10. Dispose of unwanted chemicals and reaction mixture according to local guidelines.

11. Store this approximately 100 μM X-34 stock solution at room temperature in the dark. This stock solution should remain usable for 6 mo.

X-34 Histostaining Protocol

This protocol is optimized for 4% paraformaldehyde-fixed brain tissue infiltrated with increasing sucrose concentrations (15%–30%), sectioned at 40-μm thickness on a freezing sliding microtome and mounted onto gelatin-coated histological slides. The procedure works equally well with all commonly used fixatives, embedding, and sectioning procedures, including formalin-fixed, paraffin embedded tissue and various tissue thicknesses (Styren *et al.*, 2000; see also Fig. 6). The staining is easily performed using Coplin jars and slide racks that can handle a large number of slides simultaneously.

1. Wash 5 min in 0.01 M phosphate-buffered saline (PBS), pH 7.4 (0.076% NaCl, 0.0152% K_2HPO_4, 0.002% $K_2H_2PO_4$ in distilled H_2O).

2. Incubate 20 min in $KMnO_4$ (0.25% in PBS) (See note 12a).

3. Wash 2 min in PBS twice.

4. Incubate 1–6 min in 1% $K_2S_2O_5$ made in 1% $C_2H_2O_4 \cdot 2H_2O$ in PBS (See note 12a).

5. Wash 2 min in PBS three times.

6. Incubate 10 min in X-34 (100 μM). Both incubation time and X-34 concentration can be varied to achieve optimal results for various applications (see "Concentration-dependent Detection of Selective Amyloid Pathology with X-34 and X-34-Dimethoxy Fluorescent Histostaining").

7. Wash five times by 1-sec dips in distilled H_2O.

8. Incubate 2 min in 0.2 g% NaOH made in 80% unbuffered EtOH.

9. Soak 10 min in distilled H_2O.

10. Coverslip with Fluoromount-G (See note 12b).

11. View X-34 staining using a fluorescence microscope equipped with either a violet filter set (e.g., excites 400–410 nm, dichroic mirror DM455, 455-nm longpass filter) or an ultraviolet filter set (e.g., excites 360–370 nm, dichroic mirror DM400, 420-nm longpass filter).

12. Technical notes:

    a. Steps 2 and 4 are oxidative methods for lipofuscin modification ("autofluorescence quenching"; Barden, 1983, 1984; Guntern *et al.*, 1989). They are necessary to reduce the endogenous fluorescence because of the presence of lipofuscin pigment inside neurons in aged human brain. These steps can be omitted when staining rodent brain tissue sections. Note that step 2 will stain the sections brown. Sections should be incubated in step 4 solution until the brown precipitate is removed from the tissue. This will take slightly longer in thicker tissue sections.

    b. Once the staining is complete, the sections can be coverslipped directly from water using Fluoromount-G as described in step 10. Sections can also be dehydrated in ascending series of ethanol concentrations, delipidated in xylene, and coverslipped with Permount or Krystalon.

    c. X-34 staining is easy to combine with immunofluorescence in studies examining colocalization/codistribution of proteins of interest (see "Concentration-dependent Detection of Selective

Amyloid Pathology with X-34 and X-34-dimethoxy Fluorescent Histostaining" and "Combining X-34 Histostaining with IHC"). Fluorophores with emission spectra overlapping the X-34 emission spectrum should be avoided. The fluorescence spectrum of X-34 shows an excitation maximum at 367 nm and an emission maximum at 497 nm (aqueous solution at pH 7.0). Note that in AD research, certain antibodies generated against Aβ proteins (e.g., clone 6E10, Signet; Aβ_{40} and Aβ_{42}, Chemicon) require formic acid (FA) pretreatment to dissemble β-sheets, which would prevent X-34 staining (Styren *et al.*, 2000). In such experiments, X-34 staining must be performed first. Sections should then be imaged for X-34 and slides de-coverslipped before FA treatment and immunofluorescence.

X-34 Is a Marker of the Full Spectrum of Amyloid Pathology in AD Brains

To define the X-34 staining pattern in various brain regions affected with AD, we used our X-34 protocol to study neuropathological changes in superior frontal cortex, middle temporal cortex, hippocampus, caudate nucleus, and cerebellum, obtained postmortem from six subjects with clinical and neuropathological diagnosis of end-stage AD (mean age 77 $\pm$ 6.8; mean MMSE $=$ 9), enrolled in the Alzheimer's Disease Research Center at the University of Pittsburgh. In addition, frontal cortex tissue was obtained postmortem from 15 subjects with no cognitive impairment (NCI, $n = 5$, MMSE $= 27.2 \pm 2.17$), mild cognitive impairment (MCI, $n = 5$, MMSE $= 25.8 \pm 2.68$), and mild AD ($n = 5$, MMSE $= 21 \pm 1.12$), who were participants in the Religious Order Study (ROS), a longitudinal clinicopathological study of aging and AD in older Catholic nuns, priests, and brothers (Bennett *et al.*, 2002). Clinical diagnoses of MCI and AD were determined as described previously (DeKosky *et al.*, 2002; Ikonomovic *et al.*, 2005; Kordower *et al.*, 2001; Lopez *et al.*, 2000; Mufson *et al.*, 2000). Brain tissue was immersion-fixed in 4% paraformaldehyde for 48 h, cryoprotected in 15% and 30% sucrose solutions, cut at 40 μm on a freezing sliding microtome, and mounted on gelatinized glass slides. Although the X-34 staining reveals the full extent of AD pathology including SP, NFT, DN, and NT, all of which can be visualized simultaneously, there was a substantial regional variability in the morphology, staining intensity, and distribution of X-34 positive pathological structures. When describing X-34 positive SP, the terms "diffuse" and "compact," "cored," and "neuritic" refer strictly to their morphological appearance and are not intended to infer their ultrastructure (e.g., "fibrillar" versus "nonfibrillar" Aβ). For example, "diffuse" SP appeared as amorphous extracellular deposits of dispersed X-34-positive material without involvement of neuritic elements, whereas "compact" SP were well-delineated, circumscribed deposits of more compact X-34-positive material. Some compact SP were termed "cored" if they had a compact center core and a surrounding halo of less dense X-34 staining, whereas "neuritic" SP were defined by the presence of X-34 positive DN inside SP.

Overall Pattern and Regional Distribution of X-34 Labeling in NCI, MCI, Mild AD, and End-stage AD Brains

Frontal and Temporal Cortex. X-34 staining of the frontal cortex revealed substantial variability in the degree of amyloid pathology between clinical groups and within groups. NCI subjects either lacked AD pathology or had modest numbers of SP and NFT stained with X-34. Semiquantitative analysis of X-34 positive SP densities in the frontal cortex, performed as described previously (Ikonomovic *et al.*, 2004), showed that MCI and mild AD subjects had more SP compared with NCI subjects, with the MCI group showing an overlap with both other ROS diagnostic groups (Fig. 1A–C). Mild AD subjects had more X-34 positive SP and NFT than the other two ROS diagnostic groups. The most abundant X-34 positive SP and NFT were observed in the frontal cortex of end-stage AD cases, where the majority of SP were neuritic. In some instances, SP were completely filled with intensely fluorescent DN (Figs. 2C, E, and 3A).

X-34 staining of severe neocortical AD pathology is illustrated in the temporal cortex from an end-stage AD case (Fig. 1D). This was a Braak stage V/VI case, because of the extensive severity of NFT pathology (see Braak and Braak, 1991 for a description of Braak staging of NFT changes). Numerous X-34 positive SP were observed, predominantly in layers II, IV, and VI. In addition, many X-34 positive NFT were present predominantly in layers III and V/VI, whereas DN and NT formed an intricate neuritic web across all cortical layers. Large blood vessels in cases with congophilic amyloid angiopathy (CAA) displayed particularly prominent X-34 staining (Fig. 2A). Diffuse, compact, cored, and neuritic SP were observed in the temporal cortex and were easily distinguished morphologically and by the color/intensity of X-34 fluorescence (see Fig. 2).

Hippocampus. The full spectrum and regional specificity of AD pathology as revealed with X-34 labeling is best illustrated in the hippocampus of an end-stage (Braak stage V/VI) AD case (Fig. 1E). In the CA1 pyramidal layer and extending into the subiculum, the vast majority of pyramidal neurons contained NFT and showed bright cyan fluorescent staining with X-34. Numerous neuritic SP were dispersed among the NFTs in subiculum/CA1–CA4;

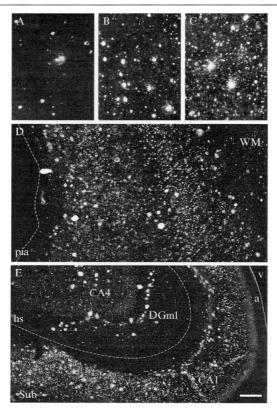

FIG. 1. X-34 fluorescence in medial frontal cortex from ROS subjects with no cognitive impairment (A; NCI), MCI (B), and mild AD (C). The MCI subject shows considerable numbers of X-34–positive SP in the frontal cortex; qualitatively, SP density is intermediate between that of the NCI and mild AD cases. In the medial temporal cortex from an end-stage AD patient (D), brightly labeled SP, blood vessels, NFT, DN, and NT are present throughout the cortical laminae. Diffuse-appearing (lighter fluorescence), compact, or cored and neuritic (brighter fluorescence) SP are most dense in layers II and IV and extend into the white matter. NFT predominate in pyramidal cell layers III and V–VI. Higher magnification images are presented in Fig. 2 to better illustrate the morphological details of the labeled components that are not visible at low power. (E) X-34 fluorescence in the hippocampus from an AD patient classified as Braak stage V/VI. The full utility of X-34 labeling can be appreciated in this overview montage, in which the hippocampal laminae and associated pathology are clearly visible. The most prominent pathology can be observed in the pyramidal cell layer of CA1 and the subiculum, CA4, and molecular layer of the dentate gyrus. Neuropil threads are seen extending from CA4 into the polymorphic cell layer. a, alveus; CA1, CA4, cornu ammonis subregions; DGml, molecular layer of dentate gyrus; hs, approximate level of hippocampal sulcus; Sub, subiculum; v, ventricular cavity; WM, white matter. Dotted line demarcates cortical pial surface (D), and the hippocampal sulcus or the separation between surface areas of hippocampus proper (CA) and dentate gyrus (E). Scale bar = 200 μm (A–C), 500 μm (D), and 750 μm (E).

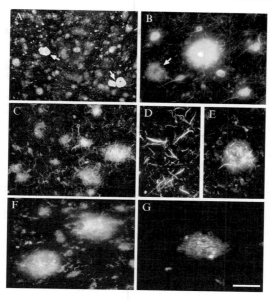

Fig. 2. Fluorescent photomicrographs illustrate region-specific morphological details of X-34 labeled structures and the spectrum of X-34 fluorescence color (Note: for color image please refer to the end of this volume). (A) Low-power photomicrograph of X-34 fluorescence in the middle temporal cortex of an AD case reveals prominent staining of large cyan-labeled blood vessels (arrows) and numerous SP and NFT. (B) Higher magnification of a classic, cored, SP in the medial temporal cortex shows bright cyan fluorescence in the central core region (bright white in the grayscale image) composed of highly aggregated amyloid. The surrounding halo of X-34 fluorescence is less prominently cyan-stained, whereas an adjacent diffuse-appearing SP (arrow) shows more greenish brown (gray in the gray scale image). Fibrils are easily observed in NT surrounding SP and in several NFT that also show bright cyan fluorescence and display a range of NFT morphology. (C) In the frontal cortex, numerous neuritic SP and NT are observed. (D, E) Higher power magnification of NT (D) and DN associated with one neuritic SP (E). (F) SP in the caudate nucleus stain with greenish brown fluorescence of the diffuse-appearing type, without compact amyloid cores. Only sporadic neuritic elements are seen in the surrounding neuropil. (G) An isolated diffuse-appearing SP in the molecular layer of cerebellar cortex shows cyan fluorescence of X-34 stained fibrils that appear loosely dispersed inside the amorphous SP formation. There are no associated neuritic elements in the neuropil. Scale bar = 250 μm (A), 50 μm (B, C, E–G), and 25 μm (D). (See color insert.)

they were most densely distributed in the CA1 pyramidale and radiatum layers and in dentate gyrus molecular layer. In the CA4 region, intensely fluorescent DN and NT were abundant, in addition to many large SP.

Caudate Nucleus and Cerebellum. The caudate nucleus contained numerous X-34 positive diffuse SP; NFT were absent (Fig. 2F). Small neuritic elements were observed infrequently in the neuropil, but DN were absent from

SP. The cerebellum showed few X-34-positive SP and other amyloid-containing structures. Small numbers of X-34-positive SP were detected in the molecular layer of cerebellar cortex, and they were almost exclusively diffuse (Fig. 2G), only very rarely containing more compact material. NFT and DN were absent from the cerebellum, and delicate X-34 positive NT were observed traversing the Purkinje cell layer (not shown).

Morphological Details of X-34 Labeled Structures in AD

Senile Plaques and CAA. Regional variations in the abundance of amyloid structures were revealed with X-34. Different SP types were easily distinguished because of morphological differences and variations in the color and intensity of X-34 fluorescence. More compact amyloid (e.g., in large blood vessel walls and cored SP) yielded a bright cyan fluorescence, whereas more diffuse-appearing SP fluoresced greenish brown and considerably less intensely (Fig. 2). In the temporal cortex, large blood vessels and cored SP (having a compact core of densely aggregated amyloid fibrils, and a diffuse-appearing "halo" of distinct amyloid fibrils) had the most prominent cyan fluorescence (Fig. 2A, B). SP in the frontal cortex contained many DN and were surrounded by NT (Fig. 2C–E). SP in the CA1 region of the hippocampus were compact, with large DN, and were accompanied by numerous NFT. In contrast, SP in the caudate nucleus contained very few neuritic elements (Fig. 2F). In the cerebellum, X-34 staining revealed isolated diffuse SP that were larger and more amorphous than in other brain regions examined (Fig. 2G).

Neurofibrillary Tangles. Besides vascular amyloidosis and cored/neuritic SP, NFT were the most brightly fluorescent structures in AD tissue stained with X-34. They were numerous in the temporal cortex (Figs. 1D, 2A, B) and most abundant in the subiculum/CA1 region of the hippocampus (Fig. 1E) and layer II of the entorhinal cortex (not shown). X-34 labeled both intracellular (flame-shaped) NFT and less compact extracellular tangles. As was the case with SP, individual bundles of fibrils were clearly resolved in all X-34-labeled NFT at high magnification (Fig. 2B).

Dystrophic Neurites and Neuropil Threads. DN and NT were particularly abundant in AD brains scored as Braak stage V/VI and displayed bright-cyan fluorescence when stained with X-34 (Fig. 2D, E). DN were present in neuritic SP and in the neuropil. NT were abundant especially in areas of high NFT formation (e.g., hippocampus, cortical layer V). X-34-positive NT varied in size from very small and delicate to large (Fig. 2D). NT were sparse in the caudate nucleus and cerebellum.

Comparative Analysis: X-34, Thioflavin S (ThioS) and
Aβ Immunohistochemistry (IHC)

We compared the labeling pattern of X-34 with that of another commonly used marker of amyloid structures in AD brain (ThioS) and immunohistochemical staining using an antibody against Aβ (Aβ IHC). Tissue sections obtained from the temporal cortex of an AD subject were processed consecutively for Aβ IHC and X-34 histochemistry. Sections were first immunolabeled using a monoclonal antibody against the NH$_2$-terminus of Aβ (10D5, Elan Pharm., 1:3000) as described in detail previously (Ikonomovic *et al.*, 2004), photographed using light microscopy, and then counterstained for X-34 and photographed using fluorescent microscopy. There was a complete overlap between the distribution of amyloid plaques labeled with X-34 and Aβ plaques revealed by IHC (Fig. 3A, B), with a direct correlation of compact Aβ immunoreactive plaques, with very bright and cyan/white-colored plaques labeled with X-34. In addition to amyloid plaques, X-34 labeled NFT, DN, and NT, which were not detectable with Aβ IHC (Fig. 3B), and provided clear distinction between neuritic and nonneuritic plaques.

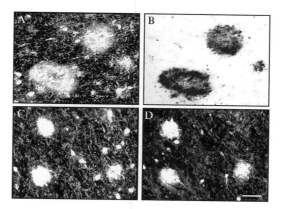

Fig. 3. Comparison of X-34 histochemistry with Aβ immunohistochemistry (A and B; single tissue section processing) and ThioS histochemistry (C and D; immediately adjacent tissue sections processing) in temporal cortex tissue sections from an end-stage AD patient. (A and B) X-34 staining in amyloid plaques completely overlaps 10D5 immunostaining and also reveals extensive neurofibrillary pathology inside and outside plaques. X-34 labeling of neuritic elements appears bright white and overlays the light gray staining of extracellular amyloid fibrils. Note the absence of NFT, DN, and NT staining in tissue section processed for Aβ IHC (B). (C and D) X-34 and ThioS both provide intense labeling of neuritic plaques and NFT, but NT staining is far more prominent with the X-34 stain. Scale bar = 150 μm.

ThioS histochemistry was performed using the modification of Guntern *et al.* (1992), which revealed close overlap with Aβ IHC (Vallet *et al.*, 1992). This procedure resulted in clear fluorescent labeling of cored/neuritic plaques, NFT, DN, and NT in AD temporal cortex (Fig. 3D). In serial sections, ThioS-labeled plaques were also labeled with X-34 (Fig. 3C), which resulted in considerably brighter fluorescent staining, especially in DN and NT. Thus, although ThioS and X-34 label the same spectrum of amyloid pathology in AD brain, X-34 staining results in better resolution of even the smallest caliber NT with an excellent signal-to-background ratio (see also Fig. 2D). This characteristic could facilitate automated procedures such as gray scale thresholding to determine overall amyloid load or other related measurements in X-34 labeled tissue.

X-34-Dimethoxy Compound

By replacing the two hydroxyl groups in X-34 with methoxy groups, the derivative X-34-dimethoxy is created that has slightly different physical and optical qualities (for overview of compounds' chemical structures, see Scheme 1). With a $\log P_{oct}$ (log of the octanol–water partition coefficient; a measure of lipophilicity) of -0.95, X-34-dimethoxy is more water soluble than X-34 ($\log P_{oct} = +0.42$), and this would likely reduce its brain entry (Dishino *et al.*, 1983) if injected systemically for the purpose of *in vivo* imaging. The excitation maximum of X-34-dimethoxy is very similar to that of X-34 ($\sim$370 nm), but the emission maximum is at a shorter wavelength ($\sim$450 nm). In addition, the fluorescence intensity of X-34-dimethoxy is about fivefold greater than that of X-34. However, the binding affinity of X-34-dimethoxy for Aβ fibrils is only one third that of X-34 (Mathis *et al.*, 2004). The staining protocol using X-34-dimethoxy is identical to that described for X-34. Similar to X-34, X-34-dimethoxy labels all AD-related pathological structures that contain amyloid, is considerably brighter, but does not have the color variation exhibited by X-34. Specifically, all structures labeled with X-34-dimethoxy show bright blue/cyan fluorescence (Fig. 4A, B), in contrast to X-34, which labels diffuse-appearing SP with greenish/brown fluorescence, and compact SP, NT, DN, and NFT with cyan/white fluorescence (Figs. 2 and 4).

Concentration-Dependent Detection of Selective Amyloid Pathology with X-34 and X-34-Dimethoxy Fluorescent Histostaining

As markers of amyloid (β-pleated sheet) structure, X-34 and related compounds (e.g., Pittsburgh Compound B, PIB, Klunk *et al.*, 2004) are not exclusive for the Aβ peptide in amyloid conformation, because they bind

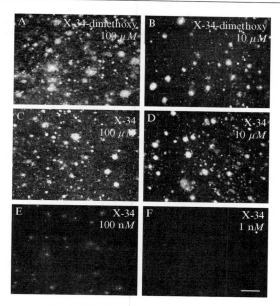

FIG. 4. Photomicrographs of X-34-dimethoxy and X-34 stainings in AD temporal cortex show that at comparable concentrations (100 μM and 10 μM), X-34-dimethoxy produces brighter and more cyan fluorescence compared with X-34 (A–D; Note: for color image please refer to the end of this volume). Dilution effect on X-34-dimethoxy (A, B) and X-34 (C–F) labeling in the temporal cortex from an AD patient is demonstrated. At the highest concentration, which shows optimal signal to background contrast (100 μM; C), X-34 labels the four main pathological structures in AD that contain the β-pleated sheet structure: SP, NFT, DN, and NT. At 10 μM (D), NT staining is significantly reduced; NFT and DN are less prominent, whereas SP remains intensely fluorescent. At 100 nM (E), NT are no longer detectable, NFT and DN are barely visible, and SP show reduced fluorescence. At the highest dilution tested (1 nM; F), compact SP are the only remaining structures exhibiting X-34 fluorescence. Scale bar = 100 μm. (See color insert.)

nonpreferentially to the amyloid structure of various protein aggregates (e.g., Aβ, phospho-tau). However, the proportion of fluorescent signal derived from dye-bound amyloid plaques can be optimized at proper dilutions. Specifically, when used at concentrations of 10–100 μM, X-34 labeling is clearly detectable in SP, NFT, DN, and NT in AD frontal and temporal cortices (Fig. 4C, D), hippocampus and caudate, in blood vessels from patients with CAA, and in diffuse SP in AD cerebellum. At this concentration, the dye is applicable for use as a simple one-step histological procedure in studies that require visualization of the entire range of AD pathology in postmortem tissue sections, obviating the need for complex and time-consuming multiple immunohistochemical labeling, histological counterstaining, or processing and analysis of multiple, serial ultrathin sections. As demonstrated here, X-34 labeled all types of SP, including cored, neuritic, and nonneuritic diffuse-appearing types, as well as the entire volume of SP (i.e., the cored SP in Fig. 2B shows X-34 fluorescence in the dense core region, as well as in the surrounding halo of dispersed amyloid fibrils). The most diffuse-appearing amyloid, detected with antibodies generated against Aβ_{42}, showed only light X-34 labeling at the high concentration (100 μM). When used at progressively more dilute concentrations (10 μM and 1 μM), X-34-labeling of NT, NFT, and DN became progressively less intense. Their staining disappeared at concentrations of 0.1 μM–0.01 μM, whereas SP and CAA vessels remained as the only detectable structures at lower concentrations (Fig. 4D, E). Thus, at low concentrations that are comparable to those used for *in vivo* PET imaging with PIB (Klunk *et al.*, 2004), X-34 amyloid labeling is detected only in SP and CAA. Note, however, that the concentrations discussed here do not translate directly to the *in vivo* PET studies, because the washes with ethanol-based solvents used in the staining procedure remove amyloid-bound X-34 more efficiently. Similar to X-34, a dilution effect was also observed with X-34-dimethoxy (Fig. 4A–B), with one order of magnitude dilution resulting in prominently labeled SP, reduced staining of NFT, and little to no labeling of DN and NT.

Combining X-34 Histostaining with IHC

X-34 can be combined with immunohistochemical procedures for light microscopy [see "Comparative Analysis: X-34, Thioflavin S (ThioS), and Aβ Immunohistochemistry (IHC)"] or with immunofluorescent staining in studies of amyloid pathology. Dual fluorescence labeling of X-34-positive structures and hyperphosphorylated tau containing NFTs was performed by processing hippocampal tissue sections immunohistochemically with a monoclonal anti-phosphorylated tau antibody (AT8, Pierce, Rockford, IL; diluted 1:5000), followed by incubation with Cy3-conjugated goat anti-mouse secondary antibody (Jackson ImmunoResearch, West Grove, PA) to visualize the antigen–antibody reaction. This immunohistochemistry protocol has been described before in detail (Desai *et al.*, 2005). After completion of AT8 immunohistochemistry, tissue sections were washed three times for 10 min in phosphate buffer (pH = 7.4) and further processed using the X-34 histochemistry as described previously. Cy3 immunofluorescence was visualized using a green filter set (e.g., excites 520–580 nm, dichroic mirror DM580, 590-nm longpass filter), and X-34 fluorescence was visualized using the violet filter set (described previously). This procedure revealed that only a subset of X-34 stained NFT contained the AT8 antigen (Fig. 5A–C). Specifically, dystrophic-appearing pyramidal neurons

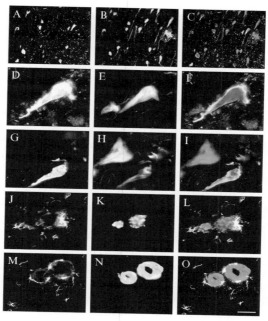

FIG. 5. Fluorescence photomicrographs of tissue sections from an AD hippocampus and transgenic mouse cortex processed for red fluorescence IHC (left) and counterstained with blue fluorescent X-34 histochemistry (center) shown with merged red/blue images (right column). For color image please refer to the end of this volume. In human hippocampus (A–C, D–F, and G–I), a subset of X-34 (blue)–stained NFT contains AT-8 immunostaining. A large extracellular NFT ("ghost tangle" in H) lacks AT-8 immunofluorescence. In transgenic mouse cortex, red GFAP immunostaining of astrocytes is observed in association with blue X-34 staining of amyloid material in plaques (J–L) and walls of large blood vessels (M–O). Scale bar = 90 μm (A–C), 15 μm (D–I), 35 μm (J–O). (See color insert.)

with "corkscrew-like" dendrites and intact cell soma, as well as some classic flame-shaped NFT, labeled prominently with both AT8 (red) and X-34 (blue; Fig. 5D–F). AT8 fluorescence also labeled some normal-appearing pyramidal neurons, and these cells showed only small amounts of X-34 staining. Extracellular ("ghost") NFT showed prominent X-34 labeling but lacked AT8 immunostaining (Fig. 5G–I). In transgenic mice, tissue sections that were immunofluorescently labeled using the astrocytic marker GFAP (Sigma, St. Louis, MO; diluted 1:1000) and a Cy3-conjugated goat anti-rabbit secondary antibody (Jackson ImmunoResearch) were counterstained using X-34 histochemistry (described previously). This double-labeling procedure revealed that Cy3-fluorescent GFAP-positive astrocytes clustered around the periphery of amyloid plaques (Fig. 5J–L) and around large blood vessels that exhibited X-34-stained amyloid material inside vascular walls (Fig. 5M–O).

X-34 Histostaining of Peripheral Amyloidosis

The binding of X-34 seems to be dependent on the β-sheet secondary structure inherent in all fibrillar amyloid deposits. Thus, as has already been seen by the ability of X-34 to stain both Aβ (SP and CAA) and tau (NFT, DN, and NT) deposits, X-34 can label many proteins that deposit as β-sheet amyloid fibrils. There are many such amyloidogenic proteins, including immunoglobulin light chain, immunoglobulin heavy chain, transthyretin, β2-microglobulin, (Apo)serum AA, apolipoprotein AI, apolipoprotein AII, gelsolin, lysozyme, fibrinogen α-chain, cystatin C, ABriPP, ADanPP, prion protein, (Pro)calcitonin, islet amyloid polypeptide, atrial natriuretic factor, prolactin, insulin, lactadherin, kerato-epithelin, Pindborg tumor associated precursor protein (tbn), and lactoferrin (Buxbaum, 2004). Besides brain amyloidosis, X-34 stained peripheral amyloid structures, including those in heart, lung, bladder, and bone (Fig. 6, tissue courtesy of Drs. Carl O'Hara and David Seldin, Boston University Medical Center), and could be useful in pathological analysis of other organs/peripheral tissues and in studies of systemic familial and nonfamilial amyloidosis.

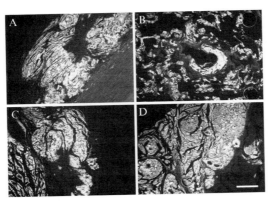

FIG. 6. Photomicrographs of X-34 staining in formalin-fixed, paraffin embedded histological sections of autopsy tissue from a patient with amyloid light-chain (AL) amyloidosis. The bright X-34 fluorescence marks amyloid aggregates in the heart (A), lung (B), bladder (C), and bone (D). For color image please refer to the end of this volume. Scale bar = 200 μm. (See color insert.)

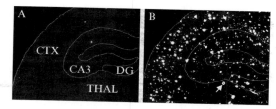

FIG. 7. X-34 fluorescent labeling of amyloid material in coronal brain tissue sections through dorsal hippocampus in (A) young (3 months) and (B) aged (24 months) PS1/APP mice (Tg 2576 APP-overexpressing mice bearing the presenilin-1 mutation used as a model of familial AD-related SP pathology). Dotted lines outline cortical surface and hippocampal formation. In young animals, small, sporadic SP are present in the cortex and the CA1 of the hippocampus. At 24 months, X-34 labeled plaques and blood vessels (arrow in B) are abundant in the cortex and hippocampus and also appear throughout subcortical regions. CA1 and DG, CA1 fields and dentate gyrus of the hippocampus; CTX, cortex; THAL, thalamus.

X-34 Histostaining in Transgenic Mouse Models of AD: Utility in Evaluation of Amyloid Plaque Development and Therapeutic Efficacy

Sections through the hippocampus of young (3 mo) and aged (24 mo) Tg 2576 mice overexpressing human APP (Swe) with the presenilin [PS1 (M146V)] mutation (PS1/APP; Holcomb et al., 1998) were stained with X-34 and demonstrated the time-dependent progression of amyloid plaque development in this animal model of AD. In PS1/APP brains, X-34 revealed only extracellular amyloid. Although these mice display some X-34-stained cerebral vascular amyloid, the X-34 staining confirms that they do not develop amyloid-containing NFT, NT, or DN. In young mice, X-34 labeled very small plaques that were present in low numbers in the hippocampus and cerebral cortex (Fig. 7A). In aged PS1/APP mice, X-34 labeling was markedly increased, with numerous plaques appearing in all cortical areas and hippocampal regions (Fig. 7B). The comparison with adjacent tissue sections immunostained with antibodies specific to Aβ peptide indicated that all Aβ-immunostained plaques were labeled with X-34 (not shown).

Discussion

X-34 is a highly fluorescent, lipophilic derivative of Congo red that intensely stains aggregated proteins with the β-pleated sheet conformation (amyloid). It is particularly useful in histochemical studies of AD, because it labels the full spectrum of known AD pathological structures that contain amyloid. This is particularly evident when comparing brain regions that contain SP in different stages of development with those that contain only a specific type of SP. For example, the hippocampus and temporal cortex contain large numbers of X-34-positive compact neuritic SP and diffuse SP (sometimes referred to as "preamyloid", Dickson, 1997), whereas the cerebellum shows X-34 staining almost exclusively in diffuse-appearing nonneuritic plaques. The fluorescence color variation inherent in X-34 bound to diffuse versus compact amyloid allows easy distinction of different SP types. X-34 appears bright cyan in SP cores and dystrophic neurites in compact and neuritic SP, and yellow-green/brown in diffuse-appearing SP (possibly the transition from preamyloid to amyloid SP). Thus, studies examining the role of proteins (e.g., apolipoproteins), cells, or other factors in SP formation and/or clearance could use the X-34 staining procedure to assess the extent of SP progression from early to mature forms.

The determining characteristic of AD pathology stained by X-34 is β-sheet fibril structure, because this staining is abolished by formic acid pretreatment (Styren *et al.*, 2000). X-34 is, therefore, specific for the secondary protein structure, not the primary amino acid sequence, and in studies of AD brain all pathological structures containing protein aggregates in β-sheet conformation are labeled with X-34. This indicates that X-34 has affinity not only for the Aβ peptide in amyloid form but also neurofibrillary pathology composed of hyperphosphorylated tau aggregated into β-sheet structure (e.g., NFT, DN, NT). Such a wide spectrum of AD pathology can be revealed also by other related and more traditionally used compounds such as ThioS; however, the clarity with which X-34 reveals the most subtle neurofibrillary lesions (NT) is exceptional (see Fig. 3). This is of particular importance considering that recent work implicated NT as the earliest form of neurofibrillary degeneration and a marker of regional vulnerability in AD brains (Ghoshal *et al.*, 2002). Despite the predisposition for its "nonspecific" labeling of a variety of pathological lesions, X-34 (and X-34-dimethoxy) staining can be manipulated to predominantly label amyloid deposits in SP and CAA in AD tissue by simply decreasing the concentration in the staining bath to the nanomolar range. This dilution effect may reflect the higher β-sheet fibril content or packing density in structures that remain labeled at nanomolar concentrations, or it could be due to different binding affinity of X-34/X-34-dimethoxy for Aβ fibrils in SP compared with hyperphosphorylated tau in neurofibrillary lesions. This is particularly relevant to *in vivo* neuroimaging studies, which use markedly lower concentrations of amyloid imaging agents (Klunk *et al.*, 2004). A related amyloid binding dye, Thioflavin T, has also yielded a derivative, PIB, which, when radiolabeled and used in positron emission tomography (PET) studies, binds amyloid in AD brain with high sensitivity (Klunk *et al.*, 2004) at concentrations comparable to (and lower than) those used in this study. These observations underscore the usefulness of compounds like PIB and its derivatives as diagnostic aids for detecting Aβ plaques in AD and even preclinical AD (pathological aging and mild cognitive impairment) and in the development of anti-Aβ therapeutics. This is supported by the good correlation between extracellular plaque deposits stained with X-34 histochemistry and Aβ IHC (see Fig. 3). Nevertheless, X-34 binds many amyloid deposits that have predominant β-sheet secondary structure and thus is not specific for the Aβ peptide. Histological studies of aged and AD brains should be designed with this characteristic in mind. In a transgenic mouse model of AD, X-34 intensely labeled plaques in the aged (24 mo) PS1/APP mouse with a pattern identical to that seen with Aβ antibodies. NFT, DN, and NT were not observed, confirming that this mouse model either lacks them altogether, or they contained negligible levels of protein aggregation into β-pleated sheet fibrils. X-34 will be particularly useful in future studies of transgenic mice that display neuritic pathology (APP[V717I], Sturchler-Pierrat, 1997), as well as mouse models that display tangles (PS1/APP with tau, P301L; Oddo et al., 2003).

X-34 staining (or any other amyloid detection procedure) cannot be used to determine clinical diagnosis of AD with certainty, because of substantial variations across diagnostic groups. As observed in previous postmortem studies of MCI using different immunohistochemical markers of AD pathology (Kordower et al., 2001; Mitchell et al., 2002; Mufson et al., 1999; Price and Morris, 1999; Price et al., 1991), the severity of neuropathological changes revealed with X-34 staining in MCI subjects is between that of nondemented controls and early AD subjects. However, correlation of clinical diagnoses with postmortem pathology must take into account that: (1) in any elderly control group, there will be some individuals who have significant AD pathology before the onset of clinical symptoms, (2) many clinically diagnosed MCI subjects have extensive plaque pathology at very advanced ages, and some may even have other disorders without AD changes (e.g., hippocampal sclerosis causing amnestic syndrome), and (3) in any clinically diagnosed AD group, there will be some individuals who have dementia caused by something other than AD and will not have AD pathology. Thus, one cannot expect distinct pathological differences among these groups along the lines drawn by clinical diagnosis (for review see DeKosky et al., 2005).

In conclusion, the X-34 histostaining is a rapid, simple, and highly reproducible procedure that is applicable to many types of tissue processing, including fresh-frozen cryostat sections fixed in ethanol or methanol; paraformaldehyde-fixed free floating sections cut on a freezing microtome; and formalin-fixed, paraffin-embedded tissue sections. In addition, it is easily combined with single or multiple fluorescent immunolabeling or with chromogen-based immunohistochemistry in studies of neuropathological structures of pathological aging, MCI, and AD. Furthermore, the affinity of X-34 for all structures rich in β-sheet fibril structure suggests that this compound could be useful not only in clinical-pathologic studies of AD but also in other studies of neurodegenerative diseases that are characterized by other abnormal protein (e.g., synuclein) folding/aggregation. Current studies in our laboratory are underway examining the X-34 staining pattern in Lewy body disease and the Lewy body variant of AD.

Acknowledgments

We thank Drs. Carl O'Hara and David Seldin (Boston University Medical Center) for providing the tissue from a subject with AL amyloidosis. Supported by NIA grants AG020226, AG14449, and AG05133.

References

Barden, H. (1983). The presence of ethylenic bonds and vic-glycol groups in neuromelanin and lipofuscin in the human brain. *J. Histochem. Cytochem.* **31,** 849–858.

Barden, H. (1984). The oxidative generation of sulfonic acid groups in neuromelanin and lipofuscin in the human brain. *J. Histochem. Cytochem.* **32,** 329–336.

Bennett, D. A., Wilson, R. S., Schneider, J. A., Evans, D. A., Beckett, L. A., Aggarwal, N. T., Barnes, L. L., Fox, J. H., and Bach, J. (2002). Natural history of mild cognitive impairment in older persons. *Neurology* **59,** 198–205.

Bennhold, H. (1922). Specific staining of amyloid by Congo red. *Munchen Med. Wochnschr.* **69,** 1537–1538.

Braak, H. and Braak, E. (1991). Neuropathological stageing of Alzheimer-related changes. *Acta Neuropathol.* **82,** 239–259.

Buxbaum, J. N. (2004). The systemic amyloidoses. *Curr. Opin. Rheumatol.* **16,** 67–75.

DeKosky, S. T., Ikonomovic, M. D., Styren, S. D., Beckett, L., Wisniewski, S., Bennett, D. A., Cochran, E. J., Kordower, J. H., and Mufson, E. J. (2002). Upregulation of choline acetyltransferase activity in hippocampus and frontal cortex of elderly subjects with mild cognitive impairment. *Ann. Neurol.* **51,** 145–155.

DeKosky, S. T., Ikonomovic, M. D., Hamilton, R. L., Bennett, D. A., and Mufson, E. J. (2005). Neuropathology of mild cognitive impairment in the elderly. *In* "Alzheimer's Disease and Related Disorders Annual," Vol. 5 (S. Gauthier, P. Scheltens, and J. L. Cummings, eds.), pp. 1–16. Martin Dunitz & Parthenon Publishing, London.

Desai, P. P., Ikonomovic, M. D., Abrahamson, E. E., Hamilton, R. L., Isanski, B. A., Hope, C. E., Klunk, W. E., DeKosky, S. T., and Kamboh, M. I. (2005). Apolipoprotein D is a component of compact but not diffuse amyloid-beta plaques in Alzheimer's disease temporal cortex. *Neurobiol. Dis.* **20,** 574–582.

Dickson, D. W. (2005). Required techniques and useful molecular markers in the neuropathological diagnosis of neurodegenerative disorders. *Acta Neuropath.* **109,** 14–24.

Dickson, D. W. (1997). The pathogenesis of senile plaques. *J. Neuropathol. Exp. Neurol.* **56,** 321–339.

Dishino, D. D., Welch, M. J., Kilbourn, M. R., and Raichle, M. E. (1983). Relationship between lipophilicity and brain extraction of C-11-labeled radiopharmaceuticals. *J. Nucl. Med.* **24,** 1030–1038.

Ghoshal, N., Garcia-Sierra, F., Wuu, J., Leurgans, S., Bennett, D. A., Berry, R. W., and Binder, L. I. (2002). Tau conformational changes correspond to impairments of episodic memory in mild cognitive impairment and Alzheimer's disease. *Exp. Neurol.* **177,** 475–493.

Guntern, R., Vallet, P. G., Bouras, C., and Constantinidis, J. (1989). An improved immunohistostaining procedure for peptides in human brain. *Experientia* **45,** 159–161.

Guntern, R., Bouras, C., Hof, P. R., and Vallet, P. G. (1992). An improved thioflavine S method for staining neurofibrillary tangles and senile plaques in Alzheimer's disease. *Experientia* **48,** 8–10.

Holcomb, L., Gordon, M. N., McGowan, E., Yu, X., Benkovic, S., Jantzen, P., Wright, K., Saad, I., Mueller, R., Morgan, D., Sanders, S., Zehr, C., O'Campo, K., Hardy, J., Prada, C. M., Eckman, C., Younkin, S., Hsiao, K., and Duff, K. (1998). Accelerated Alzheimer-type phenotype in transgenic mice carrying both mutant amyloid precursor protein and presenilin 1 transgenes. *Nat. Med.* **4,** 97–100.

Ikonomovic, M. D., Uryu, K., Abrahamson, E. E., Ciallella, J. R., Trojanowski, J. Q., Lee, V. M-Y., Clark, R. S., Marion, D. W., Wisniewski, S. R., and DeKosky, S. T. (2004). Alzheimer's pathology in human temporal cortex surgically excised after severe brain injury. *Exp. Neurol.* **190,** 192–203.

Ikonomovic, M. D., Mufson, E. J., Wuu, J., Bennett, D. A., and DeKosky, S. T. (2005). Reduction of choline acetyltransferase activity in primary visual cortex in mild-moderate Alzheimer's disease. *Arch. Neurol.* **62,** 425–430.

Klunk, W. E., Engler, H., Nordberg, A., Wang, Y., Blomqvist, G., Holt, D. P., Bergstrom, M., Savitcheva, I., Huang, G. F., Estrada, S., Ausen, B., Debnath, M. L., Barletta, J., Price, J. C., Sandell, J., Lopresti, B. J., Wall, A., Koivisto, P., Antoni, G., Mathis, C. A., and Langstrom, B. (2004). Imaging brain amyloid in Alzheimer's disease with Pittsburgh Compound-B. *Ann. Neurol.* **55,** 306–319.

Kordower, J. H., Chu, Y., Stebbins, G. T., DeKosky, S. T., Cochran, E. J., Bennett, D., and Mufson, E. J. (2001). Loss and atrophy of layer II entorhinal cortex neurons in elderly people with mild cognitive impairment. *Ann. Neurol.* **49,** 202–213.

Lopez, O. L., Becker, J. T., Klunk, W., Saxton, J., Hamilton, R. L., Kaufer, D. I., Sweet, R. A., Cidis Meltzer, C., Wisniewski, S., Kamboh, M. I., and DeKosky, S. T. (2000). Research evaluation and diagnosis of probable Alzheimer's disease over the last two decades: I. *Neurology* **55,** 1854–1862.

Mathis, C. A., Klunk, W. E., Price, J. C., and DeKosky, S. T. (2005). Imaging technology for neurodegenerative diseases. *Arch. Neurol.* **62,** 196–200.

Mathis, C. A., Wang, Y., and Klunk, W. E. (2004). Imaging beta-amyloid plaques and neurofibrillary tangles in the aging human brain. *Curr. Pharm. Des.* **10,** 1469–1492.

Mitchell, T. W., Mufson, E. J., Schneider, J. A., Cochran, E. J., Nissanov, J., Han, L. Y., Bienias, J. L., Lee, V. M., Trojanowski, J. Q., Bennett, D. A., and Arnold, S. E. (2002). Parahippocampal tau pathology in healthy aging, mild cognitive impairment, and early Alzheimer's disease. *Ann. Neurol.* **51,** 182–189.

Mufson, E. J., Chen, E. Y., Cochran, E. J., Beckett, L. A., Bennett, D. A., and Kordower, J. H. (1999). Entorhinal cortex beta-amyloid load in individuals with mild cognitive impairment. *Exp. Neurol.* **158,** 469–490.

Mufson, E. J., Ma, S. Y., Cochran, E. J., Bennett, D. A., Beckett, L. A., Jaffar, S., Saragovi, H. U., and Kordower, J. H. (2000). Loss of nucleus basalis neurons containing trkA immunoreactivity in individuals with mild cognitive impairment and early Alzheimer's disease. *J. Comp. Neurol.* **427,** 19–30.

National Institute on Aging and Reagan Institute working group on diagnosis criteria for the neuropathological assessment of Alzheimer's disease (1997). Consensus recommendations for the postmortem diagnosis of AD. *Neurobiol. Aging* **18,** S1–S3.

Oddo, S., Caccamo, A., Shepherd, J. D., Murphy, M. P., Golde, T. E., Kayed, R., Metherate, R., Mattson, M. P., Akbari, Y., and LaFerla, F. M. (2003). Triple-transgenic model of Alzheimer's disease with plaques and tangles: Intracellular Abeta and synaptic dysfunction. *Neuron* **39,** 409–421.

Price, J. L., Davis, P. B., Morris, J. C., and White, D. L. (1991). The distribution of tangles, plaques and related immunohistochemical markers in healthy aging and Alzheimer's disease. *Neurobiol. Aging* **12,** 295–312.

Price, J. L., and Morris, J. C. (1999). Tangles and plaques in nondemented aging and "preclinical" Alzheimer's disease. *Ann. Neurol.* **45,** 358–368.

Puchtler, H., Sweat, F., and Levine, M. (1962). On the binding of Congo red by amyloid. *J. Histochem. Cytochem.* **10,** 355–364.

Selkoe, D. J. (2003). Folding proteins in fatal ways. *Nature* **426,** 900–904.

Sturchler-Pierrat, C., Abramowski, D., Duke, M., Wiederhold, K. H., Mistl, C., Rothacher, S., Ledermann, B., Burki, K., Frey, P., Paganetti, P. A., Waridel, C., Calhoun, M. E., Jucker, M., Probst, A., Staufenbiel, M., and Sommer, B. (1997). Two amyloid precursor protein transgenic mouse models with Alzheimer disease-like pathology. *Proc. Natl. Acad. Sci. USA* **94,** 13287–13292.

Styren, S. D., Hamilton, R. L., Styren, G. C., and Klunk, W. E. (2000). X-34, a fluorescent derivative of Congo red: A novel histochemical stain for Alzheimer's disease pathology. *J. Histochem. Cytochem.* **48,** 1223–1232.

Vallet, P. G., Guntern, R., Hof, P. R., Golaz, J., Delacourte, A., Robakis, N. K., and Bouras, C. (1992). A comparative study of histological and immunohistochemical methods for neurofibrillary tangles and senile plaques in Alzheimer's disease. *Acta Neuropathol. (Berl).* **83,** 170–178.

[10] Visualizing Pathology Deposits in the Living Brain of Patients with Alzheimer's Disease

By Vladimir Kepe, Sung-Cheng Huang, Gary W. Small, Nagichettiar Satyamurthy, and Jorge R. Barrio

Abstract

One of the major neuropathological changes characteristic of Alzheimer's disease (AD) are deposits of β-amyloid plaques and neurofibrillary tangles in neocortical and subcortical regions of the AD brain. The histochemical detection of these lesions in *postmortem* brain tissue is necessary for definitive diagnosis of AD. Methods for their *in vivo* detection would greatly aid the diagnosis of AD in early stages when neuronal loss and related functional impairment are still limited and also open the opportunity for effective therapeutic interventions. Positron emission tomography (PET) using an appropriate radiolabeled imaging probe with high binding affinity for these lesions is one of such techniques. We have developed 2-(1-{6-[(2-[F-18]fluoroethyl)(methyl)amino]-2-naphthyl}

METHODS IN ENZYMOLOGY, VOL. 412   0076-6879/06 $35.00
Copyright 2006, Elsevier Inc. All rights reserved.   DOI: 10.1016/S0076-6879(06)12010-8

ethylidene)malononitrile ([F-18]FDDNP), a naphthalene-based radio-fluorinated PET imaging probe with binding affinity for amyloid and amyloid-like structures, and applied it for *in vivo* brain imaging of patients with Alzheimer's disease and cognitively normal controls with PET. Analysis of *in vivo* [F-18]FDDNP imaging data using Logan plot graphical analysis with the cerebellum as a reference region was performed, and the binding levels in several areas of neocortex were determined. We observed increased levels of [F-18]FDDNP binding in patients in several neocortical regions in Alzheimer's disease compared with the cerebellum. In contrast, control subjects have uniformly low levels of [F-18]FDDNP binding in all areas, which is comparable to that of cerebellum.

Introduction

Alzheimer's disease (AD) is the most common neurodegenerative disease of advanced age. The increasing life expectancy in humans, because of significant improvements in prevention and treatment of cardiovascular disease and cancer, has, ironically, increased AD incidence and its subsequent impact on patients and families. It has also increased its social significance because of the increasing burden it imposes on an already strained health system in industrialized nations. Alzheimer's disease is characterized by a progressive loss of cognitive function with neuronal loss and with β-amyloid plaques (senile plaques, SPs) and neurofibrillary tangles (NFTs) as the pathological hallmarks of the disease (Vickers *et al.*, 2000). In 1990, more than 4 million Americans were affected with AD, making it the most common form of dementia (Evans, 1990), and assuming a cure is not found, the number of AD patients has been extrapolated to be quintupled by 2040 (Iqbal, 1991).

AD manifests itself clinically as impairment of a broad spectrum of cognitive processes, including verbal and nonverbal memory; language and semantic knowledge; attention and executive functions; and visuoperceptual and spatial abilities; all of which can be assessed with neuropsychological measures (Salmon and Lange, 2001). However, clinical diagnosis of AD has moderate reliability and, despite being most accurate in advanced stages of the disease, still has limited sensitivity and specificity (Knopman *et al.*, 2001). It is important to note that the definitive diagnosis of AD can only be made on the basis of postmortem histopathological examination of brain tissue and detection of NFTs and SPs (Ball *et al.*, 1997).

Major constituents of SPs are fibrillar aggregates of Aβ(1–40) peptide (Selkoe, 1994; Teplow, 1998). Aβ peptides are formed by the cleavage of the larger amyloid precursor protein (APP) by β- and γ-secretases (Wisniewski *et al.*, 1997). The high-resolution atomic structure of Aβ fibrils has not been explained, despite the repeated attempts by conventional X-ray crystallography, because of the inability of insoluble Aβ fibrils to form single crystals (Lansbury, 1996). However, information about the overall fibrillar structure of *ex vivo* Aβ fibrils is available from x-ray diffraction (Kirschner *et al.*, 1986; Serpell, 2000). X-ray diffraction patterns of Aβ fibrils and other amyloids led to the model of cross-β sheets as a secondary structure element in Aβ protofilaments (Teplow, 1998). In a typical aggregate, multiple protofilaments twist around each other to form a straight rigid fibril, as observed with electron microscopy (Malinchik *et al.*, 1998). Fibrils of synthetic Aβ(1–40), formed *in vitro*, resemble *in vivo* fibrils in terms of neurotoxic properties (Seilheimer *et al.*, 1997) and also ultrastructurally (Kirschner *et al.*, 1987; Miyakawa *et al.*, 1986), but this may not hold true for tau aggregates.

NFTs are cytoskeletal lesions largely composed of paired helical (PHF) and straight filaments (SF), which are intraneuronal fibrillar aggregates of the hyperphosphorylated microtubule-associated protein tau (Lee *et al.*, 2001). After the NFT-laden neurons die, extracellular "tombstone" tangles remain visible (Braak *et al.*, 1999). Hyperphosphorylated microtubule-associated protein aggregates into paired helical filaments by forming cross-β sheet with its repeat domain and with the rest of the peptide forming largely unstructured "fuzzy" coat in which the cross-β-sheet structure is embedded (Barghorn *et al.*, 2004). Because of their fibrillar nature that is based on cross-β sheet polymerization, the tau fibrils can be considered as amyloid in a broad sense. The presence of cross-β sheet in NFTs and in the *in vitro* assembled tau polymers has also been confirmed by x-ray diffraction experiments (Kirchner *et al.*, 1986; von Bergen *et al.*, 2005).

The spatial pattern of SP and NFT distribution in neocortex depends on the severity of disease. Braak and Braak (1991) have proposed six stages of disease progression on the basis of the spatial pattern of NFT distribution in the brain. In the initial stages ("transentorhinal" stages I and II), NFTs are observed in limited regions of entorhinal cortex, in the "limbic" stages III and IV dense deposits of NFTs can be observed throughout the entire entorhinal cortex as well as hippocampal formation with associated areas of temporal lobe, cingulate gyrus, and orbitofrontal cortex having some deposits. In more advanced "cortical" stages V and VI, NFTs finally spread throughout the neocortex leaving only the sensory motor cortex relatively spared. The deposition of β-amyloid SPs also follows distinctive spatial and temporal patterns for which Braak and Braak (1991) have proposed three stages. In the stage A, neocortical areas of the temporal lobe and orbitofrontal cortex first develop deposits of SPs, which in stage B become denser in the same regions and also spread into the rest of frontal lobe and into the parietal lobe. Finally, in stage C, SPs can be found throughout the entire neocortex. This indicates that both pathology load (SPs and NFTs) in a specific region, as well as the pattern of pathology distribution in the brain, are important factors to consider for diagnostic work with the *in vivo* imaging techniques targeting these pathologies.

In advanced stages of the disease, when a diagnosis of probable AD can be made, the death toll of CNS neurons is already heavy and has spread to regions beyond the hippocampus, one of the critical sites of neuronal damage (Braak and Braak, 1991; Price and Morris, 1999). These neuronal losses seem strongly correlated with the presence of abundant intraneuronal NFTs in the same areas but not well correlated with amyloid deposition (Braak and Braak, 1991; Giannakopoulos *et al.*, 2003). The large glutamatergic pyramidal neurons, mainly associated with corticocortical and cortico-hippocampal projections, are particularly vulnerable in AD (Hof, 1997; Morrison and Hof, 1997).

It is essential to develop new tools for early diagnosis that can target pathological changes well before significant neuronal loss, and this is a key for effective therapeutic interventions. 2-Deoxy-2-[F-18]fluoro-*D*-glucose ([F-18]FDG) PET has been investigated as a diagnostic tool for AD for an extended period of time (Silverman *et al.*, 2001, 2003). It provides an *in vivo*, noninvasive method for determination of neuronal function in patients with AD (Silverman *et al.*, 2001; Small *et al.*, 2000). The sensitivity of [F-18]FDG-PET diagnosis is 93–95%, and its specificity was established at 89–92%. The typical glucose metabolic rates observed in the brains of patients with AD may be the consequence of reduced neurite densities resulting from dysfunctional neurons or neuronal death. Therefore, [F-18] FDG provides metabolic (functional) but no information about specific brain pathologies or processes likely to be drug targets in the brain of these patients.

Noninvasive In Vivo Visualization of Neuropathological Deposits

SPs and NFTs are structures largely lacking functional activity connected to the aggregated fibrils, with the exception of the polymerization reaction. This poses a serious challenge for the development of molecular probes with specific binding to these structures. The most obvious molecular imaging probes would be the monomeric peptides themselves (e.g., radiolabeled β-amyloid peptides [Kurihara and Pardridge, 2000; Lee *et al.*, 2002]) or the monoclonal antibodies directed against them (see for example Friedland *et al.*, 1994). These probes have very low capacity to cross the blood–brain barrier because of their size and protein nature, which puts in serious doubt their usefulness for *in vivo* imaging of pathological brain deposits.

Many laboratories have attempted the development of β-amyloid–specific small molecule imaging agents with improved brain entry following various approaches. Cross-β sheet structure in the core of the fibril with its highly ordered arrangement of peptide monomers, anchored together with the hydrogen bonds formed between the peptide backbones and with π-π stacking of the aromatic amino acid residues, as well as glutamic acid–lysine electrostatic interactions, are with high probability the target of these ligands (Makin *et al.*, 2005). In most cases these compounds have structural features similar to Congo red or to Thioflavin T, two histological dyes used as "gold standards" for *in vitro* detection of amyloid-like structures. Examples of Congo red–related structures include X-34 (Styren *et al.*, 2000), methoxy-X04 (Klunk *et al.*, 2002), and BSB (Lee *et al.*, 2001). Examples of molecular probes with Thioflavin T–related structures are 2-[(4′-methylamino)phenyl]-6-hydroxybenzothiazole (PIB) (Mathis *et al.*, 2003), imidazo[1,2-a]pyridine derivatives (Cai *et al.*, 2004; Kung *et al.*, 2002), benzoxazole derivatives such as IBOX (Zhuang *et al.*, 2001), styrylbenzoxazoles (Okamura *et al.*, 2004), and 4-methylamino-4′-hydroxystylbene (Ono *et al.*, 2003). Another type of probe is based on other aromatic or heteroaromatic polycyclic cores such as fluorene (Lee *et al.*, 2003), acridine in BF-108 (Suemoto *et al.*, 2004), or naphthalene with internal dipole formation in 2-(1-{6-[(2-[F-18] fluoroethyl)(methyl)amino]-2-naphthyl}ethylidene)malononitrile ([F-18] FDDNP) (Agdeppa *et al.*, 2003a).

Only three of these molecular imaging probes have been applied to *in vivo* imaging of the Alzheimer's disease brain with PET. [F-18]FDDNP was the first molecular imaging probe reported to be effective in the visualization of neuropathology in the living brain of patients with AD (Agdeppa *et al.*, 2001a; Barrio *et al.*, 1999; Shoghi-Jadid *et al.*, 2002). Specifically, [F-18] FDDNP labels β-amyloid and NFTs both *in vitro* and *in vivo* (Agdeppa *et al.*, 2003b) and has proven useful to reliably follow the neuropathological progression of the disease in the living brain. Initial kinetic analysis of human data was performed using a novel analysis procedure (relative residence time, RRT) of the cerebral region of interest, using the pons as a reference tissue (Shoghi-Jadid *et al.*, 2002). RRT is significantly higher in patients with AD than controls and also significantly correlated with a broad range of memory MMSE scores (8–30). To verify whether flow and other parameters would influence RRT determinations we used additional kinetic analysis, including standardized uptake values (SUV) at equilibrium and Logan distribution volume ratios with the cerebellum as a reference region with equivalent results (Kepe *et al.*, 2004). [F-18]FDDNP accumulates significantly in several cortical areas (medial and lateral temporal lobe, parietal lobe, frontal lobe), with the highest increases in the medial temporal lobe of patients with AD as a result of $A\beta$ and NFT deposition in this area. Because the medial temporal lobe is associated with initial pathology formation (Braak and Braak, 1991), [F-18]FDDNP offers an excellent opportunity for early detection (i.e., patients at risk and patients with mild cognitive impairment). The second successful attempt to image $A\beta$ aggregates in living human subjects was reported by Klunk *et al.* (2004) using the hydroxylated benzothiazole aniline derivative ([C-11] 2-(4-methylamino-phenyl)-6-hydroxybenzothiazole) ([C-11] 6-OH-BTA or [C-11] PIB). In patients with AD, [C-11]PIB retention was most prominently increased in the frontal cortex (SUV = 1.56) and parietal areas (1.45). Temporal accumulation (1.26) is low, similar to that of the pons (1.31), known to lack $A\beta$ aggregates (Klunk *et al.*, 2004). [C-11]PIB retention was equivalent in patients with AD and control subjects for areas known to have no or minimum amyloid deposition (e.g., white matter, pons, and cerebellum). More recently, Verhoeff *et al.* (2004) reported the first human PET data with a novel hydroxylated stilbene derivative, 4-[C-11]methylamino-4'-hydroxystylbene ([C-11]SB-13), with very promising results supporting its *in vivo* binding to $A\beta$-aggregates.

The availability of animal models of $A\beta$-deposition in the brain has also opened a new avenue for research on *in vivo* imaging of β-amyloid deposition with microPET. They offer an invaluable opportunity for testing new, potential molecular imaging probes for amyloid aggregates and also for evaluating new anti-aggregation therapies. For example, [F-18]FDDNP binds to the β-amyloid–rich areas of the rat brain (frontal cortex and hippocampus) in triple transgenic rat model of β-amyloid deposition (Kepe *et al.*, 2005). In contrast, microPET imaging of β-amyloid deposits in Tg2576 β-amyloid transgenic mice with [C-11]PIB did not show any difference between plaque-rich areas (frontal cortex) and the cerebellum (Toyama *et al.*, 2005).

Characterization of [F-18]FDDNP Binding to Aβ Fibrils and SPs

The binding constant of [F-18]FDDNP to synthetic $A\beta$(1–40) fibrils was initially determined by fluorescence titration of nonradioactive [F-18] FDDNP (Agdeppa *et al.*, 2001b). The binding of the $A\beta$(1–40) fibrils to [F-18]FDDNP yielded apparent K_D values of 0.12 nM and 1.86 nM in 0.25% ethanol in PBS. The fluorescence enhancement observed with [F-18]FDDNP bound to $A\beta$(1–40) fibrils indicates probe binding to hydrophobic surface clefts on the basis of fluorescence emission determinations in solvents of different polarity. Furthermore, the fluorescence emission maxima of [F-18]FDDNP bound to $A\beta(1\text{--}40)$ fibrils ($\lambda_{max} = 500$ nm) indicate that the local microenvironment in these fibrils is remarkably hydrophobic [e.g., comparable to that of dichloromethane, $E_T(30) = 39.1$] (Agdeppa *et al.*, 2001b). Confocal fluorescence microscopy and immuno-histochemistry were used to correlate the distribution of radiofluorinated [F-18]FDDNP in digital autoradiograms of AD brain specimens. Digital autoradiography of AD brain specimens using [F-18]FDDNP in 1% ethanol in saline (Agdeppa *et al.*, 2003b) revealed its binding in the temporal and parietal cortices matching the immunohistochemistry of adjacent slices. Fluorescence microscopy revealed that the pattern in the autoradiograms and immunostained tissue originated from SPs and NFTs. Radioactive binding assays with brain homogenates from AD and normal control patients, performed in 1% ethanol in PBS, confirm the high affinity binding of [F-18]FDDNP to *ex vivo* SPs and NFTs. The resulting Scatchard plot of [F-18]FDDNP binding in AD homogenates yielded a K_D value of 0.75 nM and a B_{max} value of 144 nM with the brain sample studied (Agdeppa *et al.*, 2003b). There was no appreciable binding of [F-18]FDDNP to homogenates from age-matched control brains. The high-affinity binding and B_{max} for [F-18]FDDNP sites in AD brain homogenates fulfills the requirement for *in vivo* PET visualization of probe binding to brain receptor sites if it is assumed that the binding sites on SPs are analogous to the receptor model of binding (Shoghi-Jadid *et al.*, 2002). The apparent K_D value for [F-18] FDDNP in the low nanomolar range is also consistent with the specific labeling of SPs and NFTs as microscopically evident by the fluorescence images and the gross pattern of binding observed with digital autoradiography, wide-field fluorescence microscopy, and immunostaining.

A note of caution needs to be made regarding *in vitro* binding experiments with these aggregates. All *in vitro* experiments with FDDNP (see preceding) were performed in appropriate solvents (PBS, normal saline) with a maximal concentration of 1% ethanol (ethanol is used to facilitate solubilization of the molecular probe) to minimize interference of the organic solvent and to mimic as closely as possible *in vivo* conditions. Because binding of molecular imaging probes to amyloid aggregates is attributed to π-π stacking as well as electrostatic interactions with lysine–glutamate residues in the peptide backbone (Makin *et al.*, 2005), increased levels of organic solvents (i.e., 10% ethanol) will disrupt normal binding and produce results—typically weaker binding affinities—that are difficult to interpret. Moreover, *in vitro* conditions with high solvent content (i.e., 10% ethanol in PBS) are significantly different from *in vivo* conditions, and only limited extrapolations can be made from these data (Ye *et al.*, 2005).

Methods

General

All solvents and chemicals were used as received from commercial vendors. HPLC columns [Waters SymmetryPrep C_{18} semi-preparative column (7 μm, 7.8 × 300 mm)] and Waters Symmetry C_{18} analytical column (5 μm, 4.6 × 150 mm) and SepPak cartridges (C_{18}, silica, $^tC_{18}$) were purchased from Waters, Milford, MA. Sterile filters Millex-GV (0.22 μm) were purchased from Fisher, Pittsburgh, PA.

Chemistry—Preparation of Precursor

2-[[6-(2,2-Dicyano-1-methylvinyl)-2-naphthyl](methyl)amino]ethyl 4-methylbenzenesulfonate, the precursor for preparation of the radiofluorinated tracer 2-(1-{6-[(2-[F-18]fluoroethyl)(methyl)amino]-2-naphthyl}ethylidene)malononitrile ([F-18]FDDNP), has been prepared as described elsewhere (Shoghi-Jadid *et al.*, 2002). The Bucherer reaction of 1-(6-hydroxy-2-naphthyl)-1-ethanone with 2-(methylamino)ethanol yields 1-{6-[(2-hydroxyethyl)(methyl)amino]-2-naphthyl}-1-ethanone, which reacts with malononitrile under the conditions of Knoevenagel reaction forming 2-(1-{6-[(hydroxyethyl)(methyl)amino]-2-naphthyl}ethylidene)malononitrile. In the final step, this compound is converted to 2-[[6-(2,2-dicyano-1-methylvinyl)-2-naphthyl](methyl)amino]ethyl 4-methylbenzenesulfonate by reacting it with 4-methylbenzenesulfonyl chloride (tosyl chloride) under basic conditions.

Preparation of [F-18]FDDNP

[F-18]FDDNP is routinely prepared by nucleophilic radiofluorination of 2-[[6-(2,2-dicyano-1-methylvinyl)-2-naphthyl](methyl)amino]ethyl 4-methylbenzenesulfonate with K[F-18]F/Kryptofix 2.2.2 in acetonitrile at 95° for 15 min. The reaction mixture is diluted with 0.1 N HCl and the crude material extracted with Waters C_{18} SepPak solid phase extraction cartridge, preactivated with 5 ml of ethanol and equilibrated with 10 ml of water. The cartridge is washed with copious amount of water and the extracted organic material eluted with dichloromethane. The eluate is passed through a Waters silica SepPak cartridge to remove the major portion of polar side products. After solvent evaporation, the crude product is dissolved in 1 ml of a mixture of tetrahydrofuran and methanol (1:1) and diluted with an equal amount of water. The resulting solution is separated on Waters Symmetry Prep C_{18} semi-preparative HPLC column

(7 μm, 7.8 × 300 mm) with a mixture of water, tetrahydrofuran, and methanol (47:31:22, respectively) at the flow rate of 3 ml/min. The radioactive product is identified by both gamma detection and UV detection at 254 nm. Under the described conditions, the retention time of [F-18]FDDNP is 18 min. The fractions containing [F-18]FDDNP are combined and diluted 10-fold with water. [F-18]FDDNP is extracted with Waters tC$_{18}$ Vac 3cc (200 mg) SepPak solid-phase extraction cartridge (trifunctional C$_{18}$), preactivated with 5 ml of ethanol, and equilibrated with 10 ml of water. The injectable solution of [F-18]FDDNP for human application is prepared as follows: extracted [F-18]FDDNP is eluted from the cartridge with 0.5 ml of absolute ethanol, mixed with 4.5 ml of normal saline (0.9% sodium chloride), and the resulting mixture added to 5 ml of 25% human serum albumin. The injectable solution of [F-18]FDDNP in 12.5% of human serum albumin is then sterilized by filtration through a sterile 0.22-mm Millex-GV filter.

The chemical and radiochemical purities of the radiotracer in the ethanol/saline solution are determined by analytical HPLC on Waters Symmetry C$_{18}$ analytical HPLC column (5 μm, 4.6 × 150 mm) with tetrahydrofuran, methanol, and water as an eluent (40:20:40, respectively) at flow rate of 0.5 ml/min. [F-18]FDDNP has retention time of 9 min under the described conditions. It is detected using gamma detection and UV detection at 400 nm. The radiochemical and chemical purity of the tracer is high (>99%), and the specific activity is also high, ranging between 2 and 8 Ci/μmol at the end of synthesis (EOS). Large quantities of activity (>100 mCi) are produced from a single radiolabeling reaction after ~90 min of total preparation time on regular basis with radiochemical yields ranging between 42 and 63% (EOS).

PET Scanning Protocol

Human subjects were scanned after informed written consent was obtained in accordance with the procedures set by the Human Subjects Protection Committee, University of California at Los Angeles. The scanning protocol and data reconstruction methods used were described elsewhere (Shoghi-Jadid *et al.*, 2002). All head scans are performed with EXACT HR + PET tomograph (Siemens-CTI, Knoxville, TN), with the subjects positioned supine with the imaging plane parallel to the orbito-meatal line. All scans are performed in a dimly lit room with low ambient noise, and the subjects do not have their ears occluded. A transmission measurement is performed before each PET scan for the purpose of attenuation correction. Venous catheterization is performed followed by administration of [F-18]FDDNP (318–430 MBq) as a bolus injection.

Sequential [F-18]FDDNP emission head scans are obtained in 3D mode over a period of 125 min starting at the time of the tracer injection with the following scan sequence: six 30-sec scans, four 3-min scans, five 10-min scans, and three 20-min scans.

Image Reconstruction and Manipulation

All PET images were reconstructed using filtered backprojection (FBP) with measured attenuation correction and were corrected for scatter. Dynamic images were reconstructed with a Hann filter (0.3 Nyquist cutoff) resulting in an in-plane resolution of 5.5-mm full-width half-maximum (FWHM). The obtained images contained 63 contiguous slices with the plane-to-plane separation of 2.42 mm. Summed PET images were coregistered to anatomical MRI images (Lin *et al.*, 1994). Subsequently, the dynamic PET data sets were analyzed using Logan plot graphical analysis with the cerebellum as reference region (Logan *et al.*, 1996), and the relative distribution volume (DVR) parametric images were generated. PET-to-MRI registration parameters were used to reslice the parametric image. Regions of interest (ROIs) were drawn on anatomical MRI images bilaterally on the cerebellum, the medial temporal lobe, and several neo-cortical regions on the lateral temporal lobe (Brodmann area 21), the frontal lobe (Brodmann area 10), the parietal lobe (Brodmann area 7), and a single ROI on the posterior cingulated gyrus. This set of ROIs was applied to the coregistered parametric image, and the DVR values for all ROIs were extracted and used for comparative analysis.

Logan Plot Graphical Analysis of Kinetic PET Data

PET cameras measure radioactivity concentration in the body after the injection of a tracer. Serial imaging over time after tracer injection would give regional kinetics of the tracer that is related to physiological or pathophysiological processes in the regional tissue. To convert the tracer kinetics to physiological or biological information in terms of absolute units, compartmental models are commonly used. In this process, model parameters are estimated using nonlinear regression that fits the measured tissue kinetics with the convolution of the metabolite-corrected time activity curve in plasma (the input function) and the response function of the model. However, when data noise is high, the estimated model parameters have large variability, thus model fitting is computationally impractical for pixel-wise kinetics. Graphical and simplified methods have been used, instead, to give more robust results that are suitable for generating parametric images of biological information (e.g., of receptor densities of irreversible [Patlak *et al.*, 1983] and reversible [Logan *et al.*, 1990] ligands).

These methods still require use of the input function. In some instances, the time activity curve from a reference region can be used as an input function in these graphical analyses, if the reference region is virtually devoid of specific types of binding sites (Logan *et al.*, 1996). In such cases, the equation used for the analysis is

$$\frac{\int_0^T C_{roi}(t)dt}{C_{roi}(T)} = DVR\left\{\frac{\int_0^T C_{ref}(t)dt + C_{ref}(T)/(k_2)_{ref}}{C_{roi}(T)}\right\} + y_{\text{intercept}}$$

where C_{roi} and C_{ref} are, respectively, the time activity curves in the region of interest and in the reference region, and $(k_2)_{ref}$ is the clearance rate constant of tracer in the reference region. The term with $(k_2)_{ref}$ can usually be neglected, because in most cases it does not affect significantly the slope of the plot. The slope of the linear portion of this Logan plot is thus the relative distribution volume (DVR), which is equal to DV_{roi}/DV_{ref} (i.e., the distribution volume of the tracer in a region of interest divided by that in the reference region). DVR is a useful measure for comparison of data sets.

Steps in preparation of parametric images include:

1. Coregistration of the [F-18]FDDNP PET dynamic data set with subject's MRI image.
2. Drawing of an ROI on cerebellum and one ROI on temporal lobe on the MRI image.
3. Extraction of the tissue time activity curves for both ROI from the [F-18]FDDNP PET dynamic data set.
4. Determination of the Logan plot linearity for the time frame 15–125 min for the temporal lobe ROI with cerebellar tissue time activity curve as the input function.
5. If Logan plot is linear, then DVR values are calculated for every voxel using in-house developed software with the cerebellar time activity curve as the input function; parametric image is prepared containing the DVR value for each voxel.

Example of Logan Plot Graphical Analysis of [F-18]FDDNP PET Data

A negligibly low level of brain pathology in the cerebellum of patients with AD—at least in moderate AD—makes the cerebellum a good candidate as a reference tissue in Logan plot graphical analysis of human [F-18] FDDNP PET determinations in Alzheimer's disease. After IV injection, [F-18]FDDNP is delivered to tissue by blood flow and crosses the blood–brain barrier rapidly. The tracer then clears with different rates from different brain regions depending on the distribution volume (related to the density of binding sites) of the tracer in tissue. Logan plot of the kinetics of various tissue regions that used the cerebellum as the reference region showed linearity for data points 15 min after tracer injection.

The [F-18]FDDNP-PET scans were performed on 20 AD patients (age: 72 ± 9 y, MMSE: 18.6 ± 6.8), 20 MCIs (age: 72 ± 10 y, MMSE: 27.4 ± 2.0), and 20 control subjects (age: 70 ± 9 y, MMSE: 29.3 ± 0.9). Analysis of [F-18] FDDNP-PET data was performed using conventional Logan graphical analysis with the cerebellum as reference region using time points between 15 and 125 min to generate DVR parametric images. A set of ROI's, drawn on the MRI, was applied to the coregistered DVR parametric images, and the DVR values were determined for the posterior cingulated gyrus, the medial temporal lobe, the lateral temporal lobe, the parietal lobe, and the frontal lobe bilaterally. An average of left and right DVR values from the same region was used for the comparison between the subjects. Comparison of the [F-18]FDDNP DVR values shows that the binding of the probe to all areas analyzed (medial and lateral temporal lobe, parietal lobe, frontal lobe, posterior cingulated gyrus) is significantly elevated in patients with AD compared with the same areas in the subjects from the control group. Representative examples of [F-18]FDDNP DVR parametric images are shown in Fig. 1. The three subject groups were compared on their global and regional [F-18]FDDNP binding values (DVR) using ANOVAs. The Spearman rank correlation, a nonparametric measure of association between ranked pairs of observations, was used to test the strength of

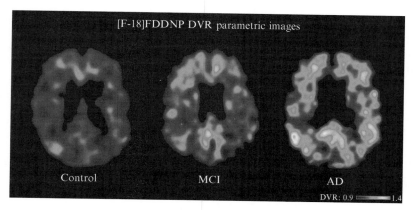

Fig. 1. Representative examples of brain [F-18]FDDNP DVR parametric images from a control subject (left), an MCI (middle), and an AD subject (right).

association between [F-18]FDDNP DVR values and neuropsychological test score (Mini-Mental State Exam, MMSE).

The average of all five regional DVR values was used as a measure of global [F-18]FDDNP binding. Results for all regional and for global [F-18]FDDNP DVR values are as follows. The mean [F-18]FDDNP DVR values differed significantly among the three diagnostic groups: values were lower for the control group than the MCI group, which had lower mean FDDNP binding than the Alzheimer's disease group for global DVR [control: 1.07 ± 0.03; MCI: 1.12 ± 0.01; AD: 1.16 ± 0.02; ANOVA: $F(2,57) = 233.8, P < 0.001$] as well as for the regional values: frontal lobe DVR [control: 1.03 ± 0.02, MCI: 1.07 ± 0.03; AD: 1.11 ± 0.02; ANOVA: $F(2,57) = 41.33, P < 0.001$], parietal lobe DVR [controls: 1.05 ± 0.03; MCI: 1.10 ± 0.03; AD: 1.15 ± 0.02; ANOVA: $F(2,57) = 70.81, P < 0.001$], lateral temporal lobe DVR [control: 1.06 ± 0.03; MCI: 1.12 ± 0.03; AD: 1.16 ± 0.02; ANOVA: $F(2,57) = 72.49, P < 0.001$], medial temporal lobe DVR [control: 1.11 ± 0.03; MCI: 1.16 ± 0.01; AD: 1.19 ± 0.02; ANOVA: $F(2,57) = 41.33, P < 0.001$]. and posterior cingulated gyrus DVR [control: 1.09 ± 0.04; MCI: 1.13 ± 0.04; AD: 1.18 ± 0.03; ANOVA: $F(2,57) = 37.97, P < 0.001$].

Global [F-18]FDDNP DVR values are also negatively correlated with the level of cognitive impairment experienced by the subjects with AD as measured by Mini Mental State Exam (Spearman correlation coefficient $r_s = -0.76, P < 0.001$).

Conclusions

The regional pattern of [F-18]FDDNP neocortical accumulation follows the pattern of disease-specific SP and NFT pathology distribution proposed by Braak and Braak (1991), with the medial temporal lobe being the central area of earliest pathological changes. Because these pathological aggregates induce neuronal death—the earliest neuronal death in hippocampus (Price and Morris, 1999)—their presence can be demonstrated before cognitive symptoms become evident. With the appropriate molecular imaging probe and PET, early diagnosis could then be possible. Also, the degree of [F-18]FDDNP brain deposition consistently correlates well with behavioral measures in these patients (e.g., MMSE).

Acknowledgments

The authors thank Dr. Linda Ercoli for performing the neuropsychological tests and Dr. Prabha Siddarth for performing the statistical analysis. Financial support from the Department of Energy (grant DE-FC03-02ER63420) is gratefully acknowledged.

References

Agdeppa, E. D., Kepe, V., Petrič, A., Satyamurthy, N., Liu, J., Huang, S. C., Small, G. W., Cole, G. M., and Barrio, J. R. (2003a). *In vitro* detection of (S)-naproxen and ibuprofen binding to plaques in the Alzheimer's brain using the positron emission tomography molecular imaging probe 2-(1-[6-[(2-[(18)F]fluoroethyl)(methyl)amino]-2-naphthyl] ethylidene)malononitrile. *Neuroscience* **117**, 723–730.

Agdeppa, E. D., Kepe, V., Liu, J., Shoghi-Jadid, K., Small, G. W., Huang, S.-C., Petrič, A., Satyamurthy, N., and Barrio, J. R. (2003b). 2-Dialkylamino-6-acylmalononitrile substituted naphtalenes (DDNP Analogs): Novel diagnostic and therapeutic tools in Alzheimer's disease. *Mol. Imaging Biol.* **4**, 404–417.

Agdeppa, E. D., Kepe, V., Liu, J., Flores-Torres, S., Satyamurthy, N., Petrič, A., Cole, G. M., Huang, S.-C., and Barrio, J. R. (2001a). Binding characteristics of radiofluorinated 6-dialkylamino-2-naphthylethylidene derivatives as positron emission tomography imaging probes for β-amyloid plaques in Alzheimer's disease. *J. Neurosci.* **21**, RC189(1–5).

Agdeppa, E. D., Kepe, V., Shoghi-Jadid, K., Satyamurthy, N., Small, G. W., Petrič, A., Vinters, H. V., Huang, S.-C., and Barrio, J. R. (2001b). *In vivo* and *in vitro* labeling of plaques and tangles in the brain of an Alzheimer's disease patient: A case study. *J. Nucl. Med.* **42**(Suppl. 5), 65.

Ball, M., Braak, H., Coleman, P., Dickson, D., Duyckaerts, C., Gambetti, P., Hansen, L., Hyman, B., Jellinger, K., Markesbery, W., Perl, D., Powers, J., Price, J., Trojanowski, J. Q., Wisniewski, H., Phelps, C., and Khachaturian, Z. (1997). Consensus recommendations for the postmortem diagnosis of Alzheimer's disease. *Neurobiol. Aging* **18**(Suppl. 4), S1–S2.

Barghorn, S., Davies, P., and Mandelkow, E. (2004). Tau paired helical filaments from Alzheimer's disease brain and assembled *in vitro* are based on β-structure in the core domain. *Biochemistry* **43**, 1694–1703.

Barrio, J. R., Huang, S.-C., Cole, G. M., Satyamurthy, N. M., Petrič, A., Phelps, M. E., and Small, G. W. (1999). PET imaging of tangles and plaques in Alzheimer disease. *J. Nucl. Med.* **40**(Suppl S), 284.

Braak, H., and Braak, E. (1991). Neuropathological staging of Alzheimer-related changes. *Acta Neuropathol.* **82**, 239–259.

Braak, E., Griffing, K., Arai, K., Bohl, J., Bratzke, H., and Braak, H. (1999). Neuropathology of Alzheimer's disease: What is new since A. Alzheimer? *Eur. Arch. Psychiatry Clin. Neurosci.* **249**(Suppl. 3), 14–22.

Cai, L., Chin, F. T., Pike, V. W., Toyama, H., Liow, J.-S., Zoghbi, S. S., Modell, K., Briard, E., Shetty, H. U., Sinclair, K., Donohue, S., Tipre, D., Kung, M.-P., Dagostin, C., Widdowson, D. A., Green, M., Gao, W., Herman, M. M., Ichise, M., and Innis, R. B. (2004). Synthesis and evaluation of two 18F-labeled 6-iodo-2-(4'-N,N-dimethylamino)phenylimidazo[1,2-a] pyridine derivatives as prospective radioligands for β-amyloid in Alzheimer's disease. *J. Med. Chem.* **47**, 2208–2218.

Evans, D. A. (1990). Estimated prevalence of Alzheimer's disease in the United States. *Milbank Q* **68**, 267–289.

Friedland, R. P., Majocha, R. E., Reno, J. M., Lyle, L. R., and Marotta, C. A. (1994). Development of an anti-A beta monoclonal antibody for in vivo imaging of amyloid angiopathy in Alzheimer's disease. *Mol. Neurobiol.* **9**, 107–113.

Giannakopoulos, P., Herrmann, F. R., Bussiere, T., Bouras, C., Kovari, E., Perl, D. P., Morrison, J. H., Gold, G., and Hof, P. R. (2003). Tangle and neuron numbers, but not amyloid load, predict cognitive status in Alzheimer's disease. *Neurology* **60**, 1495–1500.

Hof, P. R. (1997). Morphology and neurochemical characteristics of the vulnerable neurons in brain aging and Alzheimer's disease. *Eur. Neurol.* **37**, 71–81.

Iqbal, K. (1991). "Alzheimer's Disease: Basic Mechanisms, Diagnosis, and Therapeutic Strategies." Wiley; Chichester, New York.

Kepe, V., Shoghi-Jadid, K., Wu, H.-M., Huang, S.-C., Small, G. W., Satyamurthy, N., Petrič, A., Phelps, M. E., and Barrio, J. R. (2004). Global and regional [F-18]FDDNP binding as *in vivo* measure of Alzheimer's disease neuropathology. *J. Nucl. Med.* **45**(Suppl.), 126.

Kepe, V., Cole, G. M., Liu, J., Shoghi-Jadid, K., Flood, D. G., Trusko, S. P., Satyamurthy, N., Huang, S.-C., Small, G. W., Petrič, A., Phelps, M. E., and Barrio, J. R. (2005). [F-18] MicroPET imaging of β-amyloid deposits in the living brain of triple transgenic rat model of β-amyloid deposition. *Mol. Imaging Biol.* **7**(2), 105.

Kirschner, D. A., Abraham, C., and Selkoe, D. J. (1986). X-ray diffraction from intraneuronal paired helical filaments and extraneuronal amyloid fibers in Alzheimer disease indicates cross-beta conformation. *Proc. Natl. Acad. Sci. USA* **83**, 503–507.

Kirschner, D. A., Inouye, H., Duffy, L. K., Sinclair, A., Lind, M., and Selkoe, D. J. (1987). Synthetic peptide homologous to beta protein from Alzheimer disease forms amyloid-like fibrils *in vitro*. *Proc. Natl. Acad. Sci. USA* **84**, 6953–6957.

Klunk, W., Bacskai, B. J., Mathis, C. A., Kajdasz, S. T., McLellan, M. E., Frosch, M. P., Debnath, M. L., Holt, D. P., Wang, Y., and Hyman, B. P. (2002). Imaging Aβ plaques in living transgenic mice with multiphoton microscopy and methoxy-X04, a systemically administered Congo Red derivative. *J. Neuropathol. Exp. Neurol.* **61**, 797–805.

Klunk, W. E., Engler, H., Nordberg, A., Wang, Y., Blomqvist, G., Holt, D. P., Bergstrom, M., Savitcheva, I., Huang, G. F., Estrada, S., Ausen, B., Debnath, M. L., Barletta, J., Price, J. C., Sandell, J., Lopresti, B. J., Wall, A., Koivisto, P., Antoni, G., Mathis, C. A., and Langstrom, B. (2004). Imaging brain amyloid in Alzheimer's disease with Pittsburgh Compound-B. *Ann. Neurol.* **55**, 306–319.

Knopman, D. S., DeKosky, S. T., Cummings, J. L., Chui, H., Corey-Bloom, J., Relkin, N., Small, G. W., Miller, B., and Stevens, J. C. (2001). Practice parameter: Diagnosis of dementia (an evidence-based review). Report of the Quality Standards Subcommittee of the American Academy of Neurology. *Neurology* **56**, 1143–1153.

Kung, M.-P., Hou, C., Zhuang, Z.-P., Zhang, B., Skovronsky, D., Trojanowski, J. Q., Lee, V. M.-Y, and Kung, H. F. (2002). IMPY: An improved thioflavin-T derivative for *in vivo* labeling of β-amyloid plaques. *Brain Res.* **956**, 202–210.

Kurihara, A., and Pardridge, W. M. (2000). Abeta(1-40) peptide radiopharmaceuticals for brain amyloid imaging: (111) In chelation, conjugation to poly(ethylene glycol)-biotin linkers, and autoradiography with Alzheimer's disease brain sections. *Bioconjug. Chem.* **11**, 380–386.

Lansbury, P. T. (1996). A reductionist view of Alzheimer's disease. *Accounts Chem. Res.* **29**, 317–321.

Lee, C.-W., Kung, Zhunag, Z.-P., Kung,, M.-P., Plossl, K., Skovronsky, D., Gur, T., Hou, C., Trojanowski, J. Q., Lee, V. M.-Y., and Kung, H. F. (2001). Isomerization of (Z,Z) to (E,E) 1-bromo-2,5-bis-(3-hydroxycarbonyl-4-hydroxy)styrylbenzene in strong base: Probes for amyloid plaques in the brain. *J. Med. Chem.* **44**, 2270–2275.

Lee, C.-W., Kung, M.-P., Hou, C., and Kung, H. K. (2003). Dimethylamino-fluorenes: Ligands for detecting β-amyloid plaques in the brain. *Nucl. Ned. Biol.* **30**, 573–580.

Lee, V. M., Goedert, M., and Trojanowski, J. Q. (2001). Neurodegenerative tauopathies. *Annu. Rev. Neurosci.* **24**, 1121–1159.

Lin, K. P., Huang, S. C., Baxter, L. R., and Phelps, M. E. (1994). A general technique for interstudy registration of multifunction and multimodality images. *IEEE Trans. Nucl. Sci.* **41**, 2850–2855.

Logan, J., Fowler, J., Volkow, N., Wolf, A., Dewey, S., Schlyer, D., MacGregor, R. R., Hitzemann, R., Bendriem, R., Gatley, R., *et al.* (1990). Graphical analysis of reversible radioligand binding from time-activity measurements applied to [N-11C-methyl]-(-)-cocaine PET studies in human subjects. *J. Cereb. Blood Flow Metab.* **10,** 740–747.

Logan, J., Fowler, J., Volkow, N., Wang, G., Ding, Y., and Alexoff, D. (1996). Distribution volume ratios without blood sampling from graphical analysis of PET data. *J. Cereb. Blood Flow Metab.* **16,** 834–840.

Makin, O. S., Atkins, E., Sikorsky, P., Johansson, J., and Serpell, L. C. (2005). Molecular basis for amyloid fibril formation and stability. *Proc. Natl. Acad. Sci. USA* **102,** 315–320.

Malinchik, S. B., Inouye, H., Szumowski, K. E., and Kirschner, D. A. (1998). Structural analysis of Alzheimer's $\beta(1\text{--}40)$ amyloid: Protofilament assembly of tubular fibrils. *Biophys. J.* **74,** 537–545.

Mathis, C. A., Wang, Y., Holt, D. P., Huang, G.-F., Debnath, M. L., and Klunk, W. E. (2003). Synthesis and evaluation of ^{11}C-labeled 6-substituted 2-arylbenzothiazoles as amyloid imaging agents. *J. Med. Chem.* **46,** 2740–2754.

Miyakawa, T., Katsuragi, S., Watanabe, K., Shimoji, A., and Ikeuchi, Y. (1986). Ultrastructural studies of amyloid fibrils and senile plaques in human brain. *Acta Neuropathol.* **70,** 202–208.

Morisson, J. H., and Hof, P. R. (1997). Life and death of neurons in the aging brain. *Science* **278,** 412–419.

Okamura, N., Suemoto, T., Shimadzu, H., Suzuki, M., Shiomitsu, T., Akatsu, H., Yamamoto, T., Staufenbiel, M., Yanai, K., Arai, H., Sasaki, H., Kudo, Y., and Sawada, T. (2004). Styrylbenzoxazole derivatives for *in vivo* imaging of amyloid plaques in the brain. *J. Neurosci.* **24,** 2535–2541.

Ono, M., Wilson, A., Nobrega, J., Westaway, D., Verhoeff, P., Zhuang, Z.-P., Kung, M.-P., and Kung, H. F. (2003). 11C-labeled stilbene derivatives as Aβ-aggregate-specific PET imaging agents for Alzheimer's disease. *Nucl. Med. Biol.* **30,** 565–571.

Patlak, C. S., Blasberg, R. G., and Fenstermacher, J. D. (1983). Graphical evaluation of blood-to-brain transfer constants from multiple-time uptake data. *J. Cereb. Blood Flow Metab.* **3,** 1–7.

Price, J. L., and Morris, J. C. (1999). Tangles and plaques in nondemented aging and "preclinical" Alzheimer's disease. *Ann. Neurol.* **45,** 358–368.

Salmon, D. P., and Lange, K. L. (2001). Cognitive screening and neuropsychological assessment in early Alzheimer's disease. *Clin. Geriatr. Med.* **17,** 229–254.

Selkoe, D. J. (1994). Cell biology of the amyloid beta-protein precursor and the mechanism of Alzheimer's disease. *Annu. Rev. Cell Biol.* **10,** 373–403.

Seilheimer, B., Bohrmann, B., Nondolfi, B., Muller, F., Stuber, D., and Dobeli, H. (1997). The toxicity of the Alzheimer's beta-amyloid peptide correlates with a distinct fiber morphology. *J. Struct. Biol.* **119,** 59–71.

Serpell, L. C. (2000). Alzheimer's amyloid fibrils: Structure and assembly. *Biochim. Biophys. Acta* **1502,** 16–30.

Shoghi-Jadid, K., Small, G. W., Agdeppa, E. D., Kepe, V., Ercoli, L. M., Siddarth, P., Read, S., Satyamurthy, N., Petrič, A., Huang, S.-C., and Barrio, J. R. (2002). Localization of neurofibrillary tangles and beta-amyloid plaques in the brains of living patients with Alzheimer disease. *Am. J. Geriatr. Psychiatry* **10,** 24–35.

Silverman, D. H., Small, G. W., Chang, C. Y., Lu, C. S., Kung De Aburto, M. A., Chen, W., Czernin, J., Rapoport, S. I., Pietrini, P., Alexander, G. E., Schapiro, M. B., Jagust, W. J., Hoffman, J. M., Welsh-Bohmer, K. A., Alavi, A., Clark, C. M., Salmon, E., de Leon, M. J., Mielke, R., Cummings, J. L., Kowell, A. P., Gambhir, S. S., Hoh, C. K., and Phelps, M. E. (2001). Positron emission tomography in evaluation of dementia: Regional brain metabolism and long-term outcome. *J. Am. Med. Assoc.* **286,** 2120–2127.

Silverman, D. H., Truong, C. T., Kim, S. K., Chang, C. Y., Chen, W., Kowell, A. P., Cummings, J. L., Czernin, J., Small, G. W., and Phelps, M. E. (2003). Prognostic value of regional cerebral metabolism in patients undergoing dementia evaluation: Comparison to a quantifying parameter of subsequent cognitive performance and to prognostic assessment without PET. *Molec. Genetics Metab.* **80,** 350–355.

Small, G. W., Ercoli, L. M., Silverman, D. H., Huang, S.-C., Komo, S., Bookheimer, S. Y., Lavretsky, H., Miller, K., Siddarth, P., Rasgon, N. L., Mazziotta, J. C., Saxena, S., Wu, H. M., Mega, M. S., Cummings, J. L., Saunders, A. M., Pericak-Vance, M. A., Roses, A. D., Barrio, J. R., and Phelps, M. E. (2000). Cerebral metabolic and cognitive decline in persons at genetic risk for Alzheimer's disease. *Proc. Natl. Acad. Sci. USA* **97,** 6037–6042.

Styren, S. D., Hamilton, R. L., Styren, G. C., and Klunk, W. E. (2000). X-34, a fluorescent derivative of Congo Red: A novel histochemical stain for Alzheimer's disease pathology. *J. Histochem. Cytochem.* **48,** 1223–1232.

Suemoto, T., Okamura, N., Shiomitsu, T., Suzuki, M., Shimadzu, H., Akatsu, H., Yamamoto, T., Kudo, Y., and Sawada, T. (2004). In vivo labeling of amyloid with BF-108. *Neurosci. Res.* **48,** 65–74.

Teplow, D. B. (1998). Structural and kinetic features of amyloid beta-protein fibrillogenesis. *Amyloid* **5,** 121–142.

Toyama, H., Ye, D., Ichise, M., Liow, J.-S., Cai, L., Jacobowitz, D., Musachio, J. L., Hong, J., Crescenzo, M., Tipre, D., Lu, J.-Q., Zoghbi, S., Vines, D. C., Seidel, J., Katada, K., Green, M. V., Pike, V. W., Cohen, R. M., and Innis, R. B. (2005). Pet imaging of brain with β-amyloid probe, [11C]6-OH-BTA-1, in a transgenic mouse model of Alzheimer's disease. *Eur. J. Nucl. Med. Mol. Imaging* **32,** 593–600.

Verhoeff, N. P., Wilson, A. A., Takeshita, S., Trop, L., Hussey, D., Singh, K., Kung, H. F., Kung, M. P., and Houle, S. (2004). *In vivo* imaging of Alzheimer disease beta-amyloid with [11C]SB-13 PET. *Am. J. Geriatr. Psychiatry* **12,** 584–595.

Vickers, J. C., Dickson, T. C., Adlard, P. A., Saunders, H. L., King, C. E., and McCormack, G. (2000). The cause of neuronal degeneration in Alzheimer's disease. *Prog. Neurobiol.* **60,** 139–165.

von Bergen, M., Barghorn, S., Biernat, J., Mandelkow, E. M., and Mandelkow, E. (2005). Tau aggregation is driven by a transition from random coil to beta sheet structure. *Biochim. Biophys. Acta* **1739,** 158–166.

Wisniewski, T., Ghiso, J., and Frangione, B. (1997). Biology of Aβ amyloid in Alzheimer's disease. *Neurobiol. Dis.* **4,** 313–328.

Ye, L., Morgenstern, J. L., Gee, A. D., Hong, G., Brown, J., and Lockhart, A. (2005). Delineation of PET imaging agent binding sites on β-amyloid peptide fibrils. *J. Biol. Chem.* **280,** 23599–23604.

Zhuang, Z.-P., Kung, M.-P., Hou, C., Plössl, K., Skovronsky, D., Gur, T. L., Trojanowski, J. Q., Lee, V. M.-Y., and Kung, H. F. (2001). IBOX (2-(4′-dimethylaminophenyl)-6-iodobenzoxazole): A ligand for imaging amyloid plaques in the brain. *Nucl. Med. Biol.* **28,** 887–894.

[11] Micro-Imaging of Amyloid in Mice

By JONATHAN S. WALL, MICHAEL J. PAULUS, SHAUN GLEASON,
JENS GREGOR, ALAN SOLOMON, and STEPHEN J. KENNEL

Abstract

Scintigraphic imaging of radioiodinated serum amyloid P-component is a proven method for the clinical detection of peripheral amyloid deposits (Hawkins *et al.*, 1990). However, the inability to perform comparably high-resolution studies in experimental animal models of amyloid disease has impacted not only basic studies into the pathogenesis of amyloidosis but also in the preclinical *in vivo* evaluation of potential anti-amyloid therapeutic agents. We have developed microimaging technologies, implemented novel computational methods, and established protocols to generate high-resolution images of amyloid deposits in mice. ^{125}I-labeled serum amyloid P component (SAP) and an amyloid-fibril reactive murine monoclonal antibody (designated 11-1F4) have been used successfully to acquire high-resolution single photon emission computed tomographic (SPECT) images that, when fused with x-ray computed tomographic (CT) data, have provided precise anatomical localization of secondary (AA) and primary (AL) amyloid deposits in mouse models of these diseases.

This chapter will provide detailed protocols for the radioiodination and purification of amyloidophilic proteins and the generation of mouse models of AA and AL amyloidosis. A brief description of the available hardware and the parameters used to acquire high-resolution microSPECT and CT images is presented, and the tools used to perform image reconstruction and visualization that permit the analysis and presentation of image data are discussed. Finally, we provide established methods for measuring organ- and tissue-specific activities with which to corroborate the microSPECT and CT images.

Introduction

The amyloidoses represent a growing number of diverse, insidious, and devastating diseases that comprise cerebral forms such as Alzheimer's disease (Bishop and Robinson, 2004; Cummings, 2004) and spongiform encephalopathies (Ghetti *et al.*, 2003), as well as peripheral disorders including type II diabetes (Hull *et al.*, 2004), primary (AL) (Bellotti *et al.*, 2000; Merlini and Bellotti, 2003), and secondary (AA) amyloidosis

METHODS IN ENZYMOLOGY, VOL. 412   0076-6879/06 $35.00
                    DOI: 10.1016/S0076-6879(06)12011-X

(Rocken and Shakespeare, 2002). All are characterized by the aggregation of normally soluble, natural proteins or their fragments into highly ordered amyloid fibrils that deposit within vital organs and tissues leading to dysfunction and, if unremitting, death.

Monitoring the pathogenesis of these diseases and their response or resistance to treatments and preventative interventions has relied historically on the histological evaluation of tissue biopsies, surveys of surrogate physiological markers, such as organ function, and postmortem histological examination. This is also true in the fields of preclinical and basic research that use experimental animal models of amyloidosis (Kisilevsky *et al.*, 1995, 2004; Mihara *et al.*, 2004; Stenstad and Husby, 1996; Wall *et al.*, 2001; Zhu *et al.*, 2001). Within the past two decades, however, advances in clinical medical practices have been made possible, by whole body scintigraphic examination (planar 2-D imaging of a bodily radiation source), the detection of amyloid deposits in patients using radioiodinated serum amyloid P-component (SAP) (Hawkins *et al.*, 1988a,c; MacRaild *et al.*, 2004; Pepys *et al.*, 1997). To date, radioiodinated SAP scintigraphy has been performed in more than 3000 patients and is currently the most universally used clinical modality for determining the scope and biodistribution of amyloid deposits in patients with non-cerebral amyloidosis (Hawkins, 2002).

Because of its high affinity for fibrils and short lifetime in the circulation, human SAP is an excellent agent for imaging amyloid *in vivo*. Indeed, human SAP exhibits a higher affinity for mouse AA-amyloid fibrils *in vivo* than the endogenous murine SAP, making it an exceptional preclinical tracer for imaging amyloid in murine models (Hawkins *et al.*, 1988b). We have used human ^{125}I-SAP to generate quantitative microSPECT/CT images of hepatic and splenic AA amyloid deposits in a transgenic murine model.

In addition to SAP, we have successfully imaged, in mice, localized human AL amyloidomas using the amyloid-reactive monoclonal antibody (mAb) 11-1F4 (Hrncic *et al.*, 2000; O'Nuallain *et al.*, 2005; Solomon *et al.*, 2003a,b) as an imaging agent. The ^{125}I-labeled 11–1F4 mAb, when injected IV into mice bearing amyloidomas composed of human AL fibrils, was seen in the coregistered SPECT/CT images to localize almost exclusively within the amyloid (Wall *et al.*, 2005b).

Microimaging Apparatus

A number of microCT and microSPECT small animal imaging machines are commercially available; however, at present there are four manufacturers of dual (combined) microSPECT/CT apparati. For a list of preclinical microimaging systems with links to the relevant manufacturers, the reader is referred to the molecular imaging site maintained by Stanford University (www.mi-central.org).

For our experiments we have exclusively used microSPECT and CT systems, designed and fabricated by Mike Paulus and Shaun Gleason while at Oak Ridge National Laboratory (Oak Ridge, TN) (Paulus *et al.*, 1999, 2001, 2000a,b). The microCT and recently developed dual-modality micro-SPECT/CT systems are now commercially available (Siemens Medical Solutions, Knoxville, TN).

The microCT system contains a 20–80 kVp X-ray source with 50 micron focal spot and a 2048×3072 charge coupled device (CCD) array detector, optically coupled to a Gadox [$Gd_2O_2S(Tb)$] phosphor screen by means of a fiberoptic taper. The source and detector are mounted on a circular gantry allowing it to rotate 360° around the subject (mouse) positioned on a stationary bed (Fig. 1). A number of detectors are now commercially available; in our studies, the detector can capture a 90 mm × 60 mm field of view, more than adequate for a whole-body mouse study. A CT data set, composed of 360 projections, can be acquired in approximately 6 min.

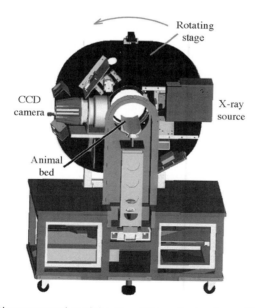

FIG. 1. Schematic representation of the microCT imaging hardware. The x-ray source and CCD camera are fixed to the stage that can rotate 360° around the stationary mouse placed on the bed.

The microSPECT apparatus used to collect the ^{125}I–labeled SAP and 11-1F4 mAb images was designed specifically for small animal imaging and fabricated at Oak Ridge National Laboratory and situated immediately next to the microCT machine. The SPECT detector consists of an array of multianode photomultiplier tubes coupled to a pixilated NaI(Tl) crystal array (Jefferson Laboratory, Newport News, VA) that revolves around the mouse subject in typically 6° increments. The detector has a 10 × 20 cm active area that is discretized into a 50 × 100 array. The resolution is approximately 2.5 mm when coupled with a high-resolution parallel-hole collimator. MicroSPECT datasets are composed of 60 projections that are collected over the course of 30–60 min, depending on the specific activity of the tracer and its concentration at the site of the amyloid.

Protein Tracers

The human SAP used in our studies was kindly provided by Prof. Philip Hawkins (Department of Medicine, Royal Free and University College Medical School, London, UK) and was of clinical grade. The protein was isolated from heat-treated human serum as previously described, by calcium-dependent affinity chromatography (De Beer and Pepys, 1982). The protein was stored frozen until used.

Murine 11-1F4 mAb was generated in our laboratory by inoculating mice with an intraperitoneal injection of 1×10^6 11-1F4–producing hybridoma cells followed by isolation of the antibody from ascitic fluid by size exclusion and ion-exchange chromatography (Abe *et al.*, 1993a,b). Certain murine 11-1F4 preparations were produced by the National Cancer Institute (Biopharmaceutical Development Program, NCI-Frederick) under the Rapid Access to Intervention Development (RAID) program. All 11-1F4 samples were stored frozen until used.

Preparation of Radioiodinated Proteins

Although unsuitable for the clinic, ^{125}I has been used in our small animal imaging studies because of its relatively long half-life (60 days), ease of labeling, and availability—the long half-life also facilitates postmortem autoradiographic analyses of tissue sections. In the clinical arena, the low energy of the ^{125}I gamma ray is unsatisfactory; the attenuation length of a 35-keV gamma ray in soft tissue is about 3.3 cm—small relative to the diameter of a human. For mouse studies, however, the attenuation length of ^{125}I is greater than the largest axial diameter of a typical adult mouse and, therefore, more than adequate for providing high-resolution images.

Indeed, the attenuation of the 35-keV gamma rays in mice is substantially less than that of the gamma rays emitted by standard clinical imaging isotopes (e.g., ^{201}Tl, ^{99m}Tc) in humans.

Our imaging experiments have used radioiodinated proteins. Whether proteins are labeled electrochemically, enzymatically, or by chemical oxidation (ICl, iodogen, or chloramine T), the iodide atoms are attached predominately to either ortho or para positions on tyrosine rings. However, iodotyrosines are subject to dehalogenation *in vivo*, particularly within the liver (Auf dem Brinke *et al.*, 1980; Kohrle *et al.*, 1990; Spanka *et al.*, 1990). The loss of iodide from the targeting molecule does not necessarily negate imaging and biodistribution studies particularly for relatively short time frames and for targets in organs other than liver. We have not found fundamental differences in the performance of targeting proteins radioiodinated by any of the oxidative methods mentioned previously. For simplicity, we chose to conduct radioiodination with limiting amounts of chloramine T, as described in the following.

In the event that dehalogenation proves problematic, several chemical approaches have been designed to alleviate the problem (De Nardo *et al.*, 1988; Hnatowich, 1990; Vaidyanathan and Zalutsky, 1990a,b; Vaidyanathan *et al.*, 2000). These involve attachment of the iodide to benzoyl rings that are subsequently coupled to proteins by means of *N*-hydroxy-succinimide (NHS) reactions with lysine amino groups. Note that the Bolton Hunter reagent, although it uses NHS coupling is *not* resistant to dehalogenation reactions, because the iodide is still conjugated to the ortho and para positions on a tyrosyl moiety (Bolton and Hunter, 1973). As an alternative to the chloramine T radioiodination, we describe a procedure for radioiodination with succinimidyl-4-iodo benzoate (SIB) (Vaidyanathan and Zalutsky, 1990a). This reaction results in stably labeled compounds but generally yields probes with lower specific activity because of the relatively inefficient coupling by the NHS reaction.

Procedure for the Chloramine T Radioiodination of Proteins

Many methods for radioiodination have been published previously (Bailey, 1994; Behr *et al.*, 2002; Garg *et al.*, 1993; Karonen, 1990; Vaidyanathan and Zalutsky, 1990a,b; Wilbur, 1992). We prefer a modification of the method of McConahey and Dixon (McConahey and Dixon, 1966). Proteins of any size can be radioiodinated by these methods if they contain an available tyrosine. Purification of the final product free of aggregates and free iodide is normally achieved by size exclusion gel filtration on media suitable for separating the radiolabeled protein from contaminants. A stepwise procedure follows:

1. In a monitored radioactive work zone, assemble Microfuge tube, micropipettes, buffers, protein, radioiodide, and chemicals; see following. Always use appropriate protective clothing, gloves, and eyewear.

2. Transfer 5 μl of 0.5 M sodium phosphate, pH 7.6, to a clean Microfuge tube (1.5 ml capacity).

3. Pipette radioactive iodide (Perkin Elmer protein iodination grade ^{125}I) into the buffer. Up to 50 μl of radioiodide can be used in this reaction. If more of the iodide solution is required, increase the amount of sodium phosphate buffer accordingly.

4. Add protein as a concentrated solution freshly dialyzed into 0.01 M sodium phosphate, 0.15 M NaCl, pH 7.6 (PBS). Proteins can also be radiolabeled in Tris-HCl buffered saline or other nonreducing buffered solutions.

5. Add 5 μl of 1 mg/ml freshly prepared chloramine T (Sigma, St. Louis, MO); dissolve 1 or 2 mg of dry crystals in the appropriate volume of water immediately before addition to the labeling mix. For reactions with more than 0.5 mg of protein, an additional 5 μl of chloramine T solution should be added for each additional 0.5 mg protein.

6. The reaction is complete within 1 min; however, residual reaction should be quenched by addition of 5 μl of 2 mg/ml $Na_2S_2O_5$ (sodium metabisulfite, [J. T. Baker, Phillipsburg, NJ]) freshly dissolved in water immediately before use.

7. For size exclusion gel filtration, a dedicated HPLC can be used; however, low yields of protein may result. For small amounts of protein, we recommend an open, disposable column method with media eluting under gravity. Pour a column with a 5-ml bed volume in a disposable plastic 5-ml pipette plugged with glass wool. For whole IgGs or SAP, use Ultragel AcA 34 resin (Sigma, St. Louis, MO) equilibrated with PBS containing 5.0 mg/ml BSA or 1.0 mg/ml gelatin as carrier protein. For smaller proteins such as scFv, Ultragel AcA 44 can be used.

8. Add 50 μl of PBS with 5 mg/ml BSA and 20 μl of 10 mg/ml blue dextran (Pharmacia Fine Chemicals) to the quenched radioiodination reaction mixture and load onto the gel filtration column.

9. Manually collect 1-min fractions (approximately 200 μl) in plastic tubes. Count a small volume of each fraction ($\sim$1–2 μl) in a gamma scintillation counter using the ^{125}I program. Pool three or four fractions with the highest activity and test for activity and radiochemical purity.

10. For immunoreactivity of radioiodinated antibodies, specific antigen tests should be done. Such a test for reactivity of 11-1F4 antibody is described in the following.

11. Radiochemical purity of proteins is checked on SDS-PAGE gels. Small fractions of the pooled material (often as little as 0.1 μl) are

separated in reducing conditions on 10% acrylamide SDS-PAGE. The gels are not fixed or washed but are covered with plastic wrap and placed on PhosphorImager screens (e.g., Cyclone Storage Phosphor System, Perkin Elmer, Shelton, CT). The resultant images are analyzed by region of interest quantitation. The preparations should be >95% pure.

Comments

Highly concentrated reactions are more efficient. The addition of large volumes of iodide solution or of proteins not adequately dialyzed can introduce agents (presumably reductants) that inhibit the reaction. Optimization of the methods described previously can provide yields of up to 50% radioiodide incorporation in the final product. By use of these techniques we have successfully radioiodinated protein amounts from 0.1 μg to >10 mg.

Immunoreactivity Protocol for Radiolabeled 11-1F4 mAb. This assay describes a bead-conjugated antigen-binding assay that has been used to confirm the immunoreactivity of the 11-1F4 mAb. This is a large format assay to measure the binding of radiolabeled 11-1F4 to antigen-coated beads and is appropriate for various radioiodinated forms. It is recommended that duplicate or triplicate samples be tested at two antibody concentrations.

1. Bead preparation: Polystyrene beads (0.91 μm-diameter, conjugated with amino groups) are obtained from Spherotech Inc. (Libertyville, IL). Approximately 0.5 ml of bead solution is washed three times in PBS by centrifugation for 2 min at 10,000g and resuspended by pipetting using a 200 μl micropipette with a small bore tip. The washed beads are then treated with 4.0 ml of 0.5% glutaraldehyde (freshly diluted from a 25% stock solution) in PBS for 5 min at RT and then washed three times as previously. After the final wash, the beads are resuspended in 4 ml PBS containing 4 mg of peptide antigen (approximately 1.2 μmoles) and tumbled end-over-end for 18 h at RT. To block free glutaraldehyde sites, sterile glycine (0.5 ml of 1 M stock in PBS) is added and the beads mixed thoroughly for 1 h. The antigen-conjugated beads are then washed three times and suspended in a final volume of 1 ml PBS. Beads can be stored as a suspension at 4°.

2. Binding assay: Add 200 μl of PBS containing 5 mg/ml BSA into a 1.5-ml Microfuge tube. To this add 5.0 μl of antigen-coated bead suspension followed by 500 μl of BSA/PBS solution containing up to 100 ng of the radiolabeled antibody. Pipette an equal amount of radiolabeled antibody into gamma counter vials to serve as a standard.

3. Incubate the mixture on a tube rotator for 1 h at RT.

4. To quantify the amount of bound mAb, the beads are isolated by centrifugation at 10,000g for 2 min and the supernatant aspirated and discarded in an appropriate radiation waste stream.

5. Resuspend the beads thoroughly in 500 μl of PBS containing 0.1% Tween 20 by pipetting. Repeat steps 4 and 5.

6. The specific activity associated with the beads can be measured in the Microfuge tube by directly inserting it into a scintillation vial, or they can be resuspended and pipetted into an appropriate counting vial.

Active preparations of the 11-1F4 mAb yield binding of >60% to the beads. By use of this protocol and a peptide antigen of 30 amino acids, the bead capacity seems to be much greater (theoretically 1000-fold) than that required to bind 100 ng of 11-1F4. We have not observed any change in % bound fraction over the range of 11-1F4 mAb concentrations that we have tested (up to 1 μg mAb). One sample of ^{125}I-labeled antibody stored at 4° for more than 2 wks retained 65% binding in this assay.

Radiolabeling using the Succinimidyl-iodobenzoate (SIB) Method. This method produces radioiodinated peptides or proteins that are resistant to dehalogenases found in the serum and liver. Compared with oxidative methods, SIB labeling is slightly slower and usually produces proteins with 5–10-fold lower specific activities. A detailed procedure follows:

1. The starting material is not commercially available at this time. Trimethyl Sn 4 succinimidyl benzoate, MeATE, was synthesized using the method of Kozoriowski (Koziorowski *et al.*, 1998). The product was dried extensively and stored desiccated at –20°. Product stored at room temperature loses significant activity for iodide incorporation within months even if it is completely dry.

2. Prepare a 10 mg/ml stock solution of MeATE in methanol with 5% acetic acid (v/v). This stock solution can be stored for about a month at –20°.

3. Immediately before labeling, dilute a small amount of the stock solution 1:50 into methanol with 5% acetic acid and transfer 5 μl (~1.0 μg) to a clean Microfuge tube. Add radioactive iodide in dilute NaOH (up to 30 μl or 3 mCi). Add 5 μl of chloramine T (freshly prepared as described previously).

4. To remove unbound iodide, add 0.5 ml of freshly prepared sodium metabisulfite solution and transfer to the barrel of a 5-ml plastic syringe with a C18 light SEPAK cartridge (Waters, Milford, MA) attached. It is prudent to prime the cartridge by rinsing with methanol and then water before use.

5. Insert the syringe plunger and slowly push the iodinated mixture through. Remove the cartridge (carefully) and then remove the plunger from the syringe barrel. Add 3 ml of pure water to the reassembled cartridge barrel and wash it through slowly.

6. Remove the cartridge and plunger again and add 0.5 ml 100% methanol to the cartridge assembly. Slowly push the solution through and collect the iodinated tracer in a clean glass tube. Dry the eluant under a stream of nitrogen.

7. Redissolve the material with 100–200 μl of analytical grade acetonitrile, transfer to a plastic Microfuge tube, and again evaporate to dryness. A quick pass with a Geiger-Müller detector should reveal >50% of the activity in this fraction.

8. To the dried, iodinated SIB, add 5 μl of 1 M sodium borate solution pH 8.6–9.0. Add protein as a concentrated solution in PBS and allow the reaction to proceed for 15 min at RT. This step is limiting, and low concentration proteins give lower yields.

9. Quench any remaining reaction by addition of 5 μl of 1 M glycine in sterile-filtered PBS.

10. The coupled material can then be purified in open size exclusion gel filtration columns as described previously. Note: Gelatin must be used as the carrier protein, because BSA will bind contaminating glycyl-SIB and carry it through the gel filtration process into the final product.

Injections. The most reproducible route of injection is intravenous (IV); in mice, we use the lateral tail vein. Mice are exposed to a heat lamp for 5 min before restraining them in a plastic tube that allows exposure of the tail. One-milliliter plastic syringes are used with 27-gauge needles. Volumes up to 0.3 ml of neutral buffered, standard ionic strength solutions (such as PBS), can be injected without complications. Care must be taken to avoid air bubbles in the delivered volume. For our 11-1F4 mAb and SAP studies, 0.2 ml of radioiodinated antibody (10–50 μg, ~300 μCi) is injected in about 5 sec. A gauze pad is used to apply pressure at the injection site and blot return blood before the animals are returned to their cages. In the rare instances that the tail vein is not accessible, intraperitoneal (IP) injections can be used. We have found that small volumes (0.2–0.5 ml) of radioiodinated antibodies injected IP equilibrate in the circulation within approximately 2 h. Thus IP injection may be used for imaging in some cases; however, biodistribution of IV- and IP-injected animals cannot be compared in the same groups even at later time points.

Generating Murine Models of AA and AL Amyloid Disease

AL Amyloidoma

There are presently no transgenic murine models of systemic AL amyloidosis, and attempts to recapitulate the disease in mice have been successful, but costly, in terms of the substantial amounts of human light chain protein required. We have, therefore, developed a model of localized AL amyloidosis in mice that can be used to assess the efficacy of novel therapeutic agents and amyloidophilic radiotracers *in vivo*. Human (hu)AL amyloid was extracted and purified from amyloid-laden organs harvested at autopsy using the water floatation method of Pras (Pras *et al.*, 1968). To prepare the amyloid for injection, lyophilized human AL amyloid extract was suspended in 25 ml of sterile PBS and homogenized with a PCU-2 Polytron apparatus at the highest setting for 3 min (Brinkman, Lucerne, Switzerland). The fibrils were sedimented by centrifugation at 6° for 30 min at ~12,000g, and 50–100 mg of the resultant pellet was resuspended in 1 ml of sterile PBS and rehomogenized using the PCU-2 Polytron as previously. Typically, two groups of three 8-wk-old Balb/c mice (Charles Rivers, Wilmington, MA) are injected subcutaneously with amyloid solution using an 18-gauge needle. All six mice receive 40–100 mg of amyloid between the scapulae in a 1–2 ml volume of sterile phosphate-buffered saline (PBS: 150 mM NaCl, 5 mM NaHPO$_4$, Na$_2$PO$_4$). The amyloidoma, which was readily palpable and visible by microCT imaging, remained resident at the site of injection and became vascularized within 7 days after injection (Fig. 2). The first group of mice received ^{125}I-11-1F4 mAb, and the second group,

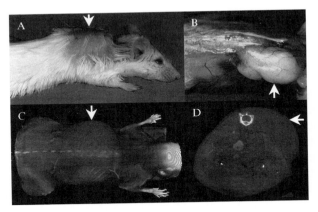

FIG. 2. The murine model of localized AL amyloidoma. (A) The human AL amyloidoma is introduced between the scapulae of Balb/c mice and is readily palpable (mouse is shaved to show the position of the amyloidoma). (B) The amyloidoma appears vascularized within 7 days after injection (skin is removed to show subcutaneous vascularization around amyloid. The amyloidoma is readily visible by microCT imaging. (C) Three-dimensional volume rendering of a mouse with a subcutaneous amyloidoma. (D) Axial CT slice through mouse shown in (B) at the level of the upper abdomen and amyloidoma. The mouse imaged in (C) and (D) was administered blood pool contrast medium (Fenestra VC, Alerion Biomedical, San Diego, CA) IV before scanning. Arrows indicate the amyloidoma.

serving as controls, was injected with ^{125}I-MOPC-31c, an isotype-matched murine IgG1κ mAb.

Induction of AA-amyloidosis in hIL-6 Transgenic Mice (TRIAD)

Systemic AA amyloidosis induced in mice has been successfully imaged by whole-body planar scintigraphy after administration of [123]I- or ^{131}I-labeled SAP (Caspi et al., 1987; Hawkins et al., 1988b) and with ^{111}In or ^{125}I-labeled AA-reactive rat monoclonal antibodies (Marshall et al., 1986). Although these studies demonstrated highly specific tracer uptake in hepatosplenic AA-amyloid deposits, the anatomical resolution was insufficient to discern liver and spleen independently. Advances in micro-imaging hardware and software capabilities now enable dual-modality (SPECT/CT), high-resolution images of amyloid-laden viscera in mice to be acquired, using ^{125}I-labeled SAP. By use of these techniques, we have demonstrated an excellent correlation between the biodistribution and quantitation of the amyloid deposits using ^{125}I-SAP imaging and standard histological evaluation (Wall et al., 2005a).

For imaging studies, AA amyloidosis is induced in mice carrying the human interleukin-6 (hIL-6) transgene, expressed constitutively under control of the murine major histocompatibility complex (H-2$^{a'}$) promoter or metallothionein (MT) promoter [furnished by Dr. Michael Potter (Kovalchuk et al., 2002) and Dr. Cilliberto (Fattori et al., 1994), respectively]. AA-amyloidosis is induced in 8–12-wk-old MT or H2/IL-6 mice by administering 10 μg of amyloid enhancing factor (AEF), prepared as previously described (Axelrad et al., 1982; Kindy and de Beer, 1999) intravenously in the tail vein, in a volume of 100 μl sterile filtered PBS. At 6 wks after AEF-injection, the amyloid deposits were visualized by dual SPECT/CT imaging using ^{125}I-labeled SAP as the tracer. AA amyloidosis can be induced in wild-type mice by injection of AEF in conjunction with a proinflammatory stimulus such as silver nitrate solution (Snow and Kisilevsky, 1985), casein (Kindy and de Beer, 1999), or lipopolysaccharide (Inoue and Kisilevsky, 1999). Note: Subcutaneous silver nitrate solution can attenuate x-rays in the CT scan and would presumably also diminish the gamma emission from the ^{125}I-SAP or 11-1F4 mAb, although the potential effects of this on SPECT image quality have not been determined.

MicroSPECT and CT Image Acquisition

Thyroid uptake of free iodide in the AA and AL amyloid mice is restricted by adding Lugol's solution (1% v/v) to the drinking water 48 h before the administration of ^{125}I-labeled tracer. Mice with systemic AA amyloidosis receive an IV injection of 200 μl of ^{125}I-SAP solution in PBS (10–15 μCi/μg protein) in the lateral tail vein. The mice are sacrificed 24 h after injection by an overdose of isoflurane inhalation or carbon dioxide asphyxiation and placed in 50-ml centrifuge tubes; they are then frozen before SPECT and CT scanning. Freezing is not necessary, but when many mice are to be scanned, it is prudent to freeze them as the acquisition of SPECT and CT data can take ~1 h per animal.

For mice-bearing subcutaneous AL amyloidomas, ~300 μCi of ^{125}I-labeled 11-1F4 mAb (~12 μCi/μg protein) is administered in the lateral tail vein. When mAbs are used the mice are sacrificed 72 h after injection to allow adequate clearance of the unbound reagent. The mice are sacrificed and frozen before collection of SPECT and CT images.

SPECT images were acquired by positioning the animal between the SPECT detector heads near the center of rotation. For these studies, hexagonal lead foil parallel hole collimators with 1.2 mm diameter, 27-mm long holes, and 0.2-mm septa were used. These collimators are very similar to clinical low-energy high-resolution (LEHR) collimators. The collimators were placed within a few millimeters of the animal, providing approximately 2.5 mm resolution with 55 cps/MBq sensitivity. Each SPECT data set consisted of 60 projections acquired over a period of 60 sec per projection.

The animals were transferred to the microCT scanner immediately after the SPECT data acquisition. The microCT system was configured for relatively high speed (~6 min), low-resolution (~100 micron) image acquisition. For these studies, the X-ray source was biased at 80 kVp with 500 μA anode current. The X-ray detector was configured for 4×4 binning, reducing the size of each projection to 512×768 pixels. Projections were acquired in 1-degree steps over a 360-degree orbit.

Image Coregistration

Our current imaging protocol uses independent microSPECT and CT machines, which necessitates that we coregister the two image sets (i.e., align them so that the anatomical CT image is correctly positioned with reference to the SPECT image). To date, we have found that this is most readily achieved manually using a 3-D visualization and analysis software package (Amira, Version 3.1: Mercury Computer Systems, Chelmsford, MA). Each SPECT and CT image contains fiducial markers supplied by capillary tubes filled with a solution of ^{125}I. Aligning the fiducials visible in 3-D reconstructions of each image ensures correct coregistration of the nuclear and anatomical images. This process will be circumvented by the new dual-modality microSPECT/CT machine that will acquire both sets of image data without the mouse being moved; application of a simple mathematical transformation will result in accurately coregistered images in this instance.

Image Reconstruction

CT Image Reconstruction

When dealing with high-count projection data as is the case here, a volumetric CT image is conveniently reconstructed using a filtered backprojection–based method such as the algorithm of Feldkamp *et al.* (Feldkamp *et al.*, 1984). Under ideal circumstances, a ramp filter will suffice, but in practice a filter is used that contains some degree of built-in smoothing to prevent undesirable high-frequency noise from being amplified. Typical filter choices include Shepp-Logan and Generalized-Hamming (Jain, 1989). Because of the divergent nature of a circular-orbit cone-beam X-ray system, filtered backprojection methods are capable of delivering perfect reconstructions of the center slice of the image, whereas off-center slices are only approximate. However, this does not tend to be a problem unless the cone-angle is very large (i.e., in slices at the periphery of the image). To achieve reconstruction times on the order of 5–10 mins, we use parallel computing, as well as a problem reducing technique called focus of attention (Gregor *et al.*, 2002). The latter allows us to separate the projection data as well as the image data into "mouse" and "background" before reconstruction so that we subsequently concentrate the available computational resources solely on the "mouse" data.

SPECT Image Reconstruction

SPECT projection data tend to be count-poor and noisy. Use of an iterative image reconstruction method is therefore necessary. The goal is to solve a linear system of equations, $Ax = y$ where vectors x and y denote the unknown image and the acquired projection data, respectively, and A is a so-called system matrix that describes the geometry of the system plus detector efficiencies if the projection data are not precorrected. Although an algebraic method could be used to solve this system in a weighted least squares sense or otherwise, it is often better to recast it in a statistical context and use a maximum-likelihood technique such as the EM algorithm by Shepp and Vardi (Shepp and Vardi, 1982). Depending on how the system matrix is configured, count preservation can be achieved meaning that each iteration redistributes activity within the image without altering the overall photon count resulting in truly quantitative images. The main disadvantage of the EM algorithm is that it is slow to converge. Although parallel computing could be used to speed up convergence, ordered subsets constitute a much simpler solution (Hudson and Larkin, 1994). We have found that approximately 8–10 iterations using four subsets are sufficient to produce images of the quality shown in this chapter. The time required to perform these calculations is about 1 min on a standard laptop (Gregor *et al.*, 2004).

Image Visualization and Analysis

For the purposes of 3-D image visualization and analysis, the raw data output from the microCT and SPECT image reconstructions are loaded into the Amira software package. The microCT and SPECT images can be represented in 3-D space as volumetric or surface views that can be readily coregistered using a manual transformation tool. The image data can be displayed as 2-D gray-scale slices in axial, sagittal, or coronal aspects, as well as in 3-D using a number of display modules. In addition, both the 2-D and 3-D visualizations can be combined to generate complex representations of the data. Amira was used to generate the images presented in this chapter. Representative examples of microSPECT/CT images depicting the binding of ^{125}I-SAP to AA amyloid (Fig. 3) and [125]I-11-1F4 mAb to an AL amyloidoma (Fig. 4) in mice were generated using Amira.

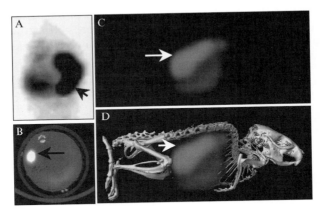

FIG. 3. Standard representations of microSPECT and microCT images obtained from ^{125}I-SAP labeling of AA-amyloid deposits in H2/IL-6 mice. (A) Planar, 2-D, scintigraphic image of a mouse with accumulation of ^{125}I-SAP in the spleen (arrow) and to a lesser degree in the liver. (B) Axial slice of a coregistered microSPECT/CT through the upper abdomen showing significant activity in the spleen (arrow) and diffuse accumulation of the SAP in the ventral liver. (C) Three-dimensional volume–texture (voltex) rendering of the microSPECT data discriminates the high amyloid content in the spleen (arrow) from the more diffuse hepatic deposits. (D) Precise anatomical localization of the amyloid is provided when a 3-D voltex rendering of the ^{125}I-SAP distribution is displayed with a surface-rendered microCT data set. (See color insert.)

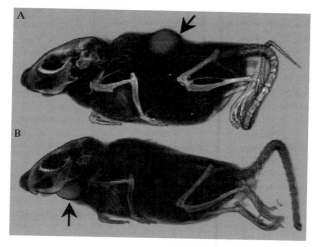

FIG. 4. Anatomical localization of a human AL amyloidoma. (A) The amyloidoma is readily visible (arrow) and anatomically pinpointed when a 3-D surface rendering of the [125]I-11–1F4 mAb distribution is displayed with the CT image. (B) When a ^{125}I-labeled isotype-matched control antibody (MOPC-31c IgG1κ; Sigma, St. Louis MO) is used, no activity is observed in the amyloidoma. However, free iodide is visible in the thyroid (arrow). In both (A) and (B), a threshold was applied to the SPECT data to display the 40% maximal activity. (See color insert.)

Corroborative Studies: Biodistribution and Autoradiography

Although microimaging can produce excellent quantitative, high-resolution data on the position of radioisotope in a live animal, studies on tissues gathered postmortem from the scanned animals can support and augment the imaging data. An effort is made to collect these data for all of our microSPECT experiments.

Biodistribution

Many methods have been published to determine the organ distribution of injected radiotracers. To collect statistically meaningful data, at least five mice per group are required. The animals should all be the same age and sex to eliminate differences caused by size and weight. The animals are sacrificed by isoflurane inhalation or carbon dioxide asphyxiation, and microSPECT/CT data are collected. The mice are either then dissected immediately after sacrifice (preferable) or after the images of animals frozen in 50-ml centrifuge tubes have been collected. Appropriate-sized pieces of organs are placed into tared, plastic scintillation vials. The vials are tightly capped and weighed soon after organ collection to minimize weight loss to evaporation. Small organ samples should be taken from mice that have been injected with relatively large amounts of radioactivity (e.g., for micro-SPECT), so that counter efficiency is not compromised by coincidence counting of samples with high activity. Note: Cuts in the liver lead to leakage of blood into the peritoneal cavity that can be collected with a 200-μl micropipette as a source of blood sampling. Alternately, blood, serum, or clotted blood can be collected from the thoracic cavity after removal of the heart lung block. Tissues should be dissected free of external blood clots and blotted on clean paper. If the mice are sacrificed using isoflurane or carbon dioxide inhalation and then frozen before the organs can be harvested, it is difficult to collect blood samples, and the mice often have lung hemorrhages that lead to anomalously high values of tracer accumulation in this organ. For most small soft tissue samples very little autoquenching is noted for gamma scintillation counting even with soft gamma emitters (I-125). Internal standards for accurate counting of large bone samples may be necessary. Values for biodistribution are usually presented as % injected dose per gram of tissue (%ID/g). These values can easily be calculated from the net weight and net recovered counts per minute (cpm) relative to standards of the injected tracer solution set aside at the onset of the experiment. Including these standard samples when measurement of the experimental tissue samples is performed eliminates the need for isotopic decay correction. Figure 5 depicts representative biodistribution data collected from mice with AA amyloid deposits primarily in the spleen and liver with minor deposits visible in the kidney and heart. A control group of healthy mice is shown for comparison. Organs were harvested from both groups 24 h after injection of ^{125}I-SAP.

Autoradiography

Autoradiographs of tissue sections yield very high-resolution microdistribution of radioactive probes (Fig. 6). Image analyses of silver grains can resolve radioactive decay (i.e., radiolabeled tracer accumulation) at the cellular level (Kennel *et al.*, 1991). Thus, measurements of probe binding at the resolution of imaging ($\sim$50 μm) or at the organ level from biodistribution studies can be refined to indicate the cell type or other target (e.g., amyloid deposits that are the actual site of tracer accumulation). For amyloid imaging in the liver, spleen, and kidneys, this is particularly useful to distinguish between actual amyloid uptake of radiolabeled probe and nonspecific accumulation that can occur in these organs because of probe interaction with the reticuloendothelial cells or during renal tubular readsorption. For this purpose it is necessary to compare the autoradiographic images with a Congo red–stained tissue section (Congo red staining of

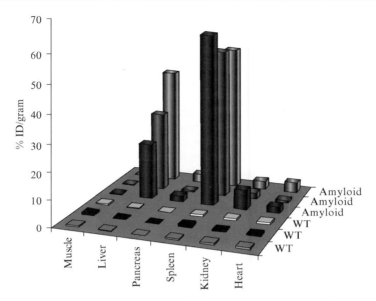

FIG. 5. Typical biodistribution data after ^{125}I-SAP injection into mice with systemic AA amyloidosis. H2/IL-6 mice (with amyloid) and wild-type, healthy mice (WT) received an IV injection of ^{125}I-SAP, were sacrificed 24 h after injection and the organs harvested for analysis. Most of the activity is observed in the spleen and liver with minor (<10% ID/g) amounts found in the kidneys and heart. No activity was associated with any of the organs evaluated from the WT mice

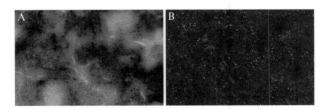

FIG. 6. Representative microautoradiographic study of splenic tissue from an AA-amyloidotic H2/IL-6 mouse administered ^{125}I-SAP for SPECT imaging. Consecutive sections were stained using photographic emulsion (A) and alkaline Congo red (B). In (A), dense areas of amyloid appear as dark punctate regions, whereas more diffuse accumulation of ^{125}I-SAP is manifest as shadowed areas. The amyloid deposits in (B) appear in cross-polarized illumination as green-blue birefringent areas. Magnification was ×40 (5× objective). (See color insert.)

amyloid is described in detail elsewhere in this series, Westermark *et al.*, 1999). Figure 6 shows consecutively cut sections of a mouse liver containing significant quantities of AA amyloid. This is evidenced by the coincident localization of ^{125}I-SAP, manifest by the appearance of black grains in the autoradiograph (Fig. 6A), with the amyloid deposits revealed by the characteristic green birefringence in the Congo red–stained section (Fig. 6B). The best autoradiographic images are obtained with [126]I. Mice injected with ~300 μCi of tracer as in the microSPECT imaging experiments are ideal tissue sources. Tissues can be collected at the time biodistribution studies are done. Small tissue samples are fixed in 10% neutral buffer formalin for 1 day. The tissues are washed from the fixative into PBS and are then processed for paraffin embedding. Sections of 5–6-μm thickness are cut and placed on Probond-coated glass slides and air dried. In absolute darkness, the slides are dipped in NTB-2 liquid emulsion (Eastman Kodak, Rochester, NY) that has been warmed to room temperature. After drying, the slides are stored in light tight boxes at room temperature for the incubation period. The time of exposure varies with the amount of radioisotope injected, the fraction accumulating at the target site, and the density of grains desired. It is useful to make several copies of a test slide and develop them at various times to gauge the amount of exposure desired for the entire set; periods from 4–7 days are generally sufficient to provide excellent images. Slides are developed with Dektol (Eastman Kodak) and fixed with Kodak fixative, then the tissues are dehydrated for hematoxylin–eosin counterstaining and coverslip mounting. Slides prepared in this manner can be stored indefinitely until analyses can be done.

Summary

Heretofore, evaluating and monitoring amyloid deposition in mouse models of disease has relied for the most part on postmortem histological examination to determine organ involvement and the extent of amyloid deposition. As an alternative, we have developed protocols for ^{125}I-labeling of SAP and the amyloid-binding mAb 11-1F4 that, in conjunction with high-resolution SPECT and CT imaging, provide a means to visualize systemic AA and localized AL amyloid deposits in mouse models of the disease. In conjunction with microautoradiographic and biodistribution studies, the SPECT and CT images, when visualized using state of the art software, can be used to quantitate and determine the extent of organ-specific amyloid deposition.

Techniques and imaging hardware and software we have described offer new opportunities not only to observe *ante mortem* and in real time the progression of amyloidosis *in vivo* but also a way to evaluate preclinically and in individual animals the efficacy of novel therapeutic agents designed to prevent or cure amyloid diseases.

Acknowledgments

The authors thank all the people who have contributed to the protocols presented herein, in particular: Rudi Hrncic, Denny Wolfenbarger, Maria Schell, Trish Smith, and Tina Richey for their tireless work with the mouse models; Trish Lankford and Justin Baba at Oak Ridge National Laboratories for their assistance with biodistribution measurements and image acquisition; Jim Wesley, Sallie Macy, and Craig Wooliver for processing and staining of the mouse tissues; and Philip Hawkins for his encouragement of this work and the gift of human SAP. A. S. is an American Cancer Society Professor. This work was supported by NIBIB Bioengineering Research Partnership Award # RO1 EB000789.

References

Abe, M., Goto, T., Kennel, S. J., Wolfenbarger, D., Macy, S. D., Weiss, D. T., and Solomon, A. (1993a). Production and immunodiagnostic applications of antihuman light chain monoclonal antibodies. *Am. J. Clin. Pathol.* **100**, 67–74.

Abe, M., Goto, T., Wolfenbarger, D., Weiss, D. T., and Solomon, A. (1993b). Novel immunization protocol and ELISA screening methods used to obtain and characterize monoclonal antibodies specific for human light chain variable-region subgroups. *Hybridoma.* **12**, 475–483.

Auf dem Brinke, D., Kohrle, J., Kodding, R., and Hesch, R. D. (1980). Subcellular localization of thyroxine-5-deiodinase in rat liver. *J. Endocrinol. Invest.* **3**, 73–76.

Axelrad, M. A., Kisilevsky, R., Willmer, J., Chen, S. J., and Skinner, M. (1982). Further characterization of amyloid-enhancing factor. *Lab. Invest.* **47**, 139–146.

Bailey, G. S. (1994). Labeling of peptides and proteins by radioiodination. *Methods Mol. Biol.* **32**, 441–448.

Behr, T. M., Gotthardt, M., Becker, W., and Behe, M. (2002). Radioiodination of monoclonal antibodies, proteins and peptides for diagnosis and therapy. A review of standardized, reliable and safe procedures for clinical grade levels kBq to GBq in the Gottingen/Marburg experience. *Nuklearmedizin.* **41**, 71–79.

Bellotti, V., Mangione, P., and Merlini, G. (2000). Review: Immunoglobulin light chain amyloidosis—the archetype of structural and pathogenic variability. *J. Struct. Biol.* **130**, 280–289.

Bishop, G. M., and Robinson, S. R. (2004). Physiological roles of amyloid-beta and implications for its removal in Alzheimer's disease. *Drugs Aging* **21**, 621–630.

Bolton, A. E., and Hunter, W. M. (1973). The labelling of proteins to high specific radioactivities by conjugation to a 125I-containing acylating agent. *Biochem. J.* **133**, 529–539.

Caspi, D., Zalzman, S., Baratz, M., Teitelbaum, Z., Yaron, M., Pras, M., Baltz, M. L., and Pepys, M. B. (1987). Imaging of experimental amyloidosis with 131I-labeled serum amyloid P component. *Arthritis Rheum.* **30**, 1303–1306.

Cummings, J. L. (2004). Alzheimer's disease. *N. Engl. J. Med.* **351**, 56–67.

De Beer, F. C., and Pepys, M. B. (1982). Isolation of human C-reactive protein and serum amyloid P component. *J. Immunol. Methods* **50**, 17–31.

De Nardo, G. L., De Nardo, S. J., Miyao, N. P., Mills, S. L., Peng, J. S., O'Grady, L. F., Epstein, A. L., and Young, W. C. (1988). Non-dehalogenation mechanisms for excretion of radioiodine after administration of labeled antibodies. *Int. J. Biol. Markers* **3**, 1–9.

Fattori, E., Della Rocca, C., Costa, P., Giorgio, M., Dente, B., Pozzi, L., and Ciliberto, G. (1994). Development of progressive kidney damage and myeloma kidney in interleukin-6 transgenic mice. *Blood* **83**, 2570–2579.

Feldkamp, L. A., Davis, L. C., and Kress, J. W. (1984). Practical cone-beam algorithm. *J. Opt. Soc. Am.* **1**, 612–619.

Garg, S., Garg, P. K., Zhao, X. G., Friedman, H. S., Bigner, D. D., and Zalutsky, M. R. (1993). Radioiodination of a monoclonal antibody using N-succinimidyl 5-iodo-3-pyridinecarboxylate. *Nucl. Med. Biol.* **20**, 835–842.

Ghetti, B., Tagliavini, F., Takao, M., Bugiani, O., and Piccardo, P. (2003). Hereditary prion protein amyloidoses. *Clin. Lab. Med.* **23**, 65–85, viii.

Gregor, J., Gleason, S. S., Kennel, S. J., Paulus, M. J., Benson, T., and Wall, J. S. (2004). Approximate volumetric system models for microSPECT. *In* "Proc. IEEE Nuclear Science Symposium and Medical Imaging Conference" p. 4, Rome, Italy.

Gregor, J., Gleason, S. S., Paulus, M. J., and Cates, J. (2002). Fast Feldkamp reconstruction based on focus of attention and distributed computing. *Int. J. Imaging Syst. Technol.* **12**, 229–234.

Hawkins, P. N. (2002). Serum amyloid P component scintigraphy for diagnosis and monitoring amyloidosis. *Curr. Opin. Nephrol. Hypertens.* **11**, 649–655.

Hawkins, P. N., Myers, M. J., Epenetos, A. A., Caspi, D., and Pepys, M. B. (1988a). Specific localization and imaging of amyloid deposits *in vivo* using 123I-labeled serum amyloid P component. *J. Exp. Med.* **167**, 903–913.

Hawkins, P. N., Myers, M. J., Epenetos, A. A., Caspi, D., and Pepys, M. B. (1988b). Specific localization and imaging of amyloid deposits *in vivo* using 123I-labeled serum amyloid P component. *J. Exp. Med.* **167**, 903–913.

Hawkins, P. N., Myers, M. J., Lavender, J. P., and Pepys, M. B. (1988c). Diagnostic radionuclide imaging of amyloid: Biological targeting by circulating human serum amyloid P component. *Lancet* **1**, 1413–1418.

Hawkins, P. N., Wooten, R., and Pepys, M. (1990). Metabolic studies of radioiodinated serum amyloid P component in normal subjects and patients with systemic amyloidosis. *J. Clinical Investigation* **89**, 1862–1869.

Hnatowich, D. J. (1990). Recent developments in the radiolabeling of antibodies with iodine, indium, and technetium. *Semin. Nucl. Med.* **20**, 80–91.

Hrncic, R., Wall, J., Wolfenbarger, D. A., Murphy, C. L., Schell, M., Weiss, D. T., and Solomon, A. (2000). Antibody-mediated resolution of light chain-associated amyloid deposits [In Process Citation]. *Am. J. Pathol.* **157**, 1239–1246.

Hudson, H. M., and Larkin, R. S. (1994). Accelerated image reconstruction using ordered subsets of projection data. *IEEE Trans. Med. Imaging* **13**, 601–609.

Hull, R. L., Westermark, G. T., Westermark, P., and Kahn, S. E. (2004). Islet amyloid: A critical entity in the pathogenesis of type 2 diabetes. *J. Clin. Endocrinol. Metab.* **89**, 3629–3643.

Inoue, S., and Kisilevsky, R. (1999). In situ electron microscopy of amyloid deposits in tissues. *Methods Enzymol.* **309**, 496–509.

Jain, A. K. (1989). "Fundamentals of Digital Image Processing." Pearson Prentice-Hall, Englewood Cliffs, N.J.

Karonen, S. L. (1990). Developments in techniques for radioiodination of peptide hormones and other proteins. *Scand. J. Clin. Lab. Invest. Suppl.* **201**, 135–138.

Kennel, S. J., Falcioni, R., and Wesley, J. W. (1991). Microdistribution of specific rat monoclonal antibodies to mouse tissues and human tumor xenografts. *Cancer Res.* **51**, 1529–1536.

Kindy, M. S., and de Beer, F. C. (1999). A mouse model for serum amyloid A amyloidosis. *Methods Enzymol.* **309**, 701–716.

Kisilevsky, R., Lemieux, L. J., Fraser, P. E., Kong, X., Hultin, P. G., and Szarek, W. A. (1995). Arresting amyloidosis *in vivo* using small-molecule anionic sulphonates or sulphates: Implications for Alzheimer's disease. *Nat. Med.* **1**, 143–148.

Kisilevsky, R., Szarek, W. A., Ancsin, J. B., Elimova, E., Marone, S., Bhat, S., and Berkin, A. (2004). Inhibition of amyloid A amyloidogenesis *in vivo* and in tissue culture by 4-deoxy analogues of peracetylated 2-acetamido-2-deoxy-alpha- and beta-d-glucose: Implications for the treatment of various amyloidoses. *Am. J. Pathol.* **164**, 2127–2137.

Kohrle, J., Rasmussen, U. B., Ekenbarger, D. M., Alex, S., Rokos, H., Hesch, R. D., and Leonard, J. L. (1990). Affinity labeling of rat liver and kidney type I 5'-deiodinase. Identification of the 27-kDa substrate binding subunit. *J. Biol. Chem.* **265**, 6155–6163.

Kovalchuk, A. L., Kim, J. S., Park, S. S., Coleman, A. E., Ward, J. M., Morse, H. C., 3rd, Kishimoto, T., Potter, M., and Janz, S. (2002). IL-6 transgenic mouse model for extraosseous plasmacytoma. *Proc. Natl. Acad. Sci. USA* **99**, 1509–1514.

Koziorowski, J., Henssen, C., and Weinrich, R. (1998). A new convenient route to radioiodinated N-succinimidyl 3- and 4-iodobenzoate, two reagents for radioiodination of proteins. *Appl. Radiat. Isot.* **49**, 955–959.

MacRaild, C. A., Stewart, C. R., Mok, Y. F., Gunzburg, M. J., Perugini, M. A., Lawrence, L. J., Tirtaatmadja, V., Cooper-White, J. J., and Howlett, G. J. (2004). Non-fibrillar components of amyloid deposits mediate the self-association and tangling of amyloid fibrils. *J. Biol. Chem.* **279**, 21038–21125.

Marshall, J., McNally, W., Muller, D., Meincken, G., Fand, I., Srivastava, S. C., Atkins, H., Wood, D. D., and Gorevic, P. D. (1986). *In vivo* radioimmunodetection of amyloid deposits in experimental amyloidosis. *In* "Amyloidosis" (G. G. Glenner, E. F. Osserman, E. P. Benditt, E. Calkins, A. S. Cohen, and D. Zucker-Franklin, eds.), pp. 163–174. Plenum Press, NY.

McConahey, P. J., and Dixon, F. J. (1966). A method of trace iodination of proteins for immunologic studies. *Int. Arch. Allergy Appl. Immunol.* **29**, 185–189.

Merlini, G., and Bellotti, V. (2003). Molecular mechanisms of amyloidosis. *N. Engl. J. Med.* **349**, 583–596.

Mihara, M., Shiina, M., Nishimoto, N., Yoshizaki, K., Kishimoto, T., and Akamatsu, K. (2004). Anti-interleukin 6 receptor antibody inhibits murine AA-amyloidosis. *J. Rheumatol.* **31**, 1132–1138.

O'Nuallain, B., Murphy, C. L., Wolfenbarger, D. A., Kennel, S. J., Solomon, A., and Wall, J. S. (2005). The amyloid-reactive monoclonal antibody 11-1F4 binds a cryptic epitope on fibrils and partially denatured immunoglobulin light chains and inhibits fibrillogenesis. *In* "Amyloid and Amyloidosis: Proceedings of the Xth International Symposium on Amyloidosis" (G. Grateau, R. A. Kyle, and M. Skinner, eds.), pp. 482–484. CRC Press, Tours, France.

Paulus, M. J., Gleason, S. S., Easterly, M. E., and Foltz, C. J. (2001). A review of high-resolution X-ray computed tomography and other imaging modalities for small animal research. *Lab. Anim. (NY)* **30**, 36–45.

Paulus, M. J., Gleason, S. S., Kennel, S. J., Hunsicker, P. R., and Johnson, D. K. (2000a). High resolution X-ray computed tomography: An emerging tool for small animal cancer research. *Neoplasia* **2**, 62–70.

Paulus, M. J., Gleason, S. S., Sari-Sarraf, H., Johnsosn, D. K., Foltz, C. J., Austin, D. W., Easterly, M. E., Michaud, E. J., Dhar, M. S., Hunsicker, P. R., Wall, J. S., and Schell, M. (2000b). High-resolution X-ray CT screening of mutant mouse models. *In* "BIOS 2000 International Biomedical Optics Symposium", San Jose, Calif.

Paulus, M. J., Sari-Sarraf, H., Gleason, S. S., Bobrek, M., Hicks, J. S., Johnson, D. B., Behel, J. K., Thompson, L. H., and Allen, W. C. (1999). A new x-ray computed tomography system for laboratory mouse imaging. *IEEE Trans. Nucl. Sci.* **46**, 558–564.

Pepys, M. B., Booth, D. R., Hutchinson, W. L., Gallimore, J. R., Collins, P. M., and Hohenester, E. (1997). Amyloid P component. A critical review. *Amyloid Int. J. Exp. Clin. Invest.* **4,** 274–295.

Pras, M., Schubert, M., Zucker-Franklin, D., Rimon, A., and Franklin, E. C. (1968). The characterization of soluble amyloid prepared in water. *J. Clin. Invest.* **47,** 924–933.

Rocken, C., and Shakespeare, A. (2002). Pathology, diagnosis and pathogenesis of AA amyloidosis. *Virchows Arch.* **440,** 111–122.

Shepp, L. A., and Vardi, Y. (1982). Maximum-likelihood reconstruction for emission tomography. *IEEE Trans. Med. Imaging* **1,** 113–122.

Snow, A. D., and Kisilevsky, R. (1985). Temporal relationship between glycosaminoglycan accumulation and amyloid deposition during experimental amyloidosis. A histochemical study. *Lab. Invest.* **53,** 37–44.

Solomon, A., Weiss, D. T., and Wall, J. S. (2003a). Immunotherapy in systemic primary (AL) amyloidosis using amyloid-reactive monoclonal antibodies. *Cancer Biother. Radiopharm.* **18,** 853–860.

Solomon, A., Weiss, D. T., and Wall, J. S. (2003b). Therapeutic potential of chimeric amyloid-reactive monoclonal antibody 11–1F4. *Clin. Cancer Res.* **9,** 3831S–3838S.

Spanka, M., Hesch, R. D., Irmscher, K., and Kohrle, J. (1990). 5'-Deiodination in rat hepatocytes: Effects of specific flavonoid inhibitors. *Endocrinology.* **126,** 1660–1667.

Stenstad, T., and Husby, G. (1996). Brefeldin A inhibits experimentally induced AA amyloidosis. *J. Rheumatol.* **23,** 93–100.

Vaidyanathan, G., Affleck, D., Welsh, P., Srinivasan, A., Schmidt, M., and Zalutsky, M. R. (2000). Radioiodination and astatination of octreotide by conjugation labeling. *Nucl. Med. Biol.* **27,** 329–337.

Vaidyanathan, G., and Zalutsky, M. R. (1990a). Protein radiohalogenation: Observations on the design of N-succinimidyl ester acylation agents. *Bioconjug. Chem.* **1,** 269–273.

Vaidyanathan, G., and Zalutsky, M. R. (1990b). Radioiodination of antibodies via N-succinimidyl 2,4-dimethoxy-3-(trialkylstannyl)benzoates. *Bioconjug. Chem.* **1,** 387–393.

Wall, J., Schell, M., Hrncic, R., Macy, S. D., Wooliver, C., Wolfenbarger, D. A., Murphy, C., Donnell, R., Weiss, D. T., and Solomon, A. (2001). Treatment of Amyloidosis using an anti-fibril monoclonal antibody: Preclinical efficacy in a murine model of AA-amyloidosis. *In* "Amyloid and Amyloidosis: Proceedings of the IXth International Symposium on Amyloidosis" (M. Bely, ed.), pp. 158–160. David Apathy, Budapest, Hungary.

Wall, J. S., Kennel, S. J., Paulus, M. J., Gleason, S., Gregor, J., Baba, J., Schell, M., Richey, T., O'Nuallain, B., Donnell, R., Hawkins, P. N., Weiss, D. T., and Solomon, A. (2005a). Quantitative high-resolution microradiographic imaging of amyloid deposits in a novel murine model of AA-amyloidosis. *Amyloid J. Protein Folding Disord.* **12,** 149–156.

Wall, J. S., Paulus, M. J., Kennel, S. J., Gleason, S., Baba, J., Gregor, J., Wolfenbarger, D. A., Weiss, D. T., and Solomon, A. (2005b). Radioimaging of primary (AL) amyloidosis with an amyloid-reactive monoclonal antibody. *In* "Amyloid and Amyloidosis: Proceedings of the Xth International Symposium on Amyloidosis" (G. Grateau, R. A. Kyle, and M. Skinner, eds.), pp. 37–39. CRC Press, Tours, France.

Westermark, G. T., Johnson, K. H., and Westermark, P. (1999). Staining methods for identification of amyloid in tissue. *Methods Enzymol.* **309,** 3–25.

Wilbur, D. S. (1992). Radiohalogenation of proteins: An overview of radionuclides, labeling methods, and reagents for conjugate labeling. *Bioconjug. Chem.* **3,** 433–470.

Zhu, H., Yu, J., and Kindy, M. S. (2001). Inhibition of amyloidosis using low-molecular-weight heparins. *Mol. Med.* **7,** 517–522.

Section II

Cell and Animal Models of Amyloid Formation and Toxicity

[12] An Efficient Protein Transformation Protocol for Introducing Prions into Yeast

By MOTOMASA TANAKA and JONATHAN S. WEISSMAN

Abstract

Although a range of robust techniques exists for transforming organisms with nucleic acids, approaches for introducing proteins into cells are far less developed. Here we describe a facile and highly efficient protein transformation protocol suitable for introducing prion particles, produced *in vitro* from pure protein or purified from an *in vivo* source, into yeast. Prion particles composed of amyloid forms of fragments of Sup35p, the protein determinant of the yeast prion state [*PSI*⁺], lead to dose-dependent *de novo* induction of [*PSI*⁺] with efficiencies approaching 100% at high protein concentrations. We also describe a procedure for generating distinct, self-propagating amyloid conformations of a prionogenic Sup35p fragment termed Sup-NM. Remarkably, infection of yeast with different Sup-NM amyloid conformations leads to distinct [*PSI*⁺] prion strains, establishing that the heritable differences in prion strain differences result directly from self-propagating differences in the conformations of the infectious protein. This protein transformation protocol can be readily adapted to the analysis of other yeast prion states, as well as to test the infectious (prion) nature of protein extracts from less well-characterized epigenetic traits. More generally, the protein transformation procedure makes it possible to bridge *in vitro* and *in vivo* studies, thus greatly facilitating efforts to explain the structural and mechanistic basis of prion inheritance.

Introduction

Infectious proteins (prions), originally postulated to explain a set of transmissible spongiform encephalopathies (TSEs), are now known to underlie a number of epigenetic elements in fungi (including the yeast [*PSI*⁺] and [*URE3*] states) (Tuite and Cox, 2003; Uptain and Lindquist, 2002; Wickner, 1994) and perhaps in higher organisms (Si *et al.*, 2003). Although mammalian PrP and fungal prion proteins are unrelated in amino acid sequence, their prion states share common structural features that place them in a class of misfolded proteins (amyloids) responsible for a broad range of noninfectious diseases. The facile genetics of yeast together with the ability to create *de novo* infectious forms of proteins from pure

METHODS IN ENZYMOLOGY, VOL. 412   0076-6879/06 $35.00

material has greatly helped to explain the principles of prion inheritance, as well as the role of cellular factors in facilitating and inhibiting prion replication.

In contrast to the mammalian case, yeast prions do not cause cell death, but rather act as epigenetic modulators of protein function. The phenotypes associated with [URE3] or [PSI+] elements are not particularly remarkable; [URE3] alters nitrogen catabolite uptake (Lacroute, 1971), whereas [PSI+] allows for the suppression of some nonsense mutations (Cox, 1965). Indeed, traditional loss-of-function mutations in the chromosomally encoded nitrogen catabolism repressor Ure2 and the translation termination factor Sup35p mimic the [URE3] and [PSI+] states, respectively. What makes [URE3] and [PSI+] remarkable are their epigenetic properties; they are inherited by all of the meiotic progeny of "heterozygous" diploid cells and can be transmitted by transfer of cytoplasm from one cell to another without the exchange of genetic material. To explain the unusual inheritance of [URE3] and [PSI+], Wickner (Wickner, 1994) proposed that these states result from the presence of self-propagating (prion) forms of the Ure2p and Sup35p proteins, respectively (Fig. 1). The prion model also explains why the phenotypes of [URE3] and [PSI+] mimic the loss of function of Ure2 and Sup35p, because conversion to the prion state inactivates the affected protein. Since then, new fungal prions and prion-like states have been identified, including the naturally occurring yeast prion [RNQ+] (also known as [PIN+]) (Sondheimer and Lindquist, 2000), the artificial prion [NU+] (Osherovich and Weissman, 2001; Santoso et al., 2000), and the *Podospora anserina* prion [Het-s] (Coustou et al., 1997); for a review, see Osherovich and Weissman (2002) and Uptain and Lindquist (2002).

The [PSI+] element is caused by self-propagating aggregates of a translation-termination factor Sup35p, which leads to a nonsense suppression phenotype (Chien et al., 2004; Tuite and Cox, 2003; Uptain and Lindquist, 2002). In yeast containing a nonsense mutation in the *ade1* gene, [PSI+] colonies are white or pink and grow on media lacking adenine, whereas nonprion [psi−] colonies are red and require adenine (Chernoff et al., 1995). [PSI+] propagation is mediated by a modular, N-terminal Gln/Asn-rich sequence and, to a lesser extent, by a highly charged middle domain (Bradley and Liebman, 2004; DePace et al., 1998; Glover et al., 1997; Liu et al., 2002; Ter-Avanesyan et al., 1994). Transient overexpression of a fusion between the Sup35p N-terminal and middle domains (Sup-NM) (residues 1–254) leads to protein aggregation and *de novo* appearance of [PSI+]. *In vitro*, the Sup-NM fragment is shown to be sufficient to form self-seeding amyloid fibers (Glover et al., 1997; King et al., 1997).

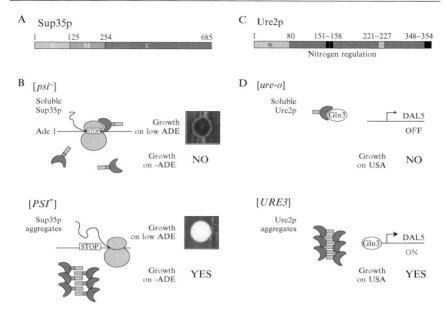

Fig. 1. The yeast states [*PSI*⁺] and [*URE3*] are due to self-propagating (prion) protein conformations. (A) Sup35p is an essential translation termination factor. The N-terminus, residues 1–125 [N], is Gln/Asn rich and mediates prion behavior. The middle domain, residues 126–254 (M; blue), is rich in charged residues. Commonly, purified NM fusions (Sup-NM) are used *in vitro*. The C-terminal domain, residues 255–685 [C], is responsible for faithful translation termination at stop codons. (B) In [*psi*⁻] yeast, Sup35p is soluble and functional. In [*PSI*⁺] yeast, Sup35p is aggregated, sequestering it from the ribosome, therefore allowing nonsense suppression. Typically, the [*PSI*⁺] nonsense suppression phenotype is monitored by read-through of a reporter gene carrying a premature stop in the *ADE1* gene, which leads to a convenient color change on low adenine media and a differential growth phenotype on media lacking adenine. (C) Ure2 is involved in regulation of nitrogen catabolism. In addition to the Gln/Asn-rich N-terminus, Ure2p also contains another region that facilitates prion behavior and portions that antagonize prion formation. The C-terminus, residues 81–354 [C], resembles glutathione-S-transferase and signals the presence of high-quality nitrogen sources through Gln3. (D) Normally, Ure2 binds the transcription factor Gln3 in the cytoplasm, keeping Gln3 from turning on a host of genes such as *DAL5*. Dal5 allows import of USA (n-carbamyl aspartate), which permits selection for [*URE3*] on USA medium in a uracil auxotroph. In a [ure-o] yeast, Ure2 is complexed with Gln3, and DAL5 is off, prohibiting growth on USA medium. For [*URE3*] yeast, Ure2 is aggregated, releasing Gln3 to the nucleus, where it activates transcription of *DAL5*, allowing USA uptake and growth on USA medium.

Studies of [PSI^+] have shed light on two of the most perplexing features of prion biology (Chien et al., 2004). One of these is the existence of transmission barriers that inhibit the passage of prions between related species (Collinge, 2001). The second remarkable feature is the existence of multiple prion strains, wherein infectious particles composed of the same protein give rise to a range of prion states that vary in incubation time, pathology, and other phenotypic aspects (Aguzzi, 2004; Prusiner, 1998).

Strain variability had been observed long before the prion hypothesis and, in fact, was originally interpreted as evidence for the existence of a nucleic acid genome in the infectious particle, with strain variation arising from mutations in this genome. To reconcile the presence of prion strains with the protein-only hypothesis, it has been proposed that a single poly-peptide can misfold into multiple infectious conformations, at least one for each phenotypic variant (Telling et al., 1996). Rather than being peculiar-ities of mammalian prions, it is now clear that both transmission barriers and strain variation are common features of amyloid-based prion elements, and both features arise from the general ability of prions to adopt multiple amyloid conformations (Chien et al., 2004). [PSI^+], in particular, exhibits a range of heritable phenotypic strain variants (Derkatch et al., 1996). These strains differ in mitotic stability (Derkatch et al., 1996), dependence on the cellular chaperone machinery (Kushnirov et al., 2000), solubility and activity of Sup35p (Derkatch et al., 1996; Kochneva-Pervukhova et al., 2001; Uptain et al., 2001; Zhou et al., 1999), and lead to differences in the ade1 color phenotype as well as in their specificity of transmission (Tanaka et al., 2005).

In addition to providing genetic systems for exploring prion biology, the fungal prions in general have proven to be highly amenable to reconstitution in vitro. Extracts from [PSI^+] yeast can induce conversion of soluble Sup35p in extracts from [psi⁻] yeast (Paushkin et al., 1997). Moreover, Sup35p, HET-s, and Rnq1, the proteins responsible for [PSI^+], [$URE3$], [Het-s], and [RNQ^+]/[PIN^+], respectively, have all been shown to form self-seeding amyloid in vitro (Baxa et al., 2003; Dos Reis et al., 2002; Glover et al., 1997; Jiang et al., 2004; King et al., 1997; Taylor et al., 1999). Nonamyloid fibrillar forms of Ure2 have also been suggested to underlie [$URE3$] (Bousset et al., 2002; Fay et al., 2003). Several lines of evidence argue that self-seeded aggregation drives prion inheritance in vivo. For example, mutations in Sup35p that affect prion formation in vivo have parallel effects on in vitro polymerization (DePace et al., 1998; Glover et al., 1997; Liu and Lindquist, 1999). More directly, for Sup35p and HET-s, it has been possible to create amyloid in vitro from recombinant proteins and to use them to convert wild-type cells to the prion state (King and Diaz-Avalos, 2004; Maddelein et al., 2002; Sparrer et al., 2000; Tanaka et al., 2004). These experiments have provided the first demonstration that a pure protein can, indeed, act as a conformationally based infectious agent. In addition to their importance for testing the prion hypothesis, the ability to create synthetic prions *de novo* from pure protein and introduce these forms into cells has enabled a wide range of mechanistic and structural investigation into the nature of prion-based inheritance that would otherwise be difficult or impossible. For example, using this protein transformation protocol, it was possible to demonstrate that amyloid is an infectious form of Sup35p and establish that infection of yeast with distinct Sup-NM amyloid conformations leads to different [*PSI*⁺] strains, thereby demonstrating that prion strain variability does, indeed, result from differences in the conformation of the infectious forms.

In the following we describe protocols for generating *de novo* distinct prion forms from recombinantly produced Sup-NM as well as for purifying *in vivo* prions from [*PSI*⁺] yeast. In addition, we describe a procedure for "infecting" live prion-free yeast with these different prion forms and monitoring the distinct [*PSI*⁺] strains that result from such infection experiments. This protocol can readily be adapted to explore other yeast prion phenomena (Brachmann *et al.*, 2005).

Purification of Bacterially Expressed Sup-NM

Sup-NM containing a polyhistidine tag at the C-terminus is expressed in *Escherichia coli* BL21(DE3) and purified as described in the following (Glover *et al.*, 1997; Santoso *et al.*, 2000). After purification, the protein is filtered through a 100-kDa filter to remove any amyloid seeds or other aggregates and can be stored frozen in one time use aliquots at $-80°$.

Purification Protocol

1. The day before expression, transform *E. coli* with the plasmid (pAED4 Sup-NM) that allows expression c-terminally polyhistidine tagged Sup-NM under control of a *T7* promoter into *E coli*. BL21(DE3).

2. Inoculate LB media containing 100 μg/ml ampicillin with transformed colonies (it is important to use fresh transformants) and culture at $37°$ until OD_{600} reaches $\sim$0.4. Add isopropyl-β-thiogalactoside (final concentration of 0.4 mM) to initiate protein expression. After 4 h, harvest bacterial cells at 5000 rpm (Sorvall SLA-3000 Rotor) for 30 min. The cell pellets can be stored at $-80°$ until use.

3. Resuspend cells in buffer A (8 M urea, 25 mM Tris, 300 mM NaCl, pH 7.8). Use 25 ml of buffer A per each liter of harvested cells. Vortex vigorously to resuspend pellets. Sonicate the suspension (Sonic Dismembrator,

Fischer Scientific, 40% intensity, tip diameter 3 mm) until the cells are fully disrupted (approximately 30 sec).

4. Incubate the protein solution with gentle agitation for 1 h at room temperature. Spin at 15,000 rpm (Sorvall SS-34 Rotor) for 30 min. Samples are prefiltered (Millex AP 20, Millipore), followed by filtration with 0.45-μm filters (Fisherbrand).

5. Pour the protein solution onto a Ni-NTA agarose column ($\sim$2 ml agarose gels per liter of cell culture) preequilibrated with buffer A. Wash the column with 4$\times$ column volumes of buffer B (8 M urea, 25 mM Tris, pH 7.8).

6. Elute bound proteins with buffer C (8 M urea, 25 mM Tris, pH 4.5). Collect 3–4-ml fractions and run them onto 4–12% SDS-PAGE. Pool fractions with highest abundance of Sup-NM and store at $-80°$.

7. Load the partially purified Sup-NM from the Ni-NTA column onto a Resource S column (6 ml, Amersham Biosciences) preequilibrated with buffer D (8 M urea, 50 mM MES, pH 6.0).

8. Elute Sup-NM with a 0–200 mM NaCl gradient in buffer D. Typically we monitor absorbances at 229 and 280 nm, and collect 1-ml fractions.

9. Analyze purity of the fractions by 4–12% SDS-PAGE. Pool fractions that contain >90% pure Sup-NM. Concentrate the fractions, exchange buffer D with 6 M guanidine hydrochloride (Gdn) containing 5 mM potassium phosphate, pH 7.4, and concentrate it again generally to more than 1 mM, using a VIVASPIN concentrator (Sartorius AG). Filter the concentrated protein with Microcon YM-100 (Millipore). Divide the filtered protein into $\sim$10-μl aliquots and store them at $-80°$ until use.

Preparation of Different Conformations of Sup-NM Amyloid

Sup-NM can adopt multiple different amyloid conformations (DePace et al., 1998), and the specific amyloid conformation that results during spontaneous polymerization of Sup-NM is highly dependent on the polymerization conditions. We have found that the simplest method for controlling the conformation of Sup-NM amyloids is to alter the temperature at which the initial polymerization occurs. For preparation of 4° or 37° Sup-NM amyloid, concentrated Sup-NM in 6 M Gdn is diluted at least 200-fold into 5 mM potassium phosphate buffer (pH 7.4) containing 150 mM NaCl in a 2-ml tube at 4° or 37°, respectively. The Sup-NM solution (typically 2.5 μM) is immediately rotated at 8 rpm in an end-over-end fashion overnight. Fiber formation can be confirmed by mixing 100 μl of polymerization reaction with 100 μl of 25 μM thioflavin-T (Sigma) in 50 mM glycine, pH 8.5, and looking for increased thioflavin-T fluorescence compared with unpolymerized reaction (Molecular Devices, 442 nm excitation and 485 nm emission) (Chien *et al.*, 2003). Once a stock of 4° and 37° fibers has been produced, it can be used as seeds (5% [wt/wt]) to amplify the different Sup-NM conformations. It is important that the differences in conformations of the Sup-NM amyloid fibers formed at the different temperatures be directly confirmed. We find that the simplest way to achieve this is to measure the melting temperature of the fibers as described in the following (Chien *et al.*, 2003; Tanaka *et al.*, 2004, 2005). Failure to robustly obtain 4° and 37° fibers using the preceding protocol has most typically in our hands resulted from high residual Gdn concentration in the reactions, highlighting the importance of using at least a 200-fold dilution of the Sup-NM stocks as described previously, or failure to preincubate the polymerization buffers at the proper temperatures.

1. Make 4 or 37° Sup-NM amyloid fibers, as described previously. Add 6× SDS sample buffer (final concentration, 1.7% SDS) to the fiber solution and aliquot 20 μl into 500-μl PCR tubes.
2. Incubate the fiber solution at increasing temperatures from 25° to 95° in 10° intervals and 100° for 5 min in a PCR thermal cycler. Transfer each tube incubated at a specific temperature into a water bath (~10°) to cool the tubes quickly.
3. Run the samples onto 4–12% SDS-PAGE. Probe thermally solubilized monomeric Sup-NM by Western blotting with a polyclonal anti-Sup-NM$_{Sc}$ antibody (Santoso *et al.*, 2000), followed by detection with chemiluminescence using FluorChem 8800 (Alpha Innotech) (Fig. 2A).
4. To estimate fiber stability, we quantitate band intensities in the Western blot with ImageJ1.3 (NIH) and fit them as a function of temperature with IgorPro5.0 (WaveMetrics Inc.), using the following equation: $y = A + B/(1 + 10^{\wedge}((C - x)/D))$, where x, y, A, B, C, D indicate temperature, band intensity, band intensity at baseline, amplitude of band intensity, melting temperature (Tm), and width of the melting transition (W), respectively (Fig. 2B); 4° fibers showed $T_m = 56 \pm 2°$ and $W = 27 \pm 2°$ compared with values for 37° fibers of $T_m = 77 \pm 2°$ and $W = 14 \pm 1°$.

Preparation of Prions from an *In Vivo* Source

For preparation of crude yeast extracts, spheroplasted yeast cells are produced with lyticase or lysed with glass beads in the presence of a protease inhibitor cocktail (Complete, Roche). The extracts were sonicated

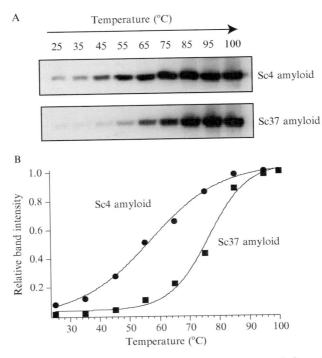

FIG. 2. (A) Thermal stability of Sup-NM amyloid fibers spontaneously formed at 4 (Sc4) or 37° (Sc37) as determined by pretreatment of the fibers at the indicated temperatures in the presence of SDS, followed by SDS-PAGE and Western blot analysis. (B) The band intensities of thermally solubilized monomeric Sup-NM in the Western blot are plotted against temperature and fit to a sigmoidal function. Modified from Tanaka *et al.* (2005).

on ice for 10 sec (Sonic Dismembrator, Fischer Scientific, 20% intensity, tip diameter 3 mm) before use.

1. For preparation of partially purified prion particles, spheroplast yeast cells with lyticase [250 μg/50 ml (OD$_{600}$ = 0.5) cultured yeast cells] in SCE-buffer (1 M sorbitol, 10 mM EDTA, 10 mM DTT, 100 mM sodium citrate, pH 5.8) buffer containing protease inhibitors (See the following section for preparation of lyticase).

2. Lyse the spheroplasts on ice by sequential addition of sodium deoxycholate to 0.5% (w/v) and after 5 min Brij-58 to 0.5% (w/v) (Uptain *et al.*, 2001). After further incubation for 15 min, spin the lysate at 10,000g for 5 min at 4° and subject the supernatant to ultra-centrifugation at 100,000g for 30 min (TLA100.3 Rotor, Beckman).

3. Resuspend the pellet with 1 M lithium acetate, incubate on ice for 30 min with gentle agitation, and further spin at 100,000g for 30 min. Resuspend the pellet with 5 mM potassium phosphate buffer (pH 7.4) containing 150 mM NaCl and sonicate on ice for 10 sec (Sonic Dismembrator, Fischer Scientific) before use. Determine the concentration of total protein in the yeast extracts and partially purified prion particles by Bradford or BCA protein assay, using BSA as a standard.

Preparation of Lyticase

1. Transform *E. coli* BL21(DE3) with a bacterial plasmid that expresses periplasmically localized lyticase under control of *T7* promoter (pUV5 lyticase) (Tanaka *et al.*, 2004).
2. Inoculate a single colony into 100 ml of LB media including 100 μg/ml ampicillin and culture it at 37° until OD_{600} reached ~0.8. Transfer 20 ml of the culture media to 1 liter of LB media with ampicillin and culture at 37°.
3. When OD_{600} reaches ~0.7, add isopropyl-β-thiogalactoside (final concentration of 0.5 mM) to initiate protein expression. After 3 h, collect cells at 5000 rpm (Sorvall SLA-3000 Rotor) for 30 min.
4. Resuspend cells in 20 ml of 25 mM Tris-HCl (pH 7.4), incubate at room temperature with gentle agitation for 30 min using a nutator, and centrifuge at 7500 rpm (Sorvall SS-34 Rotor) for 10 min.
5. Resuspend the pellet in 20 ml of 5 mM $MgCl_2$, incubate at 4° for 30 min, and spin at 15,000 rpm for 30 min (Sorvall SS-34 Rotor) to separate periplasmic components.
6. Dialyze the supernatant against 4 liters of 50 mM sodium citrate buffer (pH 5.8) overnight and store at −80°. Determine the concentration of lyticase by Bradford assay.

Protein Transformation into Yeast

Throughout, we use isogenic [*psi*⁻][*PIN*⁺] and [*PSI*⁺] derivatives of 74D-694 [MATa, *his3*, *leu2*, *trp1*, *ura3*; suppressible marker *ade1-14* (UGA)] (Santoso *et al.*, 2000), although similar results have been obtained in a W303 background. Although *de novo* prion induction after Sup35p overexpression depends on the [*PIN*] status of [*psi*⁻] yeast (Derkatch *et al.*, 1997), the infection efficiency of protein transformation does not depend on the [*PIN*] state of yeast (Tanaka *et al.*, 2004). Soluble or fibrillar Sup-NM are transformed by the following procedure (Fig. 3A). Note the presence of a *URA3* plasmid allows one to preselect for the yeasts that have successfully taken up material from the solution.

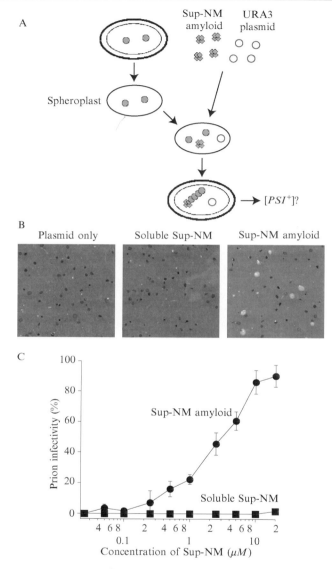

FIG. 3. Induction of the [PSI⁺] prion by *in vitro*-converted Sup-NM amyloid fibers. (A) Schematic of transformation procedure. The [PSI⁺] status is assessed by plating transformed spheroplasts on SD-URA media containing trace amounts of adenine or on SD-URA plates, followed by streaking transformants onto YEPD plates. (B) Examples of transformations using the indicated materials for infections and SD-URA trace Ade plates. Large and white colonies represent authentic [PSI⁺] cells. (C) Concentration-dependence of prion infectivity. The indicated concentration of fibrillar (circle) or soluble (square) Sup-NM was transformed into [psi⁻][PIN⁺] strains. Values with error bars are expressed as mean ± standard deviation. The prion infectivity does not depend on the [PIN] state of [psi⁻] yeast. Modified from Tanaka *et al.* (2004).

1. Grow yeast cells in YEPD media (1% yeast extract, 2% bactopeptone, 2% dextrose) (typically 50 ml) to an OD_{600} of 0.5 and successively wash with 20 ml of sterile H_2O, 1 M sorbitol, and SCE-buffer (1 M sorbitol, 10 mM EDTA, 10 mM DTT, 100 mM sodium citrate, pH 5.8).

2. Spheroplaste cells with lyticase ($\sim$250 μg for yeast cells cultured from 50 ml YEPD) in SCE-buffer at 30° for 30 min. The spheroplasts are then collected by centrifugation (400g, 5 min) and successively washed with 20 ml of 1 M sorbitol and STC-buffer (1 M sorbitol, 10 mM $CaCl_2$, 10 mM Tris, pH 7.5).

3. Resuspend pelleted spheroplasts in 1 ml of STC-buffer. Mix 100 μl of the yeast spheroplasts with sonicated Sup-NM amyloid fibers (final concentration 2.5–10 μM) or in vivo prions (final concentration 200–400 μg/mL), URA3 marked plasmid (pRS316) (20 μg/ml) and salmon sperm DNA (100 μg/ml). Incubate mixture for 30 min at room temperature.

4. Induce fusion to the spheroplasts by addition of 9 volumes of PEG-buffer (20% [w/v] PEG 8000, 10 mM $CaCl_2$, 10 mM Tris, pH 7.5) at room temperature for 30 min.

5. Collect cells by centrifugation (400g, 5 min), resuspend with 150 μl of SOS-buffer (1 M sorbitol, 7 mM $CaCl_2$, 0.25% yeast extract, 0.5% bactopeptone), incubate at 30° for 30 min, and plate on synthetic media lacking uracil overlaid with 7 ml of synthetic media containing 2.5% agar. Adenine (20 mg/L) was absent (procedure [I]) or present (procedure [II]) in the agar plate.

6. Incubate the plates at 30° for $\sim$10 days [I] or 4–6 days [II]. For procedure [II], after the incubation, streak single colonies randomly chosen from the SD-URA plates onto modified YEPD plates containing 1/4 of the standard amount of yeast extract to enhance color phenotypes of [PSI+] and [psi−] states (Osherovich et al., 2004). Include streaks of strong [PSI+] and [psi−] controls on each plate for comparison.

Determination of Prion Conversion Efficiency and Prion Strain Phenotypes

For procedure [I], [psi−] yeast cells form small and intensely red colonies whereas [PSI+] yeast form large white colonies (Fig. 3B). Transformation of [psi−] yeast with plasmid alone or with soluble Sup-NM does not lead to detectable formation of [PSI+] colonies, whereas inclusion of preformed Sup-NM amyloid seeds leads to significant production of large, white Ade+ colonies. These Ade+ convertants exhibit the hallmarks of the [PSI+] prion: the Ade+ trait is inherited in a non-Mendelian manner, is readily cured by transient growth on medium containing Gdn, and is associated with the formation of large Sup35p aggregates that are readily pelletable by high-speed centrifugation. We rarely observe Ade$^+$ revertants that were not curable by Gdn in the infection experiments. To determine the infection efficiency, measure the fraction of Ade$^+$ colonies from more than 200 total colonies on at least three different SD-URA plates containing trace amounts of adenine. Increasing the concentration of Sup-NM fibers should result in a dose-dependent increase in Ade$^+$ convertants, with the fraction of Ade$^+$ colonies among the Ura$^+$ colonies approaching 100% at high Sup-NM concentrations (Fig. 3C). As expected for a prion, the efficiency of prion conversion is sensitive to protease but not nuclease treatment (Tanaka et al., 2004).

For procedure [II], efficiency of conversion to prion state, as well as phenotypic strength of prion strains, can be determined by monitoring the color phenotype of colonies on 1/4 YEPD plates, which indicate whether transformants remain [psi$^-$] (red) or converted to strong [PSI$^+$] (white) or weak [PSI$^+$] (pink) states. After the 1/4 YEPD plates are incubated at 30° for a few days, each streak should be classified into strong [PSI$^+$] (white), weak (pink) [PSI$^+$] or [psi$^-$] (red) strains. In quantification experiments, typically at least 50 colonies from at least three independent transformations are streaked on YEPD plates. Ade$^+$ revertants rarely exist but are readily excluded from the statistics, because they show brown color on 1/4 YEPD plates.

Fibers formed at 4° should have a relatively high efficiency of infection, with the large majority of colonies showing a strong (white) [PSI$^+$] strain phenotype, whereas fibers formed at 37° have lower infectivity and produce almost exclusively weak (pink and/or sectored) [PSI$^+$] strains (Fig. 4A, B). The weak strains show increased levels of soluble Sup35p and are more readily cured by Hsp104 overexpression (Tanaka et al., 2004). The strain phenotypes do not depend on the concentration of seed or the infection efficiency; 4° fibers yield strong strains even when diluted 10-fold (infection rate ~20%), and 37° fibers induce weak strains even when the concentration is increased 10-fold (infection rate ~80%) (Tanaka et al., 2004). Thus, the [PSI$^+$] strain is determined by the conformation of the Sup-NM amyloid fiber rather than the number of seeds introduced into a cell. When distinct Sup-NM amyloids such as Sc4 and Sc37 are introduced into [psi$^-$] yeast by the protein infection protocol, physical properties of these in vitro amyloids are faithfully propagated in vivo (Tanaka et al., 2004). In addition, for the Sup-NM polymerization conditions described here, small nonamyloid oligomeric intermediates are not detectable (Collins et al., 2004). These results establish that amyloid is an infectious form of prion protein and that conformational differences in amyloid determine prion strain variations.

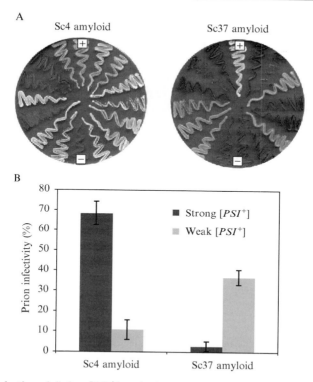

FIG. 4. Induction of distinct $[PSI^+]$ strains by Sup-NM amyloid fibers formed at different temperatures. (A) Examples of $[PSI^+]$ strains resulting from infection with Sup-NM amyloid spontaneously formed at 4 or 37°. White and pink and/or sectored colonies are strong and weak $[PSI^+]$ variants, respectively. + and − show strong $[PSI^+]$ and $[psi^-]$ controls. (B) Quantification of frequency of $[PSI^+]$ strains induced by transformation with Sc4 or Sc37 amyloid fibers. Values with error bars are expressed as mean ± standard deviation. Modified from Tanaka *et al.* (2004).

Application to the Study of Other Prion or non-Mendelian Elements

We described previously a protocol optimized for exploring the prion nature of Sup-NM amyloids as well as the role of conformation in determining the subsequent $[PSI^+]$ prion strains. More recently we have exploited the protein transformation assay to examine the mechanism of cross-species prion transmission (Tanaka *et al.*, 2005). In addition, the protein transformation procedure should be readily adaptable to the analysis of other yeast prions, with the major variable being the requirement to modify the phenotypic assay so that it is appropriate to monitor formation of the prion state under study. Finally, the protein transformation assay should in principle be broadly applicable for the biochemical characterization of any non-Mendelian trait that can be transferred by cytoplasmic transfer (cytoduction). In this case, the "infectious" material would be generated from extracts of cells propagating the non-Mendelian trait of interest. Although conceptually similar to cytoduction experiments, transfer of such elements using the preceding protein transformation protocol allows for the biochemical analysis and purification of the agent responsible for transmission. Thus, if a prion is suspected, one could test for the ability of protease (versus nuclease) treatment to inhibit transfection and ultimately couple the transfection assay with biochemical purification to allow for the identification of the prion protein.

Acknowledgments

We thank members of the Weissman laboratory for helpful comments on the manuscript. M. T. was partly supported by JSPS postdoctoral fellowships for research abroad and Uehara Memorial Foundation research fellowships. Funding was also provided by the Howard Hughes Medical Institute and the National Institutes of Health (J. S. W.).

References

Aguzzi, A. (2004). Understanding the diversity of prions. *Nat. Cell. Biol.* **6,** 290–292.

Baxa, U., Taylor, K. L., Wall, J. S., Simon, M. N., Cheng, N., Wickner, R. B., and Steven, A. C. (2003). Architecture of Ure2p prion filaments: The N-terminal domains form a central core fiber. *J. Biol. Chem.* **278,** 43717–43727.

Bousset, L., Thual, C., Belrhali, H., Morera, S., and Melki, R. (2002). Structure and assembly properties of the yeast prion Ure2p. *C. R. Biol.* **325,** 3–8.

Brachmann, A., Baxa, U., and Wickner, R. B. (2005). Prion generation *in vitro*: Amyloid of Ure2p is infectious. *EMBO J.* **24,** 3082–3092.

Bradley, M. E., and Liebman, S. W. (2004). The Sup35 domains required for maintenance of weak, strong or undifferentiated yeast [*PSI*+] prions. *Mol. Microbiol.* **51,** 1649–1659.

Chernoff, Y. O., Lindquist, S. L., Ono, B., Inge-Vechtomov, S. G., and Liebman, S. W. (1995). Role of the chaperone protein Hsp104 in propagation of the yeast prion-like factor [*PSI*+]. *Science* **268,** 880–884.

Chien, P., DePace, A. H., Collins, S., and Weissman, J. S. (2003). Generation of prion transmission barriers by mutational control of amyloid conformations. *Nature* **424,** 948–951.

Chien, P., Weissman, J. S., and DePace, A. H. (2004). Emerging principles of conformation-based prion inheritance. *Annu. Rev. Biochem.* **73,** 617–656.

Collinge, J. (2001). Prion diseases of humans and animals: Their causes and molecular basis. *Annu. Rev. Neurosci.* **24,** 519–550.

Collins, S. R., Douglass, A., Vale, R. D., and Weissman, J. S. (2004). Mechanism of prion propagation: Amyloid growth occurs by monomer addition. *PLoS Biol.* **2,** e321.

Coustou, V., Deleu, C., Saupe, S., and Begueret, J. (1997). The protein product of the het-s heterokaryon incompatibility gene of the fungus *Podospora anserina* behaves as a prion analog. *Proc. Natl. Acad. Sci. USA* **94,** 9773–9778.

Cox, B. (1965). PSI, a cytoplasmic suppressor of super-suppressor in yeast. *Heredity* **20**, 505–521.

DePace, A. H., Santoso, A., Hillner, P., and Weissman, J. S. (1998). A critical role for amino-terminal glutamine/asparagine repeats in the formation and propagation of a yeast prion. *Cell* **93**, 1241–1252.

Derkatch, I. L., Chernoff, Y. O., Kushnirov, V. V., Inge-Vechtomov, S. G., and Liebman, S. W. (1996). Genesis and variability of [*PSI*⁺] prion factors in *Saccharomyces cerevisiae*. *Genetics* **144**, 1375–1386.

Derkatch, I. L., Bradley, M. E., Zhou, P., Chernoff, Y. O., and Liebman, S. W. (1997). Genetic and environmental factors affecting the de novo appearance of the [*PSI*⁺] prion in *Saccharomyces cerevisiae*. *Genetics* **147**, 507–519.

Dos Reis, S., Coulary-Salin, B., Forge, V., Lascu, I., Begueret, J., and Saupe, S. J. (2002). The HET-s prion protein of the filamentous fungus Podospora anserina aggregates *in vitro* into amyloid-like fibrils. *J. Biol. Chem.* **277**, 5703–5706.

Fay, N., Inoue, Y., Bousset, L., Taguchi, H., and Melki, R. (2003). Assembly of the yeast prion Ure2p into protein fibrils. Thermodynamic and kinetic characterization. *J. Biol. Chem.* **278**, 30199–30205.

Glover, J. R., Kowal, A. S., Schirmer, E. C., Patino, M. M., Liu, J. J., and Lindquist, S. (1997). Self-seeded fibers formed by Sup35, the protein determinant of [*PSI*⁺], a heritable prion-like factor of S. cerevisiae. *Cell* **89**, 811–819.

Jiang, Y., Li, H., Zhu, L., Zhou, J. M., and Perrett, S. (2004). Amyloid nucleation and hierarchical assembly of Ure2p fibrils. Role of asparagine/glutamine repeat and nonrepeat regions of the prion domains. *J. Biol. Chem.* **279**, 3361–3369.

King, C. Y., Tittmann, P., Gross, H., Gebert, R., Aebi, M., and Wuthrich, K. (1997). Prion-inducing domain 2-114 of yeast Sup35 protein transforms *in vitro* into amyloid-like filaments. *Proc. Natl. Acad. Sci. USA* **94**, 6618–6622.

King, C. Y., and Diaz-Avalos, R. (2004). Protein-only transmission of three yeast prion strains. *Nature* **428**, 319–323.

Kochneva-Pervukhova, N. V., Chechenova, M. B., Valouev, I. A., Kushnirov, V. V., Smirnov, V. N., and Ter-Avanesyan, M. D. (2001). [*PSI*⁺] prion generation in yeast: Characterization of the 'strain' difference. *Yeast* **18**, 489–497.

Kushnirov, V. V., Kryndushkin, D. S., Boguta, M., Smirnov, V. N., and Ter-Avanesyan, M. D. (2000). Chaperones that cure yeast artificial [*PSI*⁺] and their prion-specific effects. *Curr. Biol.* **10**, 1443–1446.

Lacroute, F. (1971). Non-Mendelian mutation allowing ureidosuccinic acid uptake in yeast. *J. Bacteriol.* **106**, 519–522.

Liu, J. J., and Lindquist, S. (1999). Oligopeptide-repeat expansions modulate 'protein only' inheritance in yeast. *Nature* **400**, 573–576.

Liu, J. J., Sondheimer, N., and Lindquist, S. L. (2002). Changes in the middle region of Sup35 profoundly alter the nature of epigenetic inheritance for the yeast prion [*PSI*⁺]. *Proc. Natl. Acad. Sci. USA* **99**, 16446–16453.

Maddelein, M.-L., Dos Reis, S., Duvezin-Caubet, S., Coulary-Salin, B., and Saupe, S. J. (2002). Amyloid aggregates of the HET-s prion protein are infectious. *Proc. Natl. Acad. Sci. USA* **99**, 7402–7407.

Osherovich, L. Z., and Weissman, J. S. (2001). Multiple Gln/Asn-rich prion domains confer susceptibility to induction of the yeast [*PSI*⁺] prion. *Cell* **106**, 183–194.

Osherovich, L. Z., and Weissman, J. S. (2002). The utility of prions. *Dev. Cell.* **2**, 143–151.

Osherovich, L. Z., Cox, B. S., Tuite, M. F., and Weissman, J. S. (2004). Dissection and design of yeast prions. *PLoS Biol.* **2**, e86.

Paushkin, S. V., Kushnirov, V. V., Smirnov, V. N., and Ter-Avanesyan, M. D. (1997). *In vitro* propagation of the prion-like state of yeast Sup35 protein. *Science* **277**, 381–383.

Prusiner, S. B., Scott, M. R., DeArmond, S. J., and Cohen, F. E. (1998). Prion protein biology. *Cell.* **93**, 337–348.

Santoso, A., Chien, P., Osherovich, L. Z., and Weissman, J. S. (2000). Molecular basis of a yeast prion species barrier. *Cell* **100**, 277–288.

Si, K., Lindquist, S., and Kandel, E. R. (2003). A neuronal isoform of the aplasia CPEB has prion-like properties. *Cell* **115**, 879–891.

Sondheimer, and Lindquist, S. (2000). Rnq 1: An epigenetic modifier of protein function in yeast. *Mol. Cell.* **5**, 163–172.

Sparrer, H. E., Santoso, A., Szoka, F. C., Jr., and Weissman, J. S. (2000). Evidence for the prion hypothesis: Induction of the yeast [*PSI*⁺] factor by *in vitro*-converted sup35 protein. *Science* **289**, 595–599.

Taylor, K. L., Cheng, N., Williams, R. W., Steven, A. C., and Wickner, R. B. (1999). Prion domain initiation of amyloid formation *in vitro* from native Ure2p. *Science* **283**, 1339–1343.

Tanaka, M., Chien, P., Naber, N., Cooke, R., and Weissman, J. S. (2004). Conformational variations in an infectious protein determine prion strain differences. *Nature* **428**, 323–328.

Tanaka, M., Chien, P., Yonekura, K., and Weissman, J. S. (2005). Mechanism of cross-species prion transmission: An infectious conformation compatible with two highly divergent yeast prion proteins. *Cell* **121**, 49–62.

Telling, G. C., Parchi, P., DeArmond, S. J., Cortelli, P., Montagna, P., Gabizon, R., Mastrianni, J., Lugaresi, E., Gambetti, P., and Prusiner, S. B. (1996). Evidence for the conformation of the pathologic isoform of the prion protein enciphering and propagating prion diversity. *Science* **274**, 2079–2082.

Ter-Avanesyan, M. D., Dagkesamanskaya, A. R., Kushnirov, V. V., and Smirnov, V. N. (1994). The *SUP35* omnipotent suppressor gene is involved in the maintenance of the non-Mendelian determinant [*PSI*⁺] in the yeast *Saccharomyces cerevisiae*. *Genetics* **137**, 671–676.

Tuite, M. F., and Cox, B. S. (2003). Propagation of yeast prions. *Nat. Rev. Mol. Cell. Biol.* **4**, 878–890.

Uptain, S. M., Sawicki, G. J., Caughey, B., and Lindquist, S. (2001). Strains of [*PSI*⁺] are distinguished by their efficiencies of prion-mediated conformational conversion. *EMBO J.* **20**, 6236–6245.

Uptain, S. M., and Lindquist, S. (2002). Prions as protein-based genetic elements. *Annu. Rev. Microbiol.* **56**, 703–741.

Wickner, R. B. (1994). [URE3] as an altered URE2 protein: Evidence for a prion analog in *Saccharomyces cerevisiae*. *Science* **264**, 566–569.

Zhou, P., Derkatch, I. L., Uptain, S. M., Patino, M. M., Lindquist, S., and Liebman, S. W. (1999). The yeast non-Mendelian factor [*ETA*⁺] is a variant of [*PSI*⁺], a prion-like form of release factor eRF3. *EMBO J.* **18**, 1182–1191.

[13] Screening for Genetic Modifiers of Amyloid Toxicity in Yeast

By Flaviano Giorgini and Paul J. Muchowski

Abstract

In recent years the facile, yet powerful, genetics of the baker's yeast *Saccharomyces cerevisiae* has been appropriated for the study of amyloid toxicity. Several models of amyloid toxicity using this simple eukaryotic organism have been developed that faithfully recapitulate many disease-relevant phenotypes. Furthermore, these models have been exploited in genetic screens that have provided insight into conserved mechanisms of amyloid toxicity and identified potential therapeutic targets for disease. In this chapter, we discuss the strengths and weaknesses of yeast models of amyloid toxicity and how experiments with these models may be relevant to amyloid disorders. We suggest approaches for development of new yeast models of amyloid toxicity and provide an overview of screening protocols for genetic modifiers of amyloid toxicity by both random and systematic approaches.

Introduction

Saccharomyces cerevisiae has proven to be an extremely important tool for studying basic cellular processes conserved with higher eukaryotes, including cell division, replication, metabolism, protein folding, and intra-cellular transport. In fact, almost everything that is known about the fundamental properties of living cells has been learned from the study of *S. cerevisiae* and other model organisms (Fields and Johnston, 2005). The conserved cellular mechanisms explained in these organisms can play a direct role in understanding disease processes. For example, the fundamental knowledge of cell cycle regulation uncovered in yeast has been directly applied toward studies in mammalian systems on cancer biology (Hartwell, 2002; Nurse, 2002). Several factors make yeast the perfect system for biological studies, but the primary reason that work in *S. cerevisiae* has proven so fruitful is the "awesome power of yeast genetics."

Yeast is perhaps the ideal organism for classical genetics, because it can exist in both haploid and diploid states, and it is quite simple to mate haploid strains and to sporulate diploid strains. These features, along with the ease of mutagenesis and the availability of several mutant collections

METHODS IN ENZYMOLOGY, VOL. 412
0076-6879/06 $35.00
DOI: 10.1016/S0076-6879(06)12013-3

and open reading frame (ORF) libraries, allow for rapid isolation of genetic modifiers of specific processes or phenotypes. In addition, this ease of genetic manipulation allows for facile dissection of molecular pathways by epistatic analysis. Epistasis is the control of a phenotype by two or more genes by means of a genetic interaction in which one gene masks or suppresses the phenotypic effect of another. This phenomenon can be used to genetically order genes within a pathway and has proven to be a critical technique for determining the order of events in many cellular pathways, such as signaling cascades.

S. cerevisiae is one of the best characterized eukaryotic organisms; in addition to its genome being fully sequenced (Goffeau et al., 1996), Saccharomyces genes and proteins are extensively annotated in several genomic and proteomic databases, which contain detailed gene expression profiles, known protein–protein interactions, and predicted orthologs in other organisms. In addition, molecular genetic manipulations, such as DNA transformation, targeted disruption of specific genes, generation of point mutations in cloned genes, and overexpression of proteins of interest can be performed in a matter of days, compared with months or years in other model organisms. The genomic era for yeast has truly arrived with the recent development of arrayed libraries allowing for the systematized testing of deletion and overexpression of most genes in the yeast genome, which permits rapid genomic screening and identification of interacting genes.

Modeling Amyloid Toxicity in Yeast

In recent years, the aggregation and toxicity of amyloid proteins implicated in neurodegenerative disorders has been studied in several nonmammalian organisms: Drosophila, C. elegans, and S. cerevisiae. Although mouse models of disease remain the "gold standard" for the analysis of genetic and pharmacological modification of phenotype/pathology, nonmammalian models provide a useful alternative to the high cost and slow pace of murine strategies. Both Drosophila and C. elegans have well-characterized nervous systems specialized for sensory, learning, and memory functions. Because of this, several neurodegenerative diseases caused by dominant mutations in single genes have been modeled in these organisms (Link, 2001; Marsh and Thompson, 2004). Many disease-relevant phenotypes are recapitulated in these models, including progressive neurodegeneration, decreased life span, and inclusion body formation. Although S. cerevisiae is a much simpler organism, yeast models of amyloid toxicity can tackle both genetic and pharmacological screens with a rapidity and ease not possible in fly and worm models, while still identifying conserved cellular mechanisms critical to toxicity and candidate therapeutic targets (Giorgini et al., 2005; Willingham et al.,

2003; Zhang *et al.*, 2005). As with any disease model, it is critical that yeast models faithfully recapitulate many features of the disease being studied. Ultimately, candidate modifier genes or compounds identified using screens in yeast must be validated in mammalian systems, such as cell models or transgenic mouse models.

Two approaches have been taken in modeling aspects of neurodegenerative diseases in yeast: (1) directly studying the function of yeast homologs of human disease genes and (2) studying the phenotypes caused by expressing human disease genes in yeast. The first approach has been used to model aspects of Friedreich's ataxia (FRDA), amyotrophic lateral sclerosis (ALS), and prion disease. Studies with *YFH1* and *SOD1*, the yeast homologs of the human genes implicated in FRDA and ALS pathogenesis, respectively, have helped explain pathogenic mechanisms underlying these diseases (Outeiro and Muchowski, 2004; Puccio and Koenig, 2000). Work in yeast, both with yeast prions and with mammalian PrP protein, has provided insight into the mechanism of prion action (Sherman and Muchowski, 2003). Although not homologous to PrP by sequence comparison and by phenotype, yeast prions behave in an analogous manner to their human counterpart; these proteins can change conformation to form self-propagating, transmittable aggregates. An excellent review of yeast prion biology is provided by Uptain and Lindquist (Uptain and Lindquist, 2002).

As mentioned previously, a second approach to modeling aspects of neurodegenerative disease is expressing a human disease gene in yeast cells. This approach has proven very successful, with development of yeast strains that model aspects of Huntington's disease (HD) and Parkinson's disease (PD). These models recapitulate many disease-relevant phenotypes and have already provided mechanistic insight into the pathology of these diseases. HD is caused by an expansion of a polyglutamine (polyQ) stretch beyond a critical threshold of $\sim$35–40 glutamines in the protein Huntingtin (Htt). Expression of mutant Htt fragments in yeast produces phenotypes reminiscent of HD pathology, including polyQ length-dependent inclusion body formation and toxicity and modulation of this phenotype by perturbation of chaperone levels (Hughes *et al.*, 2001; Krobitsch and Lindquist, 2000; Meriin *et al.*, 2002; Muchowski *et al.*, 2000). Genetic screens using these models have uncovered genes in cellular pathways implicated in HD, including transcriptional regulation, protein folding, vesicle transport, and the kynurenine pathway (Giorgini *et al.*, 2005; Willingham *et al.*, 2003). Although most cases of PD are idiopathic, a small proportion of disease incidence is caused by missense mutations in the α-synuclein gene (Kruger *et al.*, 1998; Polymeropoulos *et al.*, 1997) or triplication of the wild-type locus (Singleton *et al.*, 2003). Consistent with studies in primary cortical neurons, expression of α-synuclein in yeast leads to formation of cytoplasmic inclusion bodies and cellular toxicity (McLean et al., 2001; Outeiro and Lindquist, 2003). Provocatively, a genome-wide screen for loss-of-function enhancers of α-synuclein toxicity (Willingham et al., 2003) has uncovered many genes involved in vesicular transport and lipid metabolism, pathways implicated in α-synuclein function in mammals (Outeiro and Muchowski, 2004). The relevance of yeast models of amyloid toxicity in regard to various amyloid diseases is discussed in greater detail elsewhere (Outeiro and Muchowski, 2004; Sherman and Muchowski, 2003); here we focus on techniques for generation of yeast models of amyloid toxicity and methods for identification of genetic modifiers of toxicity.

Thus far, the use of yeast models of amyloid toxicity has provided insight into conserved mechanisms of toxicity in several amyloid disorders. This low-cost, facile approach for identification of genetic modifiers can theoretically be applied to any amyloid disorder in which a protein has been associated with pathogenesis. How applicable are findings in yeast to studies in higher eukaryotes? The answer to this question will not be known until genetic modifiers and molecular mechanisms identified in yeast studies are validated by genetic tests in mouse models of neurodegenerative disease (and for that matter, studies in mouse models will not be validated until the proposed disease mechanisms are proven by pharmacological intervention in diseased humans). Nevertheless, there are obvious limitations to studying amyloid toxicity in yeast that are worth mentioning (as there are with any model system). First of all, genes involved in neurodegeneration may not be present in the yeast genome. Second, the toxicity observed in yeast may not be related to that involved in neurodegeneration. In this regard, it is important to consider recent work that has shown that yeast can undergo apoptosis-like cell death in response to several stimuli and that several yeast orthologs of crucial apoptotic regulators exist (Madeo et al., 2004). In addition, it has recently been shown that expression of wild-type α-synuclein or the inherited mutants (A30P, A53T) in yeast triggers several markers of apoptosis and that deletion of a yeast metacaspase gene suppresses many of these apoptosis-like phenotypes (Flower et al., 2005). It is thus possible that toxicity in yeast may be more similar to neurotoxicity than previously thought. Ultimately, any candidate modifier identified in yeast needs to be validated in more physiologically relevant models of neurodegeneration. This chapter is designed as a primer on yeast genetic screens for the amyloid researcher who wishes to exploit this simple organism for molecular genetic studies on the mechanisms of amyloid toxicity. We will not discuss the two-hybrid system, perhaps the most extensively used yeast screening technique, because this has been discussed at length elsewhere (Fields and Sternglanz, 1994).

Genetic Manipulation and Screening

Generating a Yeast Model of Amyloid Toxicity

Developing a yeast model for amyloid toxicity that recapitulates aspects of a particular disease may involve some troubleshooting and will likely require more than merely expressing the implicated protein in yeast cells. Several factors may influence toxicity of expressing your favorite gene (YFG) in yeast, including, but not limited to, promoter strength, plasmid copy number, and protein localization. Other factors to consider when generating constructs are the use of an inducible promoter for gene expression, the ploidy number of the yeast strain (haploid versus diploid), and the addition of a molecular tag (FLAG, GFP, etc.).

Two very useful and versatile expression systems for *S. cerevisiae* have been developed by Martin Funk and colleagues that allow either constitutive or regulated expression of protein over a range of two to three orders of magnitude in several genetic backgrounds (Table I) (Mumberg et al., 1994, 1995). The constitutive expression system consists of 32 expression vectors with which investigators can modulate the level of transcription of YFG by selecting among four promoters (*CYC1, ADH, TEF, GPD*), alter expression level by changing copy number of the plasmid (using either a centromeric or 2μ plasmid), and use the appropriate selectable marker (*HIS3, LEU2, TRP1,* or *URA3*). All these vectors have multiple cloning sites derived from the Bluescript vectors (Stratagene, La Jolla, CA) containing six to nine unique restriction sites for ease of cloning.

Although the preceding system is extremely versatile, expression of YFG using an inducible and regulatable system may be desirable, especially if the protein is strongly toxic to yeast or if you wish to compare growth in mutant strains with and without protein expression. The regulatable vectors mentioned previously take advantage of the *GAL1* promoter, which is tightly repressed by glucose (GLU) and is strongly induced by galactose (GAL) (Johnston and Davis, 1984). In addition, *GAL1* is an extremely strong promoter, capable of inducing protein expression over 1000-fold in GAL, such that the protein of interest can constitute up to 0.8% of total cell protein (Schneider and Guarente, 1991). Funk and colleagues generated deletion variants of the *GAL1* promoter that contain either one and one-half (GALS) or two (GALL) of the three upstream activator sequences (UAS) required for full induction of the *GAL1* promoter by galactose (Mumberg et al., 1994; West et al., 1984), allowing for inducible expression of heterologous proteins over a range of two to three orders of magnitude. As with the constitutive expression vectors, the set of GAL-inducible vectors included both centromeric and 2μ plasmids, several

TABLE I
CONSTITUTIVE AND INDUCIBLE VECTORS FOR EXPRESSION OF HETEROLOGOUS
PROTEINS IN YEAST

Origin of replication	Selectable marker	Name	Promoter type	Strength
p41X–CEN6/ARSH4				
p42X–2–micron	P4X3–HIS3	CYC1	Constitutive	Weak
	P4X4–TRP1	ADH	Constitutive	Intermediate
	P4X5–LEU2	TEF	Constitutive	Strong
	P4X6–URA3	GPD	Constitutive	Very strong
		GAL1	Inducible	Weak
		GALL	Inducible	Strong
		GALS	Inducible	Very strong

From Mumberg *et al.* (1994, 1995).

choices for selectable markers (*HIS3, LEU2, TRP1*, or *URA3)*, and a versatile multiple cloning site with six to nine unique sites. To clone YFG into the plasmid of choice, one can either use standard cloning techniques or PCR-based homologous recombination in yeast (Longtine *et al.*, 1998; Reid *et al.*, 2002). It is worth noting that several yeast gene deletion strains grow slowly in GAL media and that this must be considered when analyzing deletion strains that seem to enhance toxicity (reduce growth) of an YFG under inducing (+GAL) conditions.

In general, when constructing a yeast model of toxicity, it is desirable that expression of YFG cause toxicity; if toxicity is not observed, low expression levels may be the cause. It has previously been observed that increasing the expression levels of both α-synuclein and a mutant Htt fragment in yeast increases toxicity (Outeiro and Lindquist, 2003; Meriin *et al.*, 2002). Fortunately, the expression systems described previously allow expression of YFG over a broad range. Integration of YFG into the yeast genome may also increase toxicity of the construct while decreasing variation in expression levels (Outeiro and Lindquist, 2003); techniques for integrating YFG by homologous recombination are described in detail in a previous volume of this series: *Volume 194: Guide To Yeast Genetics and Molecular and Cell Biology—Part A* (Rothstein, 1991). Another method for obtaining or increasing toxicity of YFG is the use of signal peptides, such as nuclear localization signals or secretion signals, to modify the cellular localization of YFG. The use of such signals in yeast models of amyloid toxicity has been observed to increase the severity of the resulting phenotypes seen in yeast (Hughes *et al.*, 2001). Another factor to consider is whether to use a haploid or diploid yeast strain for the model, because ploidy of the cell may lead to differences in phenotype and toxicity because of YFP expression. Finally, it is of utmost importance to construct a valid control for toxicity experiments, because measurement of toxicity is always relative. In the case of polyQ expansion disease models, a non-expanded polyQ control is an obvious choice. For toxicity models expressing another mutant protein, expression of the wild-type protein may be a good control. In some cases, the best control may be a scrambled version of the toxic construct, thus ensuring that overexpression of a similar protein with the same localization does not produce toxicity.

A final factor to consider when generating a construct for YFG is the use of molecular tags. It is highly recommended to generate a fusion with a fluorescent protein (i.e., GFP, RFP, CFP, YFP) if possible, this greatly simplifies imaging of protein localization by microscopy, since this technique requires no additional processing in yeast. This is of special importance when dealing with protein aggregation diseases, because it is important to monitor whether the protein of interest forms inclusion bodies. It may also be desirable to fuse your heterologous protein to epitope tags (i.e., MYC, FLAG, HA), because antibodies are readily available for these epitopes that allow immunoblotting and immunoprecipitation of the protein. However, it is important to validate that the tag used does not influence various properties of *YFG*. In the end, many variables should be tested when generating a yeast model of amyloid toxicity. Proper planning and a little patience will result in the development of the optimal model for your particular needs.

Obtaining Gene Mutations in Yeast

The heart of any genetic screen is scanning a collection of mutants that provides thorough genomic coverage of the organism of interest. These gene mutations can either be generated directly by mutagenesis or can be obtained in mutant collections. In yeast, several methods are available for random generation of forward mutations (the change of a gene from the wild-type to a mutant form), including chemical mutagenesis, exposure to ultraviolet (UV) light, and insertional mutagenesis. The widely used alkylating agents ethylmethane sulfonate (EMS) and N-methyl-N'-nitro-N-nitrosoguanidine (MNNG) induce high frequencies of base-pair substitutions and low levels of lethality to yeast (Kohalmi and Kunz, 1988). Although these mutagens almost exclusively produce transition mutations at G-C base pairs, this specificity is not a problem for most applications. A wider range of mutations can be generated using treatment of cells with UV light that induces both transitions and transversions and also can produce frameshift mutations, most often by single nucleotide deletion (Kunz *et al.*, 1987; Lee *et al.*, 1988). More recently, transposon-insertion libraries have been used as mutagens in yeast. One approach is to use bacterial transposons to mutagenize genomic yeast DNA in *Escherichia coli*, and then shuttle this DNA back into yeast *en masse*. Insertion alleles of genes can then replace their chromosomal counterparts by homologous recombination (Kumar *et al.*, 2002). Insertion alleles provide the added benefit that the insertion site (gene of interest) is easy to identify by PCR amplification or plasmid rescue. Detailed protocols and excellent reviews of these mutagenesis techniques are found in a previous edition of this series, *Volume 350: Guide To Yeast Genetics and Molecular and Cell Biology—Part B* (Kumar *et al.*, 2002; Lawrence, 2002).

Although several mutant collections, or libraries, are available for yeast, the most powerful collection of mutant strains is the yeast gene knockout (YKO) collection. The YKO set was developed by an international consortium and is commercially available from several sources (Winzeler *et al.*, 1999). The YKO set is an array of ~4850 viable gene deletion strains generated by systematically deleting each predicted gene in yeast using a "knockout," or targeted disruption, method. During the gene-disruption process, a unique DNA sequence tag or "molecular bar code" was introduced into the genome for each deletion strain (Johnston, 2000). This allows rapid determination of each strain by PCR amplification of the bar code followed by DNA sequencing. In addition, because the deletion strains are arrayed in microtiter plates, not only can individual strains be identified by position in the array, but the entire library can be manipulated systematically *en masse* using robotics. This set of strains can also be pooled for growth competition assays among the strains (whole-genome parallel analysis) in response to environmental factors; the abundance of each deletion strain is quantified by determining levels of the associated molecular bar code using hybridization to an oligonucleotide array of the complementary bar-code sequences (Giaever *et al.*, 2002; Winzeler *et al.*, 1999). One drawback to the YKO deletion set is that second site mutations exist in the set; by one estimate ~6.5% of the strains contain such mutations (Grunenfelder and Winzeler, 2002). In addition, as much as 8% of the strains may be aneuploid, complicating phenotypic analysis in these strains (Hughes *et al.*, 2000). The most comprehensive selection of YKO collections is available from Open Biosystems (Huntsville, AL). The YKO collections are available as haploid strains [Mat A (BY4741 background) and Mat α (BY4742)], homozygous diploid knockouts (BY4743), and also heterozygous diploids (BY4743). The loss-of-function genetic screening techniques presented herein are based on the use of these YKO collections.

Loss-of-Function Screens

All published screens using yeast models of amyloid toxicity have been performed with loss-of-function mutant sets, specifically the YKO collection. This approach consists of transforming your construct of interest into all the strains of the deletion set, either individually or *en masse* in pools (protocols are described in the following). If the construct being used drives expression of YFG by a GAL-inducible (*GAL1*) promoter, then transformants are plated on media containing GLU (on which expression is repressed). On growth, the colonies are transferred to GAL-containing plates by replica plating that induces expression of YFG. In the wild-type strain, if YFG is toxic, little to no growth will be observed on GAL media; if on the other hand, expression of YFG is not toxic, the yeast will grow normally on GAL media (Fig. 1). In the case of a toxic construct, the gene deletion collection can be screened for deletions that eliminate, or suppress, toxicity. These loss-of-function mutants, or suppressors, are deletions of genes required for toxicity in yeast. Because genetic inhibition of a suppressor's activity alleviates toxicity, the respective gene products are excellent candidate therapeutic targets for small molecule inhibitors, assuming the gene is conserved in humans. Thus, loss-of-function suppressor screens of amyloid toxicity in yeast are predicted to isolate not only genes required for toxicity but can also help identify targets for rational drug design.

Enhancer screens with nontoxic constructs can be equally rewarding and may provide great insight into the mechanisms of amyloid toxicity. As stated previously, if YFG is not toxic to wild-type yeast, colonies will grow normally on both GLU media and GAL media. In the case of a nontoxic construct, the gene deletion collection can be screened for deletions that enhance toxicity. These enhancer strains represent deletions in genes required for suppression of toxicity and identify pathways that are sensitive to expression of the amyloid protein of interest. Therefore, genes identified in loss-of-function enhancer screens represent pathways somehow disrupted or perturbed by expression of YFG.

Overexpression Screens

An alternate approach to genetic screening relies on overexpression of gene products, an approach that has identified genes involved in many cellular functions (Rine, 1991). A gene overexpression screen is the converse of a loss-of-function screen; the object is to alter the phenotype of choice by increased gene dosage. An overexpression suppressor will rescue the original phenotype; in the case of a yeast model of amyloid toxicity, such a suppressor would relieve toxicity. Enhancement of the phenotype

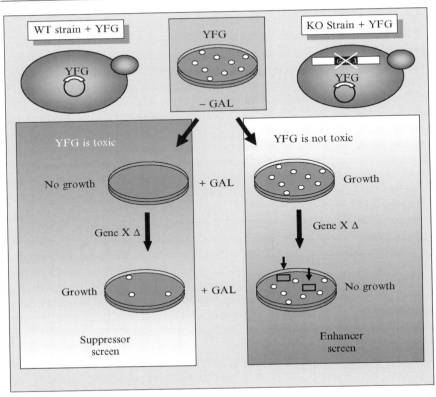

FIG. 1. An overview of a loss-of-function screen for suppressors or enhancers of amyloid toxicity in yeast. A plasmid containing YFG under the control of a GAL-inducible promoter is transformed into wild-type yeast cells and the YKO deletion set. If YFG is toxic to wild-type yeast cells, then the YKO set can be screened for gene deletions (i.e., Gene X Δ) that restore growth or suppress toxicity. If on the other hand, YFG is not toxic to wild-type cells, then the YKO set can be screened for gene deletions that enhance toxicity of YFG.

would exacerbate its severity, or in our example, increase toxicity (Forsburg, 2001). Identification of both types of genetic modifiers leads to a better understanding of the pathways disrupted by the toxic protein or activated by its presence. Although random clone libraries have been available for quite some time for overexpression studies in yeast, only recently has an array of overexpression constructs for most (>90%) of the yeast genome become available (Yeast ORF Collection [YOC], Open Biosystems, Huntsville, AL). The YOC is a collection of more than 5500 constructs each containing an individual S. cerevisiae open reading frame (ORF). Each construct is under the control of the *GAL1* promoter and contains Protein A and 6xHis domains together with a HA epitope tag. This collection is arrayed in 96-well microtiter plates, allowing systematic transformation into yeast models and individual analysis of the effect of candidate gene overexpression on amyloid toxicity. The clones may also be pooled and transformed *en masse* into the yeast model for a more traditional overexpression library screen.

The screening approach with the YOC overexpression set is quite similar to that described previously for the YKO set. A double-transformation of YFG and the pooled YOC library into a wild-type yeast strain is performed (Fig. 2). Transformants are plated on media containing GLU; once grown,

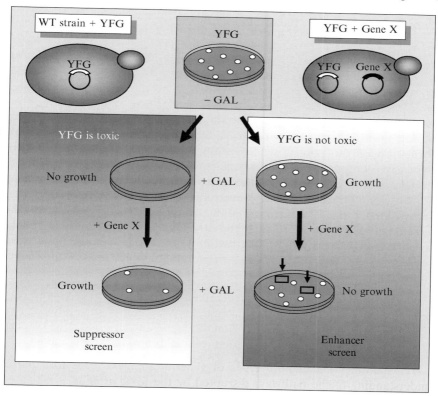

FIG. 2. An overview of an overexpression screen for suppressors or enhancers of amyloid toxicity in yeast. A plasmid containing YFG under the control of a GAL-inducible promoter is transformed into wild-type yeast cells alone or in combination with an ORF from a clone library (i.e., Gene X). If YFG is toxic to wild-type yeast cells, then the ORF library can be screened for a gene that when overexpressed restores growth or suppresses toxicity. If on the other hand, YFG is not toxic to wild-type cells, then the ORF library can be screened for a gene that when overexpressed enhances toxicity of YFG.

the colonies are transferred to GAL-containing plates by replica plating, thereby inducing expression of both YFG and the clones from the YOC library. If YFG is toxic to wild-type cells, little to no growth will be observed on GAL media, and suppressing colonies can be selected. These will represent colonies overexpressing a gene that suppresses the toxicity of YFG. If, on the other hand, expression of YFG is not toxic, the yeast will grow normally on GAL media, and enhancers of toxicity can be identified. An added benefit of overexpression screens is that mammalian cDNA libraries can be exploited. For example, in a genetic screen for suppressors of amyloid toxicity in yeast, a human cDNA library could be used, allowing investigators to directly test genetic interactions between YFG and human genes from tissues affected by the disease.

Methods

Pooling the YKO Library Strains

Screening the pooled YKO strain set offers the benefit of random screening of mutants without *a priori* bias in selection. In addition, transformation of the pooled strains is quite simple and does not require expensive equipment. One drawback is that the variable growth rates and transformation efficiencies of the deletion strains within the pools may allow certain strains to become overrepresented and others to be underrepresented in the final transformation. To reduce competition among the gene deletion strains, we recommend dividing the YKO set into five pools, four for the strains with normal growth rates (~50 plates total), and a fifth pool for all the slow-growing strains (two plates, which are grouped separately in the YKO sets).

1. Inoculate 100 μl of YPD media in 96-well microtiter plates with cells from the YKO stocks by using a multichannel pipetter or a pinning tool (see "Systematic Approaches to Genetic Screening"). Thaw YKO frozen stocks two to four plates at a time on ice, pin 1–5 μl into YPD media, and immediately return stocks to $-80°$. To ensure that cross-contamination of strains does not occur, be very careful when handling these plates.

2. Incubate the plates at $30°$ for 3–4 days, until the cultures are saturated ($OD_{600} \sim 5.0$– 10.0; 1–2 $\times 10^8$ cells/ml).

3. Pool strains using a multipipetter to remove cultures from wells and pipette into a sterile solution basin. The large pools will consist of ~12–13 plates, or approximately 125 ml of culture.

4. Centrifuge at 1500g for 5 min at $4°$, discard media.

5. Resuspend cell pellet in 1/10th volume YP-media + 5% DMSO (~12.5 ml).

6. Distribute into 250-μl aliquots in cryovials, and store at $-80°$.

High-Efficiency Yeast Transformation

This method is a modification of the lithium acetate (LiAc) TRAFO method (Gietz and Schiestl, 1996). Variations of this protocol have been used extensively for high-efficiency transformation of libraries into yeast for two-hybrid and three-hybrid screens. This protocol can be used either for transforming a pool of YKO deletion strains with a plasmid encoding a GAL-inducible YFG or transforming a parental strain with a GAL-inducible cDNA library. This protocol yields transformation efficiencies of up to 2.2×10^7 transformants/μg of plasmid DNA.

1. Inoculate 50 ml of YPD from a single colony or an aliquot of the pooled YKO strains and incubate overnight at $30°$ with shaking.

2. Harvest cells when density reaches 5×10^6 cells/ml culture by centrifugation in a table-top centrifuge at 1500g for 5 min at $4°$. A hemocytometer may be used to calculate the cell number from an appropriate dilution of the culture. When determining cell number, count cells with equal size buds as two cells; cells with an obvious larger mother cell and smaller bud, should be counted as a single cell. Although the relationship between cell number and OD_{600} is strain specific, an estimate of ~ 1.0–2.0×10^7 cells/ml at OD_{600} of 1.0 may be used.

3. Pour off media, and wash cells with 10 ml sterile water. Centrifuge at 1500g for 5 min at $4°$.

4. Resuspend in 10 ml of 100 mM LiAc (sterile), and refrigerate overnight at $4°$.

5. Centrifuge at 1500g for 5 min.

6. Pour off supernatant and resuspend in sterile water to a final volume of 1 ml.

7. Pipette 100-μl aliquots of the cells into Microfuge tubes.

8. Pellet cells by centrifugation at 1500g for 5 min at $4°$, and remove water with a micropipetter.

9. Layer on each pellet in the following order:
   a. 240 μl 50% PEG
   b. 36 μl 1.0 M LiAc
   c. 5 μl salmon sperm DNA (10 mg/ml)
   d. 0.1–1.0 μg plasmid DNA (in a total of 79 μl sterile water); 360 μl total

10. Vortex vigorously for 1 min to ensure cell pellet is resuspended.

11. Incubate at 30° for 30 min.

12. Heat shock at 42° for 30 min in a water bath.

13. Pellet cells by centrifugation at 1500g for 5 min at 4°, and remove transformation mix with a micropipetter.

14. Resuspend cells in 100 μl of sterile water, and plate each aliquot onto a large (15-cm diameter) GLU plate with appropriate selection.

15. Incubate the plates at 30° for 2–3 days.

16. Replicate plate onto GAL plates with appropriate selection.

17. For isolation of suppressors of toxicity, identify colonies that grow on both GLU and GAL plates. To isolate deletion strains that enhance toxicity, identify colonies that grow normally on GLU, but weakly on GAL.

18. Streak modifying deletion strains onto GLU plates with appropriate selection to obtain colonies for further testing.

Amplification and Sequencing of Molecular Barcode

Once individual strains of interest from the YKO set have been identified above, the identity of the strain is determined by PCR amplification of the molecular barcode or TAG, followed by DNA sequencing and identification of the gene deletion strain using the database at the *Saccharomyces* Genome Deletion Project web page (see later).

1. Pick colony of interest into 20 μl of 20 m*M* NaOH in a tube appropriate for a thermocycler.

2. Lyse colonies at 94° for 20 min using a thermocycler.

3. Centrifuge at 1500g for 5 min at 4° to pellet cell debris.

4. Use 3 μl of supernatant lysate for PCR amplification.
   Per 25-μl reaction:
   a. 3 μl yeast genomic lysate
   b. 0.5 μl Forward primer (5 pmol)
      i. 5′ GCCTCGACATCATCTGCCCAG 3′
   c. 0.5 μl Reverse primer (5 pmol)
      ii. 5′ CGGTGTCGGTCTCGTAG 3′
   d. 0.5 μl dNTP mix (10 m*M*)
   e. 0.75 μl MgCl$_2$ (50 m*M*)
   f. 2.5 μl PCR reaction buffer (10×)
   g. 0.1 μl Taq polymerase (1 unit)
   h. 17.15 μl H$_2$O

5. Amplify TAG with thermocycler as follows:
   a. 94° for 2 min
   b. 35 cycles of:

    i. 94° for 15 sec
    ii. 55° for 15 sec
    iii. 72° for 30 sec
  c. 72° for 10 min
  d. Cool to 4°
6. Purify and sequence the amplified product by standard methods.
7. Identify gene deletion strain using TAG sequence at the *Saccharomyces* Genome Deletion Project web page: http://sequence-www.stanford.edu/group/yeast_deletion_project/deletions3.html

High-Throughput Transformation Protocol

Because of the high level of false positives isolated in these screens (see "Systematic Approaches to Genetic Screening"), all candidate modifier YKO strains and YOC plasmids must be retested. In the case of the YKO strains, each strain should be freshly retrieved from the frozen stock and retransformed with the construct of interest. In the case of the YOC plasmids, genetic interactions with YFG can be confirmed by double transformation into the parental strain. The following protocol is a simple method for performing up to 96 transformations at a time, allowing for rapid retesting of candidate modifiers. Modified from the "One step transformation of yeast in stationary phase" protocol (Chen *et al.*, 1992).

One Step Buffer (prepare fresh)

0.2 M Lithium acetate
40% Polyethylene glycol (PEG) 3350
100 mM Dithiothreitol (DTT); add fresh from 1 M frozen stock

1. Add 100 μl of YPD media per well of a 96-well microtiter plate using a multichannel pipetter.
2. Inoculate media with a single colony of the strain of interest.
3. Grow cells 1–2 days at 30° until the cultures reach stationary phase (OD$_{600}$ ∼ 5.0–10.0; 1–2 × 10^8 cells/ml).
4. Centrifuge plate(s) in a table-top centrifuge for 5 min at 1200g.
5. Remove as much media as possible using a multichannel pipetter, being careful not to disturb the cell pellet.
6. Wash cell pellet by resuspending in 50 μl of sterile water using a multichannel pipetter.
7. Centrifuge plate(s) in a table-top centrifuge for 5 min at 1200g.
8. Remove as much water as possible using a multichannel pipetter, being careful not to disturb the cell pellet. Prepare transformation cocktail

by mixing 2 ml of One Step Buffer, 100 μl of 10 mg/ml sheared, boiled salmon sperm DNA (1.0 mg), and 20 μg of plasmid.

9. Add 20 μl of the transformation cocktail per well using a multichannel pipetter.

10. Resuspend cells thoroughly in the transformation cocktail by pipetting up and down 10–20 times.

11. Incubate at 45° for 30 min in a water bath.

12. Use multichannel pipetter to spot 5 μl of cells per well onto the appropriate selective plate. Allow spots to absorb into media before placing in oven. Drying plates ahead of time speeds this process.

13. Incubate at 30° for 2 days.

14. Growth spots represent a mixture of many transformed cells; to obtain colonies derived from individual transformants, streak out a sampling of the cells onto a fresh selective plate.

Spotting Assays

Spotting assays are a simple, yet powerful, method to measure relative toxicity or growth between strains expressing heterologous proteins (Fig. 3). The method that follows is designed for 96-well microtiter plates and is well suited for high-throughput testing.

1. Add 100 μl of appropriate selective media + raffinose (RAF) per well of a 96-well microtiter plate using a multichannel pipetter.

2. Inoculate with a single colony from desired transformant. Because of variation in copy number it is advisable to test several transformants per

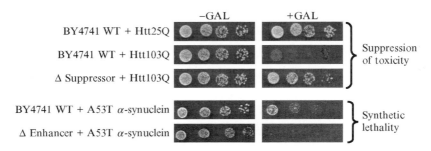

Fig. 3. Example of using spotting assays to measure the relative effect of genetic modifiers on amyloid toxicity. Assuming YFG is under the control of a GAL-inducible promoter, a sample of cultures in log phase is serially diluted and plated on both GLU and GAL plates. In the top example, a loss-of-function suppressor of toxicity is observed to relieve the growth defect caused by Htt103Q. In the bottom example, a loss-of-function enhancer is observed to increase toxicity of A53T α-synuclein.

clone. Incubate at 30° for 1–2 days until the cultures reach stationary phase ($OD_{600} \sim 5.0$–10.0; 1–2 × 10^8 cells/ml).

3. Read the optical density of the cultures using a plate reader, and then individually adjust each culture to OD_{600} 0.4.

4. Incubate at 30° for 3 h. This incubation allows the yeast cells to exit lag phase and begin actively dividing (log phase) before plating.

5. Perform serial fivefold dilutions of cultures in water (1:5, 1:25, 1:125, 1:625) in 96-well microtiter plates (diluting 20 μl of cells into 80 μl of water is convenient). Because yeast cells fall out of solution quickly, make sure that cells are fully resuspended when making dilutions.

6. Plate 5 μl of each dilution onto both GLU and GAL selective plates to compare growth of the strains with and without expression of YFG. Make sure to include a transformant containing a control construct on each plate for reference. It is critical that the plates are extremely dry or else the spots may run. Dry the plates at room temperature for 2–4 days, or 30 min in a sterile vacuum hood.

7. Incubate at 30° for 2 days (GLU plates) or 3 days (GAL plates). Relative toxicity is determined by comparisons of growth between the clones of interest and controls on GAL and GLU media (under inducing and repressed conditions).

Systematic Approaches to Genetic Screening

Although pooled approaches to screening are simple to perform and allow random isolation of genetic modifiers without *a priori* bias, this approach has several drawbacks and limitations that array approaches can eliminate. Perhaps the largest obstacle to this screening method is the large number of false positives isolated during transformation. During a recent loss-of-function screen, we found that >80% of modifiers isolated were false positives, likely because of second site mutations not related to the bar-coded gene knockout (Giorgini *et al.*, 2005). Such mutations may be due to the process of DNA transformation itself, which is mutagenic. If the promoter being used is at all "leaky," there is strong selective pressure within the pool for any second site mutation that relieves toxicity. Because the pool transformation approach requires screening individual transformants, any second site mutations affecting toxicity will confound the true effect of the deletion strain or clone being tested. On the other hand, if transformations are performed with the arrayed strains or ORF, each "spot" of transformants derived from the 96-well microtiter plates will represent dozens, perhaps hundreds, of transformants. By testing these transformants *en masse*, the confounding effect of second site mutations is greatly reduced. Furthermore, analysis of transformants *en masse* also reduces clone to clone differences in expression level of YFG because of variability in plasmid copy number, which can also lead to false positives.

Another major benefit of keeping the strains and ORFs ordered in arrays during testing is that the clones are readily identifiable by their position in the 96-well plate, which eliminates the need for PCR amplification of the bar code or plasmid isolation of the ORF and subsequent DNA sequencing, which can become quite costly when done on a large scale. Systematic testing of arrayed strains or ORFs can be performed either manually using 96- or 384-pin manual pinning tools (V&P Scientific, Inc., San Diego, CA) or with the aid of robotics, such as the Biomek FX Laboratory Automation Workstation (Beckman Coulter, Fullerton, CA). Even without the benefit of automation, the availability of the YKO and YOC sets, combined with manual pinning tools, allows for rapid systematic genomic screening in yeast.

Automated screening has been combined with several new genomic techniques in yeast, including construction of protein–protein interaction (PPI) networks in yeast using the two-hybrid system (Ito *et al.*, 2000; Uetz *et al.*, 2000), synthetic genetic array (SGA) analysis (Tong *et al.*, 2001), and diploid-based synthetic lethality analysis on microarrays (dSLAM) (Pan *et al.*, 2004). The PPI network approach has been used to generate a protein interaction map for the yeast genome (Schwikowski *et al.*, 2000) and can be used to generate PPI networks for disease genes (Giorgini and Muchowski, 2005; Goehler *et al.*, 2004). SGA analysis is a method for systematic construction of double mutants, in which a query mutation (i.e., YFG integrated into the yeast genome) is mated to the YKO deletion set. Because YFG is integrated into the yeast genome, expression levels of YFG are much more consistent, reducing the number of false positives in screens for enhancers and suppressors of toxicity. In addition, bringing together the gene deletion strain of interest and YFG is far simpler by mating than by individual transformation of each strain. Finally, the dSLAM method is another array-based analysis using the YKO set in which heterozygous strains are converted to haploid YKO strains immediately before genomic profiling, eliminating variation in growth rate among the strains, making the set more amenable to manipulation as a population. All of these rapid automated techniques, and others in development, make the use of yeast genomics an extremely powerful and informative approach for explanation of the mechanisms of amyloid toxicity.

Conclusion

In this chapter, we have only scratched the surface of the plethora of genomic and proteomic strategies available in yeast that can be applied toward studies of amyloid toxicity. Approaches from simple, classical genetics (i.e., epistatic analysis to order genes in pathways or networks) to complicated, data-intensive gene profiling experiments with microarrays, can be used to exploit these yeast models to their fullest potentials. In the end, a combination of several approaches in yeast and mammalian models is required for validation of interesting genetic pathways and for subsequent application of this information toward explaining amyloid disease. It is worth noting that results from loss-of-function screens can be nicely complemented by the use of RNA interference in mammalian cell models of disease. Nonetheless, the "Golden Age" of yeast genomics is on us in earnest (Fields and Johnston, 2005; Johnston, 2000), and the application of these genomic techniques in studies of amyloid toxicity using yeast and other model organisms is leading to an explosion of exciting new information of potential relevance to amyloid disease.

Acknowledgments

P. J. M. is supported by the National Institute of Neurological Disease and Stroke (R01NS47237), by an NIH construction award (C06 RR 14571), by the Alzheimer's Disease Research Center at the University of Washington, and by the Hereditary Disease Foundation under the auspices of the "Cure Huntington's Disease Initiative." F. G. is supported by a postdoctoral fellowship from the HighQ foundation and the Royal Society. The authors thank Tiago Outeiro for permission to adapt his illustrations for Figs. 1 and 2.

References

Chen, D. C., Yang, B. C., and Kuo, T. T. (1992). One-step transformation of yeast in stationary phase. *Curr. Genet.* **21**, 83–84.

Fields, S., and Johnston, M. (2005). Cell biology. Whither model organism research? *Science* **307**, 1885–1886.

Fields, S., and Sternglanz, R. (1994). The two-hybrid system: An assay for protein-protein interactions. *Trends Genet.* **10**, 286–292.

Flower, T. R., Chesnokova, L. S., Froelich, C. A., Dixon, C., and Witt, S. N. (2005). Heat shock prevents alpha-synuclein-induced apoptosis in a yeast model of Parkinson's disease. *J. Mol. Biol.* **351**, 1081–1100.

Forsburg, S. L. (2001). The art and design of genetic screens: Yeast. *Nat. Rev. Genet.* **2**, 659–668.

Giaever, G., Chu, A. M., Ni, L., Connelly, C., Riles, L., Veronneau, S., Dow, S., Lucau-Danila, A., Anderson, K., Andre, B., Arkin, A. P., Astromoff, A., El-Bakkoury, M., Bangham, R., Benito, R., Brachat, S., Campanaro, S., Curtiss, M., Davis, K., Deutschbauer, A., Entian, K. D., Flaherty, P., Foury, F., Garfinkel, D. J., Gerstein, M., Gotte, D., Guldener, U., Hegemann, J. H., Hempel, S., Herman, Z., Jaramillo, D. F., Kelly, D. E., Kelly, S. L., Kotter, P., LaBonte, D., Lamb, D. C., Lan, N., Liang, H., Liao, H., Liu, L., Luo, C., Lussier, M., Mao, R., Menard, P., Ooi, S. L., Revuelta, J. L., Roberts, C. J., Rose, M., Ross-Macdonald, P., Scherens, B., Schimmack, G., Shafer, B., Shoemaker, D. D., Sookhai-Mahadeo, S., Storms, R. K., Strathern, J. N., Valle, G., Voet, M., Volckaert, G., Wang, C. Y., Ward, T. R., Wilhelmy, J., Winzeler, E. A., Yang, Y., Yen, G., Youngman, E., Yu, K., Bussey, H., Boeke, J. D., Snyder, M.,

Philippsen, M., Davis, R. W., and Johnston, M. (2002). Functional profiling of the *Saccharomyces cerevisiae* genome. *Nature* **418,** 387–391.

Gietz, R. D., and Schiestl, R. H. (1996). Transforming yeast with DNA. *Methods Mol. Cell. Biol.* **5,** 255–269.

Giorgini, F., Guidetti, P., Nguyen, Q., Bennett, S. C., and Muchowski, P. J. (2005). A genomic screen in yeast implicates kynurenine 3-monooxygenase as a therapeutic target for Huntington disease. *Nat. Genet.* **37**(5), 526–531.

Giorgini, F., and Muchowski, P. J. (2005). Connecting the dots in Huntington's disease with protein interaction networks. *Genome Biol.* **6,** 210.

Goehler, H., Lalowski, M., Stelzl, U., Waelter, S., Stroedicke, M., Worm, U., Droege, A., Lindenberg, K. S., Knoblich, M., Haenig, C., Herbst, M., Suopanki, J., Scherzinger, E., Abraham, C., Bauer, B., Hasenbank, R., Fritzsche, A., Ludewig, A. H., Buessow, K., Coleman, S. H., Gutekunst, C. A., Landwehrmeyer, B. G., Lehrach, H., and Wanker, E. E. (2004). A protein interaction network links GIT1, an enhancer of huntingtin aggregation, to Huntington's disease. *Mol. Cell.* **15,** 853–865.

Goffeau, A., Barrell, B. G., Bussey, H., Davis, R. W., Dujon, B., Feldmann, H., Galibert, F., Hoheisel, J. D., Jacq, C., Johnston, M., Louis, E. J., Mewes, H. W., Murakami, Y., Philippsen, P., Tettelin, H., and Oliver, S. G. (1996). Life with 6000 genes. *Science* **274,** 546, 563–567.

Grunenfelder, B., and Winzeler, E. A. (2002). Treasures and traps in genome-wide data sets: Case examples from yeast. *Nat. Rev. Genet.* **3,** 653–661.

Hartwell, L. H. (2002). Nobel Lecture. Yeast and cancer. *Biosci. Rep.* **22,** 373–394.

Hughes, R. E., Lo, R. S., Davis, C., Strand, A. D., Neal, C. L., Olson, J. M., and Fields, S. (2001). Altered transcription in yeast expressing expanded polyglutamine. *Proc. Natl. Acad. Sci. USA* **98,** 13201–13206.

Hughes, T. R., Roberts, C. J., Dai, H., Jones, A. R., Meyer, M. R., Slade, D., Burchard, J., Dow, S., Ward, T. R., Kidd, M. J., Friend, S. H., and Marton, M. J. (2000). Widespread aneuploidy revealed by DNA microarray expression profiling. *Nat. Genet.* **25,** 333–337.

Ito, T., Tashiro, K., Muta, S., Ozawa, R., Chiba, T., Nishizawa, M., Yamamoto, K., Kuhara, S., and Sakaki, Y. (2000). Toward a protein-protein interaction map of the budding yeast: A comprehensive system to examine two-hybrid interactions in all possible combinations between the yeast proteins. *Proc. Natl. Acad. Sci. USA* **97,** 1143–1147.

Johnston, M. (2000). The yeast genome: On the road to the Golden Age. *Curr. Opin. Genet. Dev.* **10,** 617–623.

Johnston, M., and Davis, R. W. (1984). Sequences that regulate the divergent GAL1-GAL10 promoter in *Saccharomyces cerevisiae. Mol. Cell. Biol.* **4,** 1440–1448.

Kohalmi, S. E., and Kunz, B. A. (1988). Role of neighbouring bases and assessment of strand specificity in ethylmethanesulphonate and N-methyl-N′-nitro-N-nitrosoguanidine mutagenesis in the SUP4-o gene of *Saccharomyces cerevisiae. J. Mol. Biol.* **204,** 561–568.

Krobitsch, S., and Lindquist, S. (2000). Aggregation of huntingtin in yeast varies with the length of the polyglutamine expansion and the expression of chaperone proteins. *Proc. Natl. Acad. Sci. USA* **97,** 1589–1594.

Kruger, R., Kuhn, W., Muller, T., Woitalla, D., Graeber, M., Kosel, S., Przuntek, H., Epplen, J. T., Schols, L., and Riess, O. (1998). Ala30Pro mutation in the gene encoding alpha-synuclein in Parkinson's disease. *Nat. Genet.* **18,** 106–108.

Kumar, A., Vidan, S., and Snyder, M. (2002). Insertional mutagenesis: Transposon-insertion libraries as mutagens in yeast. *Methods Enzymol.* **350,** 219–229.

Kunz, B. A., Pierce, M. K., Mis, J. R., and Giroux, C. N. (1987). DNA sequence analysis of the mutational specificity of u.v. light in the SUP4-o gene of yeast. *Mutagenesis* **2,** 445–453.

Lawrence, C. W. (2002). Classical mutagenesis techniques. *Methods Enzymol.* **350,** 189–199.

Lee, G. S., Savage, E. A., Ritzel, R. G., and von Borstel, R. C. (1988). The base-alteration spectrum of spontaneous and ultraviolet radiation-induced forward mutations in the URA3 locus of *Saccharomyces cerevisiae. Mol. Gen. Genet.* **214**, 396–404.

Link, C. D. (2001). Transgenic invertebrate models of age-associated neurodegenerative diseases. *Mech. Ageing Dev.* **122**, 1639–1649.

Longtine, M. S., McKenzie, A., 3rd, Demarini, D. J., Shah, N. G., Wach, A., Brachat, A., Philippsen, P., and Pringle, J. R. (1998). Additional modules for versatile and economical PCR-based gene deletion and modification in *Saccharomyces cerevisiae. Yeast* **14**, 953–961.

Madeo, F., Herker, E., Wissing, S., Jungwirth, H., Eisenberg, T., and Frohlich, K. U. (2004). Apoptosis in yeast. *Curr. Opin. Microbiol.* **7**, 655–660.

Marsh, J. L., and Thompson, L. M. (2004). Can flies help humans treat neurodegenerative diseases? *Bioessays* **26**, 485–496.

McLean, P. J., Kawamata, H., and Hyman, B. T. (2001). Alpha-synuclein-enhanced green fluorescent protein fusion proteins form proteasome sensitive inclusions in primary neurons. *Neuroscience* **104**, 901–912.

Meriin, A. B., Zhang, X., He, X., Newnam, G. P., Chernoff, Y. O., and Sherman, M. Y. (2002). Huntington toxicity in yeast model depends on polyglutamine aggregation mediated by a prion-like protein Rnq1. *J. Cell Biol.* **157**, 997–1004.

Muchowski, P. J., Schaffar, G., Sittler, A., Wanker, E. E., Hayer-Hartl, M. K., and Hartl, F. U. (2000). Hsp70 and hsp40 chaperones can inhibit self-assembly of polyglutamine proteins into amyloid-like fibrils. *Proc. Natl. Acad. Sci. USA* **97**, 7841–7846.

Mumberg, D., Muller, R., and Funk, M. (1994). Regulatable promoters of *Saccharomyces cerevisiae:* Comparison of transcriptional activity and their use for heterologous expression. *Nucleic Acids Res.* **22**, 5767–5768.

Mumberg, D., Muller, R., and Funk, M. (1995). Yeast vectors for the controlled expression of heterologous proteins in different genetic backgrounds. *Gene* **156**, 119–122.

Nurse, P. M. (2002). Nobel Lecture. Cyclin dependent kinases and cell cycle control. *Biosci. Rep.* **22**, 487–499.

Outeiro, T. F., and Lindquist, S. (2003). Yeast cells provide insight into alpha-synuclein biology and pathobiology. *Science* **302**, 1772–1775.

Outeiro, T. F., and Muchowski, P. J. (2004). Molecular genetics approaches in yeast to study amyloid diseases. *J. Mol. Neurosci.* **23**, 49–60.

Pan, X., Yuan, D. S., Xiang, D., Wang, X., Sookhai-Mahadeo, S., Bader, J. S., Hieter, P., Spencer, F., and Boeke, J. D. (2004). A robust toolkit for functional profiling of the yeast genome. *Mol. Cell.* **16**, 487–496.

Polymeropoulos, M. H., Lavedan, C., Leroy, E., Ide, S. E., Dehejia, A., Dutra, A., Pike, B., Root, H., Rubenstein, J., Boyer, R., Stenroos, E. S., Chandrasekharappa, S., Athanassiadou, A., Papapetropoulos, T., Johnson, W. G., Lazzarini, A. M., Duvoisin, R. C., Di Iorio, G., Golbe, L. I., and Nussbaum, R. L. (1997). Mutation in the alpha-synuclein gene identified in families with Parkinson's disease. *Science* **276**, 2045–2047.

Puccio, H., and Koenig, M. (2000). Recent advances in the molecular pathogenesis of Friedreich ataxia. *Hum. Mol. Genet.* **9**, 887–892.

Reid, R. J., Lisby, M., and Rothstein, R. (2002). Cloning-free genome alterations in *Saccharomyces cerevisiae* using adaptamer-mediated PCR. *Methods Enzymol.* **350**, 258–277.

Rine, J. (1991). Gene overexpression in studies of *Saccharomyces cerevisiae. Methods Enzymol.* **194**, 239–251.

Rothstein, R. (1991). Targeting, disruption, replacement, and allele rescue: Integrative DNA transformation in yeast. *Methods Enzymol.* **194**, 281–301.

Schneider, J. C., and Guarente, L. (1991). Vectors for expression of cloned genes in yeast: Regulation, overproduction, and underproduction. *Methods Enzymol.* **194,** 373–388.

Schwikowski, B., Uetz, P., and Fields, S. (2000). A network of protein-protein interactions in yeast. *Nat. Biotechnol.* **18,** 1257–1261.

Sherman, M. Y., and Muchowski, P. J. (2003). Making yeast tremble: Yeast models as tools to study neurodegenerative disorders. *Neuromolecular Med.* **4,** 133–146.

Singleton, A. B., Farrer, M., Johnson, J., Singleton, A., Hague, S., Kachergus, J., Hulihan, M., Peuralinna, T., Dutra, A., Nussbaum, R., Lincoln, S., Crawley, A., Hanson, M., Maraganore, D., Adler, C., Cookson, M. R., Muenter, M., Baptista, M., Miller, D., Blancato, J., Hardy, J., and Gwinn-Hardy, K. (2003). alpha-Synuclein locus triplication causes Parkinson's disease. *Science* **302,** 841.

Tong, A. H., Evangelista, M., Parsons, A. B., Xu, H., Bader, G. D., Page, N., Robinson, M., Raghibizadeh, S., Hogue, C. W., Bussey, H., Andrews, B., Tyers, M., and Boone, C. (2001). Systematic genetic analysis with ordered arrays of yeast deletion mutants. *Science* **294,** 2364–2368.

Uetz, P., Giot, L., Cagney, G., Mansfield, T. A., Judson, R. S., Knight, J. R., Lockshon, D., Narayan, V., Srinivasan, M., Pochart, P., Qureshi-Emili, A., Li, Y., Godwin, B., Conover, D., Kalbfleisch, T., Vijayadamodar, G., Yang, M., Johnston, M., Fields, S., and Rothberg, J. M. (2000). A comprehensive analysis of protein-protein interactions in Saccharomyces cerevisiae. *Nature* **403,** 623–627.

Uptain, S. M., and Lindquist, S. (2002). Prions as protein-based genetic elements. *Annu. Rev. Microbiol.* **56,** 703–741.

West, R. W., Jr., Yocum, R. R., and Ptashne, M. (1984). *Saccharomyces cerevisiae* GAL1-GAL10 divergent promoter region: Location and function of the upstream activating sequence UASG. *Mol. Cell. Biol.* **4,** 2467–2478.

Willingham, S., Outeiro, T. F., DeVit, M. J., Lindquist, S. L., and Muchowski, P. J. (2003). Yeast genes that enhance the toxicity of a mutant huntingtin fragment or alpha-synuclein. *Science* **302,** 1769–1772.

Winzeler, E. A., Shoemaker, D. D., Astromoff, A., Liang, H., Anderson, K., Andre, B., Bangham, R., Benito, R., Boeke, J. D., Bussey, H., Chu, A. M., Connelly, C., Davis, K., Dietrich, F., Dow, S. W., El Bakkoury, M., Foury, F., Friend, S. H., Gentalen, E., Giaever, G., Hegemann, J. H., Jones, T., Laub, M., Liao, H., Licbundguth, N., Lockhart, D. J., Lucau-Danila, A., Lussier, M., M'Rabet, N., Menard, P., Mittmann, M., Pai, C., Rebischung, C., Revuelta, J. L., Riles, L., Roberts, C. J., Ross-MacDonald, P., Scherens, B., Snyder, M., Sookhai-Mahadeo, S., Storms, R. K., Veronneau, S., Voet, M., Volckaert, G., Ward, T. R., Wysocki, R., Yen, G. S., Yu, K., Zimmermann, K., Philippsen, P., Johnston, M., and Davis, R. W. (1999). Functional characterization of the *S. cerevisiae* genome by gene deletion and parallel analysis. *Science* **285,** 901–906.

Zhang, X., Smith, D. L., Meriin, A. B., Engemann, S., Russel, D. E., Roark, M., Washington, S. L., Maxwell, M. M., Marsh, J. L., Thompson, L. M., Wanker, E. E., Young, A. B., Housman, D. E., Bates, G. P., Sherman, M. Y., and Kazantsev, A. G. (2005). A potent small molecule inhibits polyglutamine aggregation in Huntington's disease neurons and suppresses neurodegeneration *in vivo. Proc. Natl. Acad. Sci. USA* **102,** 892–897.

[14] Searching for Anti-Prion Compounds: Cell-Based High-Throughput *In Vitro* Assays and Animal Testing Strategies

By DAVID A. KOCISKO and BYRON CAUGHEY

Abstract

The transmissible spongiform encephalopathies (TSEs) or prion diseases are infectious neurodegenerative diseases of mammals. Protease-resistant prion protein (PrP-res) is only associated with TSEs and thus has been a target for therapeutic intervention. The most effective compounds known against scrapie *in vivo* are inhibitors of PrP-res in infected cells. Mouse neuroblastoma (N2a) cells have been chronically infected with several strains of mouse scrapie including RML and 22L. Also, rabbit epithelial cells that produce sheep prion protein in the presence of doxycycline (Rov9) have been infected with sheep scrapie. Here a high-throughput 96-well plate PrP-res inhibition assay is described for each of these scrapie-infected cell lines. With this dot-blot assay, thousands of compounds can easily be screened for inhibition of PrP-res formation. This assay is designed to find new PrP-res inhibitors, which may make good candidates for *in vivo* anti-scrapie testing. However, an *in vitro* assay can only suggest that a given compound might have *in vivo* anti-scrapie activity, which is typically measured as increased survival times. Methods for *in vivo* testing of compounds for anti-scrapie activity in transgenic mice, a much more lengthy and expensive process, are also discussed.

Introduction

The transmissible spongiform encephalopathies (TSEs) or prion diseases are closely related incurable infectious neurodegenerative diseases of humans and other mammals. The incubation periods of these diseases range from months to decades. Creutzfeldt-Jakob disease (CJD) is a human TSE with an incidence of about 1 case per million people per year. Bovine spongiform encephalopathy (BSE) is a well-known TSE that has caused many billions of dollars of economic damage worldwide. BSE is also most likely responsible for approximately 180 cases of human variant CJD transmitted by consumption of contaminated beef. Strict measures to stop the spread of BSE and protect the food supply have resulted in a greatly reduced incidence in cattle and seem to have reduced the

METHODS IN ENZYMOLOGY, VOL. 412

0076-6879/06 $35.00
DOI: 10.1016/S0076-6879(06)12014-5

incidence of variant CJD as well (Andrews et al., 2003; Smith and Bradley, 2003).

Prion protein (PrP) is a 33–35-kDa membrane-associated glycoprotein of unknown function. The only form of prion protein found in healthy mammals is detergent soluble and sensitive to protease-degradation (PrP^C or PrP-sen). A TSE-associated form of prion protein (PrP^{Sc} or PrP-res) is highly aggregated and resistant to protease degradation (Caughey and Lansbury, 2003). PrP-res and PrP-sen have the same amino acid sequence (Stahl et al., 1993), and PrP-res is formed from PrP-sen by a posttranslational conformational modification (Borchelt et al., 1990; Caughey and Raymond, 1991). PrP-res is the major component of purified infectivity and is postulated to be the infectious particle of the TSEs (Prusiner, 1998).

PrP-res has consequently been a target for therapeutic intervention of the TSEs (Aguzzi et al., 2001; Brown, 2002; Cashman and Caughey, 2004; Dormont, 2003). The role of PrP in TSE pathology is not well understood mechanistically, but animals lacking PrP are not susceptible to TSE infection (Bueler et al., 1993). Compounds that have demonstrated anti-scrapie activity in vivo, which is typically measured as increased survival times, are usually also inhibitors of PrP-res in cell culture. Pentosan polysulfate, perhaps the most active anti-scrapie compound in vivo (Diringer and Ehlers, 1991; Doh-ura et al., 2004; Ladogana et al., 1992), strongly inhibits PrP-res formation in cells (Caughey and Raymond, 1993). Amphotericin B (Adjou et al., 1995; Mange et al., 2000; Pocchiari et al., 1987) and a number of porphyrins (Caughey et al., 1998; Priola et al., 2000) with anti-scrapie activity also inhibit the formation of PrP-res in cell culture. Regardless of the mechanism by which these compounds work in vivo, inhibition of PrP-res in cell culture is one feature these anti-scrapie compounds share. Thus, new compounds that effectively inhibit PrP-res in cell culture are good candidates for the expensive and time-consuming process of testing against scrapie in vivo. High-throughput screening of compound libraries for PrP-res inhibitors is an efficient way to find these new candidates. In this chapter, high-throughput testing of compounds for PrP-res inhibitory activity using TSE-infected cells and a dot-blot apparatus is discussed. Demonstrating anti-TSE activity requires in vivo experimentation, and several different approaches to this testing in transgenic mice are also discussed.

Cell Lines Chronically Infected with TSEs

Cell lines chronically infected with TSEs have been useful tools for studying cellular processes of PrP-res (reviewed by Solassol et al., 2003). However, relatively few chronically infected cell lines have been developed despite the efforts of many research groups; among these are RML mouse scrapie strain (RML) (Race *et al.*, 1988), 22L mouse scrapie strain (22L) (Nishida *et al.*, 2000), and Fukuoka mouse–adapted CJD strain (Ishikawa *et al.*, 2004) infected mouse neuroblastoma cells (N2a). 22L also infects two different mouse fibroblast lines, NIH/3T3 and ψ2C2 (Vorberg *et al.*, 2004). In addition, mouse neuronal gonadotropin-releasing hormone cells have been infected with RML (Sandberg *et al.*, 2004). A rabbit epithelial cell line that produces sheep PrP in the presence of doxycycline (Rov9) has been chronically infected with sheep scrapie (Vilette *et al.*, 2001). Recently, we have developed a mule deer brain cell line persistently infected with chronic wasting disease (MDB-CWD) (Raymond *et al.*, 2005). Although scrapie-infected hamster cells (Taraboulos *et al.*, 1990) and CJD-infected human cells (Ladogana *et al.*, 1995) have been reported, they seem to have been lost. Hence, Rov9 and MDB-CWD are the only non-mouse TSE-infected cell lines that are currently available.

Compounds can be tested for the ability to inhibit PrP-res accumulation in chronically TSE-infected cell lines. An assay based on cells grown in 96-well plates with dot-blot PrP-res detection can greatly increase the throughput of such testing. A requirement for this increase in throughput is that the cell line must produce enough PrP-res from one well of a 96-well plate to be readily detected on a dot blot. Mouse N2a cells infected with RML and 22L (Kocisko *et al.*, 2003), and Rov9 cells infected with sheep scrapie produce enough PrP-res to be used with dot-blot detection and 96-well plate testing (Kocisko *et al.*, 2005). In the next sections, the use of these cells in a high-throughput assay will be discussed.

N2a Cell-Based High-Throughput PrP-res Inhibition Assay

The following description of the assay is written in the context of testing a commercially available compound library such as the Spectrum Collection (Microsource Discovery). In this case, the compounds were received as 10 m*M* DMSO solutions in 96-well format, which was convenient for this assay.

Before the addition of compounds, approximately 20,000 RML or 22L-infected N2a cells are added to each well of a Costar 3595 flat-bottom 96-well plate with a low evaporation lid (Corning) in 100 μl of OPTIMEM cell medium (Invitrogen) supplemented with 10% fetal bovine serum (FBS) (Invitrogen). The OPTIMEM and the FBS lots used are pretested for the ability to sustain RML scrapie infection in mouse N2a cells for five passes as measured by analysis of PrP-res signal on immunoblot. For unknown reasons RML scrapie infections can be rapidly lost with growth in a majority of recent individual lots of OPTIMEM and rare lots of Invitrogen certified FBS. 22L-infected cells were developed by the curing of RML-infected N2a cells by seven passages including treatment with 1 μg/ml pentosan polysulfate. The cured cells were then reinfected with 22L using the method of Nishida *et al.* (2000). The N2a cells reinfected with 22L scrapie have continuously expressed PrP-res for more than 80 passages. The cells are allowed to settle for at least 90 min in a 5% CO_2 incubator at 37° before compounds are added.

The 10-mM solutions of compounds in DMSO are diluted several times with PBS before addition to the cell medium. Typically, compounds are screened at 1 or 10 μM. From the final dilution into PBS, 5 μl is added to the 100-μl cell medium. For example, if compounds are being screened at 1 μM, then 5 μl of 21 μM compound solution is added to the 100 μl cell medium. If aqueous-soluble compounds are being tested, up to 20 μl of physiologically compatible aqueous solutions containing no DMSO or other solvent have been added to the cell medium without decreasing PrP-res production. Final DMSO concentrations in the cell media as the cells grow to confluence are never higher than 0.5% (v/v). DMSO concentrations higher than 0.5% (v/v) have caused morphological changes in the cells. After compound is added, the cells are allowed to grow for 4–6 days at 37° in a 5% CO_2 incubator before being lysed at confluence.

Immediately before cell lysis, the cells of each well are inspected for toxic effects, bacterial contamination, and density by light microscopy. Any differences in the cells compared with controls are noted. Cytotoxicity detected initially by light microscopy is confirmed with 3-[4,5-dimethylthiazol-2-yl]-2,5-diphenyltetrazolium bromide (MTT) cell viability assays (May *et al.*, 2003). However, under the conditions of growth from low density to confluence in the presence of test compounds, cytotoxicity is usually obvious by light microscopy. So far, the MTT assay results have always agreed with what was noted as cytotoxicity by light microscopy.

After removal of the cell media, 50 μl of lysis buffer is added to each well. Lysis buffer contains 150 mM NaCl, 5 mM EDTA, 0.5% (w/v) triton X-100, 0.5% (w/v) sodium deoxycholate, and 5 mM tris-HCl, pH 7.4, at 4°. At this point the plates containing cell lysates can be frozen at −20° for up to 2 weeks, thawed, and the processing continued without any loss of signal. The frozen cell lysates may be stable longer than 2 weeks, but this has not been tested. Several minutes after adding lysis buffer, 25 μl of 0.1 mg/ml proteinase K (PK) (Calbiochem) in TBS is added to each well and incubated at 37° for 50 min. The treatment with PK eliminates PrP-sen and most other proteins in the lysate but only has a limited effect on PrP-res, which can then be more easily detected; 225 μl of 1 mM Pefabloc (Boehringer Mannheim) is then added to each well to inhibit PK before dot-blot analysis.

Rov9 Cell-Based High-Throughput PrP-res Inhibition Assay

The creation and characterization of the sheep scrapie-infected Rov9 cells used in this 96-well plate assay has been reported by Vilette *et al.* (2001). Rov9 cells must be grown in the presence of 1 μg/ml (~1 μM) doxycycline to maintain expression of ovine PrP. Rov9 cells are grown at 37° in 5% CO_2 and are passaged at a 1:4 dilution weekly. We have adapted Rov9 cells from MEM supplemented with 10% FBS to OPTIMEM (Invitrogen) supplemented with 10% FBS (Invitrogen), because cells chronically infected with TSEs often maintain infection better when grown with pretested lots of OPTIMEM. This adaptation was completed over the course of three passages by increasing to 50, 75, and finally 100% OPTI-MEM and resulted in an increase in PrP-res production by the cells (Kocisko *et al.*, 2005). Rov9 cells are plated in 96-well plate wells in 100 μl medium, as was the case with N2a cells. After at least 90 min, appropriate dilutions of potential inhibitors in DMSO or PBS solutions are added and the cells allowed to grow to confluence during the next 7 days. DMSO in the medium at up to 0.5% (v/v) does not affect Rov9 cell growth or morphology. At confluence, any morphological changes or toxic-ities seen by light microscopy caused by test compounds are noted as described for N2a cells. An MTT cell viability assay (May *et al.*, 2003) is also useful with inhibitors of PrP-res in the Rov9 cells to corroborate any toxicity noted by light microscopy. The cell medium is then removed by aspiration, and 50 μl of lysing buffer is added; 25 μl of 0.2 U/μl benzonase (Sigma) is added 5 min after lysis, and the lysates are then incubated for 30 min at 37°. The benzonase treatment eliminates clumps of nucleic acids to produce more homogeneous signals in the subsequent dot blots. This treatment is critical with the Rov9 cells and is optional with the N2a cells; 25 μl of 100 μg/ml PK is added after benzonase treatment to give a final concentration of 25 μg/ml, and the plates are incubated at 37° for 50 min. Immediately after protease treatment, 200 μl of 1 mM Pefabloc is added to each well to inhibit further proteolysis.

Dot-Blot Procedure and Immunodetection of PrP-res on Membranes

The dot-blot procedure and immunodetection of PrP-res are identical for RML- and 22L-infected N2a cells and sheep-scrapie infected Rov9 cells. Each opening of the dot-blot apparatus (Minifold I dot-blot system, Schleicher and Schuell) is rinsed with 500 μl of TBS. The suction is adjusted so that 500 μl of liquid will go through the apparatus in about 30 sec. Variation in suction strength can lead to distortion of the signal. The PK-treated cell lysates are then put onto a PVDF membrane (Immobilon-P, 0.45-μm pore size, Millipore) through the dot-blot apparatus along with a second rinse of 500 μl TBS. The membrane is removed, treated with 3 M guanidine thiocyanate for 10 min, and blocked in 5% (wt/v) milk in TBS-T (TBS with 0.5% [v/v] Tween 20 added). The 3 M guanidine thiocyanate denatures PrP-res and makes it more accessible to an antibody. The membrane is then incubated with an anti-PrP monoclonal antibody, in our case 6B10 (Kocisko et al., 2003), which was effective against mouse and sheep PrP-res with low background. 6H4 antibody (Prionics) is effective and presumably others will work as well. The membrane is then incubated with an alkaline phosphatase–conjugated goat anti-mouse secondary antibody in 5% milk, and then after TBST-T rinsing, an enhanced chemifluorescence agent (Zymed) is applied. PrP-res is quantified by scanning the membrane with a Storm Scanner (Molecular Dynamics) and using ImageQuant software.

The amount of input PrP-res is virtually undetectable from RML- and 22L-infected N2a cells, and the quantified PrP-res data can be used at this point. However, because the Rov9 cells are initially plated at ~25% confluent density before addition of potential inhibitors, the amount of input PrP-res in the seeded cells needs to be subtracted from all wells for more accurate results. To measure the amount of preexisting PrP-res in seeded Rov9 cells, a cytotoxic compound such as 20 μM thiothixene is added to at least three wells per 96-well plate to prevent new PrP-res formation while cells in other wells are growing to confluency. The addition of the cytotoxic compound does not affect detection of PrP-res in the input (seeded) cells. New PrP-res accumulation during growth to confluency is calculated as the difference between the total PrP-res signal intensity and the average signal intensity from the wells containing the cytotoxic compound.

Inhibitors Found with This Assay

This assay has been useful to screen several libraries of compounds for PrP-res inhibitory activity. Many new inhibitors have been discovered through screening compounds (Kocisko et al., 2003). This assay is also useful for testing smaller numbers of compounds at a range of concentrations to determine IC$_{50}$ values. An arbitrary IC$_{50}$ value of 1 μM or less has been used as a standard for advancing a compound to animal scrapie testing, but this is only a guide, because porphyrins with in vivo activity have IC$_{50}$ values between 1 and 10 μM (Caughey et al., 1998; Priola et al., 2000). Because RML and 22L mouse strains are available as chronic infections in N2a cells, any differences in compounds' inhibitory activity between these strains can be readily detected. Many compounds have been found that are better inhibitors of RML than 22L PrP-res (Kocisko et al., 2005).

The availability of sheep scrapie–infected Rov9 cells allows the comparison of a compounds' PrP-res inhibitory activity to be extended to other species. Many compounds that are good inhibitors of RML or 22L PrP-res are not inhibitors of sheep PrP-res in the Rov9 cells (Kocisko *et al.*, 2005), and it is not clear whether this is due primarily to differences in PrP-res or cell type. Nonetheless, these examples show that PrP-res inhibitors can have striking species-, strain-, and/or cell-type specificities that should be considered as a potential confounding aspect in anti-TSE applications.

Screening Throughput

For chronically infected cell lines amenable to a 96-well plate assay with dot-blot detection, testing compounds for PrP-res inhibitory activity is much more rapid than using a Western blot–based assay. A person assaying compounds by this method should be able to screen hundreds of compounds per week, but this number depends on how batches of test compounds are received. As noted previously, receiving a library of compounds pre-solubilized in 96-well format saves considerable setup time. Quantifying PrP-res from two plates in a day at the same time is easy. Experienced personnel can increase output to four per day by processing plates in parallel batches. Culturing multiple flasks of cells that are passed on different days of the week can help increase testing output. Finally, knowing that plates of cell lysates can be frozen to process later allows more scheduling freedom. This assay may be amenable to robotics, but this has not been attempted.

The Use of Transgenic Mice for *In Vivo* Anti-Scrapie Testing

These *in vitro* assays select promising candidates for *in vivo* anti-TSE activity on the basis of inhibition of PrP-res formation in chronically infected cell culture. Unfortunately, there is no substitute for animal testing to prove that a compound actually has *in vivo* anti-TSE activity. The TSEs are known for long incubation periods, so testing compounds for anti-TSE activity *in vivo* is a lengthy and expensive process. However, transgenic mice have been developed with greatly reduced incubation periods. One such line, Tg7 (Priola *et al.*, 2000; Race *et al.*, 2000), overexpresses hamster PrP and is highly susceptible to hamster 263K scrapie (263K) infection. High doses of 263K given intracerebrally (IC) into Tg7 mice cause disease in about 44 days, whereas high doses given intraperitoneally (IP) cause disease in 80–90 days. Another transgenic mouse line, Tga20 (Fischer *et al.*, 1996), overexpresses mouse PrP and its incubation period from RML is roughly the same as the incubation period of Tg7 mice from 263K.

Compounds can be tested for either prophylaxis or postexposure activity, depending on when dosing begins relative to scrapie inoculation. A prophylaxis test has the greater chance of success, because compound is present before inoculation. Also, a prophylaxis test against an IP inoculation allows a compound to intercept infectivity before it gets established in the brain. Once infection is established in the brain, the blood–brain barrier penetration of the compound is an issue. In general, designing *in vivo* anti-scrapie experiments involves arbitrary decisions such as when compound dosing is started relative to inoculation and how long it lasts. There are many other valid experimental designs besides the schemes outlined in the following, which have been used with Tg7 mice and 263K infection (Kocisko *et al.*, 2004).

Another variable in animal testing is the amount of infectivity to deliver. Regardless of the route of inoculation, using high amounts of infectivity has the advantage of shorter and less variable incubation periods. This must be balanced with the possibility that high amounts of inoculated infectivity may make therapy or prophylaxis more difficult. Naturally occurring infections are likely to involve much lower levels of infectivity than can be dosed in a laboratory setting. A compromise approach is to use intermediate doses of infectivity that result in reasonable incubation periods. In the case of 263K dosed IC into Tg7 mice, 50 μl of 0.001% brain homogenate results in incubation periods of approximately 70 days, and this has been used in some tests (Kocisko *et al.*, 2004).

To test for treatment of an established scrapie infection in the brain, compound administration is started 2 weeks after IC scrapie inoculation and continues for 5–6 weeks. A 2-week delay after IC inoculation before starting treatment allows the disease time to progress before the compound is administered. To test for prophylaxis, compound is administered for a total of 6 weeks starting 2 weeks before and continuing for 4 weeks after IP scrapie inoculation. *In vivo* compound levels should be approaching a steady state in the mouse at the time of inoculation, enabling it to block peripheral scrapie infectivity from being established in the brain. The treatment after inoculation conceivably allows time for the animal to eliminate infectious material while the compound prevents further formation of PrP-res.

Compounds are administered either as an IP injection or in the drinking water. For IP injections, compounds are dissolved or suspended in an appropriate buffer and the dose volume is 10 ml/kg. Generally, the highest known tolerated dose of a compound in mice is given to maximize the chance of seeing an effect in all types of testing. Injections are given three times per week on Monday, Wednesday, and Friday. This dosing schedule is largely for convenience of laboratory personnel; many other dosing regimens are possible. If it is available, a compounds' pharmacokinetics can be helpful in planning a dosing regimen. Compound administration in drinking water is less labor intensive for sufficiently stable and soluble molecules that have known oral bioavailability. Solutions of compounds in drinking water are made to yield the desired dose on the basis of the average daily consumption of water by mice, 15 ml/100 g body weight. Compound dissolved in the drinking water is the mouse's only source of water during the dosing period. All 263K scrapie brain homogenates made up for inoculation are in PBS supplemented with 2% fetal bovine serum. Tg7 mice are euthanized when clinical signs of scrapie such as ruffed fur, lethargy, ataxia, and weight loss are present. Animals that die from inoculation, dosing, anesthetizing procedures, and any other non-scrapie causes are excluded from the data. In the course of experiments involving mice, there will be occasional deaths for reasons other than scrapie. Watching mice regularly for clinical signs of scrapie and testing brain homogenates for PrP-res by protease treatment and Western blot can eliminate scrapie as a cause of death.

Another way that compounds can be tested for *in vivo* activity is to mix them with infectious brain homogenates before inoculation. After 1-h incubation at 37°, 50 μl of the homogenate/compound mixture is inoculated IC to see whether infectivity in the sample has been reduced. The IC inoculation route is used, because it has the fastest incubation period and no other compound administrations are done. This method has the advantages of needing only one injection and using very little compound. Compounds have been dosed directly into the ventricle of the brain by catheter and osmotic pump to test for scrapie treatment activity (Doh-ura *et al.*, 2004), but this is a labor-intensive procedure. A "mixing" experiment as described here can test many compounds for activity with much less labor. However, a compound directly injected into the brain by a needle or osmotic pump can result in problematic toxicity. Even a compound that seems relatively nontoxic dosed IP may be toxic directly injected into the brain because the blood–brain barrier has been bypassed.

Conclusion

The high-throughput dot-blot assay is a rapid and easy way to measure the amount of PrP-res produced by chronically infected cells as they grow from low density to confluence over the course of 4–6 days. A single addition of potential inhibitors to wells of a 96-well plate, added soon after plating cells, allows for inhibition of PrP-res formation to be assayed. The output data from the assay is that a given concentration of compound added to cell medium allows accumulation of a certain amount of PrP-res in that time. As mentioned previously, the most effective known anti-scrapie compounds *in vivo* inhibit PrP-res formation in cell culture. Exactly how these compounds fight scrapie *in vivo* is not clear, and the precise role of PrP-res in disease pathology is not understood. However, on a purely practical level, screening compounds for the ability to inhibit PrP-res in cells is a rational way to seek new compounds that might be active *in vivo*. This assay on its own is not designed to discriminate between the different mechanisms that can be envisioned for inhibiting PrP-res production. For example, the assay cannot distinguish between PrP-res accumulation because of a compound binding to PrP-sen or PrP-res or inhibiting some cellular process required for PrP-res accumulation. Regardless of how inhibitors work in cell culture, testing in animals must be done to show *in vivo* anti-scrapie activity.

Cell lines infected with additional strains and species of TSEs will hopefully be available soon. As previously noted, compounds that inhibit one strain or species of PrP-res cannot be assumed to be inhibitors of all. Different activities against various mouse scrapie strains *in vivo* by the same compound have already been demonstrated (Ishikawa *et al.*, 2004). Certainly the best cell-based test for compounds effective against human TSEs will be cells infected with human TSEs, but these are currently not available.

References

Adjou, K. T., Demaimay, R., Lasmezas, C., Deslys, J. P., Seman, M., and Dormont, D. (1995). MS-8209, a new amphotericin B derivative, provides enhanced efficacy in delaying hamster scrapie. *Antimicrob. Agents Chemother.* **39,** 2810–2812.

Aguzzi, A., Glatzel, M., Montrasio, F., Prinz, M., and Heppner, F. L. (2001). Interventional strategies against prion diseases. *Nat. Rev. Neurosci.* **2,** 745–749.

Andrews, N. J., Farrington, C. P., Ward, H. J., Cousens, S. N., Smith, P. G., Molesworth, A. M., Knight, R. S., Ironside, J. W., and Will, R. G. (2003). Deaths from variant Creutzfeldt-Jakob disease in the UK. *Lancet* **361,** 751–752.

Borchelt, D. R., Scott, M., Taraboulos, A., Stahl, N., and Prusiner, S. B. (1990). Scrapie and cellular prion proteins differ in their kinetics of synthesis and topology in cultured cells. *J. Cell Biol.* **110,** 743–752.

Brown, P. (2002). Drug therapy in human and experimental transmissible spongiform encephalopathy. *Neurology* **58,** 1720–1725.

Bueler, H., Aguzzi, A., Sailer, A., Greiner, R. A., Autenried, P., Aguet, M., and Weissmann, C. (1993). Mice devoid of PrP are resistant to scrapie. *Cell* **73,** 1339–1347.

Cashman, N. R., and Caughey, B. (2004). Prion diseases—close to effective therapy? *Nat. Rev. Drug Discov.* **3,** 874–884.

Caughey, B., and Lansbury, P. T. (2003). Protofibrils, pores, fibrils, and neurodegeneration: Separating the responsible protein aggregates from the innocent bystanders. *Annu. Rev. Neurosci.* **26,** 267–298.

Caughey, B., and Raymond, G. J. (1991). The scrapie-associated form of PrP is made from a cell surface precursor that is both protease- and phospholipase-sensitive. *J. Biol. Chem.* **266**, 18217–18223.

Caughey, B., and Raymond, G. J. (1993). Sulfated polyanion inhibition of scrapie-associated PrP accumulation in cultured cells. *J. Virol.* **67**, 643–650.

Caughey, W. S., Raymond, L. D., Horiuchi, M., and Caughey, B. (1998). Inhibition of protease-resistant prion protein formation by porphyrins and phthalocyanines. *Proc. Natl. Acad. Sci. USA* **95**, 12117–12122.

Diringer, H., and Ehlers, B. (1991). Chemoprophylaxis of scrapie in mice. *J. Gen. Virol.* **72**, 457–460.

Doh-ura, K., Ishikawa, K., Murakami-Kubo, I., Sasaki, K., Mohri, S., Race, R., and Iwaki, T. (2004). Treatment of transmissible spongiform encephalopathy by intraventricular drug infusion in animal models. *J. Virol.* **78**, 4999–5006.

Dormont, D. (2003). Approaches to prophylaxis and therapy. *Br. Med. Bull.* **66**, 281–292.

Fischer, M., Rulicke, T., Raeber, A., Sailer, A., Moser, M., Oesch, B., Brandner, S., Aguzzi, A., and Weissmann, C. (1996). Prion protein (PrP) with amino-proximal deletions restoring susceptibility of PrP knockout mice to scrapie. *EMBO J.* **15**, 1255–1264.

Ishikawa, K., Doh-ura, K., Kudo, Y., Nishida, N., Murakami-Kubo, I., Ando, Y., Sawada, T., and Iwaki, T. (2004). Amyloid imaging probes are useful for detection of prion plaques and treatment of transmissible spongiform encephalopathies. *J. Gen. Virol.* **85**, 1785–1790.

Kocisko, D. A., Baron, G. S., Rubenstein, R., Chen, J., Kuizon, S., and Caughey, B. (2003). New inhibitors of scrapie-associated prion protein formation in a library of 2000 drugs and natural products. *J. Virol.* **77**, 10288–10294.

Kocisko, D. A., Morrey, J. D., Race, R. E., Chen, J., and Caughey, B. (2004). Evaluation of new cell culture inhibitors of protease-resistant prion protein against scrapie infection in mice. *J. Gen. Virol.* **85**, 2479–2483.

Kocisko, D. A., Engel, A. L., Harbuck, K., Arnold, K. M., Olsen, E., Raymond, L. D., Vilette, D., and Caughey, B. (2005). Comparison of protease-resistant prion protein inhibitors in cell cultures infected with two strains of mouse and sheep scrapie. *Neurosci. Lett.* **388**, 106–111.

Ladogana, A., Casaccia, P., Ingrosso, L., Cibati, M., Salvatore, M., Xi, Y. G., Masullo, C., and Pocchiari, M. (1992). Sulphate polyanions prolong the incubation period of scrapie-infected hamsters. *J. Gen. Virol.* **73**, 661–665.

Ladogana, A., Liu, Q., Xi, Y. G., and Pocchiari, M. (1995). Proteinase-resistant protein in human neuroblastoma cells infected with brain material from Creutzfeldt-Jakob patient. *Lancet* **345**, 594–595.

Mange, A., Nishida, N., Milhavet, O., McMahon, H. E., Casanova, D., and Lehmann, S. (2000). Amphotericin B inhibits the generation of the scrapie isoform of the prion protein in infected cultures. *J. Virol.* **74**, 3135–3140.

May, B. C., Fafarman, A. T., Hong, S. B., Rogers, M., Deady, L. W., Prusiner, S. B., and Cohen, F. E. (2003). Potent inhibition of scrapie prion replication in cultured cells by bis-acridines. *Proc. Natl. Acad. Sci. USA* **100**, 3416–3421.

Nishida, N., Harris, D. A., Vilette, D., Laude, H., Frobert, Y., Grassi, J., Casanova, D., Milhavet, O., and Lehmann, S. (2000). Successful transmission of three mouse-adapted scrapie strains to murine neuroblastoma cell lines overexpressing wild-type mouse prion protein. *J. Virol.* **74**, 320–325.

Pocchiari, M., Schmittinger, S., and Masullo, C. (1987). Amphotericin B delays the incubation period of scrapie in intracerebrally inoculated hamsters. *J. Gen. Virol.* **68**, 219–223.

Priola, S. A., Raines, A., and Caughey, W. S. (2000). Porphyrin and phthalocyanine antiscrapie compounds. *Science* **287**, 1503–1506.

Prusiner, S. B. (1998). Prions. *Proc. Natl. Acad. Sci. USA* **95**, 13363–13383.

Race, R. E., Caughey, B., Graham, K., Ernst, D., and Chesebro, B. (1988). Analyses of frequency of infection, specific infectivity, and prion protein biosynthesis in scrapie-infected neuroblastoma cell clones. *J. Virol.* **62**, 2845–2849.

Race, R., Oldstone, M., and Chesebro, B. (2000). Entry versus blockade of brain infection after oral or intraperitoneal scrapie administration: Role of prion protein expression in peripheral nerves and spleen. *J. Virol.* **74**, 828–833.

Raymond, G. J., Olsen, E. A., Raymond, L. D., Bryant III, P. K., Lee, K. S., Baron, G. S., Caughey, W. S., Kocisko, D. A., McHolland, L. E., Favara, C., Langeveld, J. P. M., van Zijderveld, F. G., Miller, M. W., Williams, E. S., and Caughey, B. (2005). Inhibition of protease-resistant prion protein formation in a transformed deer cell line infected with chronic wasting disease. *J. Virol.* **80**, 596–604.

Sandberg, M. K., Wallen, P., Wikstrom, M. A., and Kristensson, K. (2004). Scrapie-infected GT1-1 cells show impaired function of voltage-gated N-type calcium channels (Ca(v) 2.2) which is ameliorated by quinacrine treatment. *Neurobiol. Dis.* **15**, 143–151.

Smith, P. G., and Bradley, R. (2003). Bovine spongiform encephalopathy (BSE) and its epidemiology. *Br. Med. Bull.* **66**, 185–198.

Solassol, J., Crozet, C., and Lehmann, S. (2003). Prion propagation in cultured cells. *Br. Med. Bull.* **66**, 87–97.

Stahl, N., Baldwin, M. A., Teplow, D. B., Hood, L., Gibson, B. W., Burlingame, A. L., and Prusiner, S. B. (1993). Structural studies of the scrapie prion protein using mass spectrometry and amino acid sequencing. *Biochemistry* **32**, 1991–2002.

Taraboulos, A., Serban, D., and Prusiner, S. B. (1990). Scrapie prion proteins accumulate in the cytoplasm of persistently infected cultured cells. *J. Cell Biol.* **110**, 2117–2132.

Vilette, D., Andreoletti, O., Archer, F., Madelaine, M. F., Vilotte, J. L., Lehmann, S., and Laude, H. (2001). *Ex vivo* propagation of infectious sheep scrapie agent in heterologous epithelial cells expressing ovine prion protein. *Proc. Natl. Acad. Sci. USA* **98**, 4055–4059.

Vorberg, I., Raines, A., Story, B., and Priola, S. A. (2004). Susceptibility of common fibroblast cell lines to transmissible spongiform encephalopathy agents. *J. Inf. Dis.* **189**, 431–439.

[15] A *Drosophila* Model of Alzheimer's Disease

By Damian C. Crowther, Richard Page,
Dhianjali Chandraratna, and David A. Lomas

Abstract

The development of a model of Alzheimer's disease in *Drosophila* allows us to identify and dissect pathological pathways using the most powerful genetic tools available to biology. By reconstructing essential steps in Alzheimer's pathology, such as amyloid β peptide and tau overexpression, we can observe clear and rapid phenotypes that are surrogate markers for human disease. The characterization of progressive phenotypes

METHODS IN ENZYMOLOGY, VOL. 412
0076-6879/06 $35.00
DOI: 10.1016/S0076-6879(06)12015-7

by immunohistochemistry of the brain combined with longevity, climbing, and pseudopupil assays allows the investigator to generate quantitative data. Phenotypes may be modulated by changes in gene expression as part of a genetic screen or by potential therapeutic compounds.

Introduction

A consequence of the aging Western population is the increasing burden of suffering and expense posed by Alzheimer's disease (AD) (Hoyert *et al.*, 2005). The research required to understand and intervene in the disease requires a range of model systems that each have particular benefits, ranging from cell-based models through invertebrate and vertebrate models of AD. At present, murine models of AD are predominant (see Chapter 11; German and Eisch, 2004); however, because of their limitations, not least the time and expense involved, there is an opportunity for *Drosophila* systems to provide complementary data.

What Is it About Alzheimer's Disease That We Want To Model?

Alzheimer's disease is considered a disease of old age, because young people who go on to have AD are normal and because the incidence of AD increases markedly with age (Price *et al.*, 1998). To reflect the late onset and progressive nature of AD, the ideal fly model should be normal during young adulthood and subsequently develop signs of progressive neuronal dysfunction and death. For this reason, much of the review will deal with assessing and quantifying progressive phenotypes that are observed in fly models of AD. Our ideal fly model should also be pathogenetically faithful, reflecting the mechanism and processes of neuronal dysfunction and neurodegeneration as seen in the brains of patients with AD. The two classical histological features of AD are the extracellular β-amyloid plaques (Dickson, 1997) and the intracellular neurofibrillary tangles (Braak and Braak, 1994). These insoluble protein deposits are composed primarily of the peptide amyloid-β_{1-42} (Aβ_{1-42}) (Glenner and Wong, 1984a,b) and the hyperphosphorylated tau protein, respectively (Goedert *et al.*, 1988). The tau-based pathology has been shown to correlate most closely with clinical dementia (Wilcock and Esiri, 1982); however, genetic evidence points to the production of excess Aβ_{1-42} as the first pathological step in AD (Selkoe, 2000). It is becoming clear that soluble aggregates of Aβ_{1-42} are the proximal neurotoxic species in AD (Kayed *et al.*, 2003; Lambert *et al.*, 1998; Walsh *et al.*, 2002), and because this oligomeric Aβ_{1-42} provides a distinct molecular target, we should ensure that our model generates this species. Thus, the ideal fly models of AD should force the toxic expression or metabolism of Aβ_{1-42} and tau.

The great advantage of the fruit fly is the wealth of genetic tools that can be applied to identify genes that modify the disease phenotype. Screens using chemical mutagenesis, genetic deletion kits, or mobile genetic elements (P-, EP- or GS-elements) allow the researcher to find unexpected biological pathways that are required for the disease process or are able to rescue the fly from disease (St Johnston, 2002). Screens require clear phenotypes that are easy to assess because of the need to analyze large numbers of flies. The phenotypes that are most convenient for a large genetic screen are not identical to the symptoms in human patients with AD. Thus, surrogate phenotypes such as reduced longevity or rough eyes are more useful (Crowther et al., 2005; Finelli et al., 2004; Greeve et al., 2004; Iijima et al., 2004; Wittmann et al., 2001) in the context of a large screen rather than deficits in memory (Iijima et al., 2004).

How Much of the AD Machinery Do We Need to Model in the Fly?

In the human, the toxic $A\beta_{1-42}$ peptide that accumulates in plaques is generated by the sequential cleavage of the transmembrane protein amyloid precursor protein (APP) by β- and γ-secretase (Selkoe, 2000). The fly homolog of APP (Appl) is a neurone-specific protein that may have a role in axonal transport; however, the sequence homology does not extend to the part of the protein that would constitute the $A\beta$ peptide (Link, 2005). Furthermore, despite having a functional γ-secretase homolog (Fossgreen et al., 1998) that is involved in Notch signalling (Struhl and Greenwald, 1999; Ye et al., 1999), there is no equivalent to β-secretase (Fossgreen et al., 1998). For these reasons, there is no endogenous production of $A\beta_{1-42}$, and the expression of human APP alone is not sufficient to produce a model of AD on the basis of the toxicity of $A\beta_{1-42}$. The fly also has a tau-homolog that is a nonessential neurone-specific protein (Doerflinger et al., 2003). Despite the presence of an endogenous tau, investigators have opted to overexpress human tau in the fly not least because the range of phosphorylation specific monoclonal antibodies allows greater analysis of tau processing.

How to Make a Fly Model of AD

Simple Models Based on Secreted Aβ Peptides

There is a consensus that the first step in the pathogenesis of AD is the generation of excess proaggregatory $A\beta$ peptides; in most sporadic and familial cases, this means an excess of the $A\beta_{1-42}$ peptide (Crowther, 2002). Although the subcellular compartment in which $A\beta$ is generated is unknown, it is clear that it is topologically distinct from the cytoplasm.

Candidate locations include the endoplasmic reticulum, trans-Golgi network (Chyung et al., 1997; Cook et al., 1997; Thinakaran et al., 1996; Wild-Bode et al., 1997; Xu et al., 1997) or lysosomes (Koo and Squazzo, 1994; Refolo et al., 1995) or the plasma membrane (Chyung et al., 2005). The simplest model of AD, therefore, uses a secretion signal peptide to direct the synthesis of $A\beta$ peptides as secreted peptides in the nervous system of the fly. Finelli et al. (2004) used the secretion signal peptide derived from the rat enkephalin gene (MAQFLRLCIWLLALGSCLLATVQA), and Crowther et al. (2005) used the secretion signal peptide from the fly spn43Ac gene (MASKVSILLLLTVHLLAAQTFAQ) to direct the secretion of either $A\beta_{1-42}$ (DAEFRHDSGYEVHHQKLVFFAEDVGSNK-GAIIGLMVGGVVIA), the control $A\beta_{1-40}$ (truncated by two amino acids at the C-terminus), or the familial Arctic mutant of $A\beta_{1-42}$ (E22G). The coding sequence for the constructs may be derived from human cDNA using PCR, or, alternately, the sequence may be synthesized de novo allowing optimization for insect codon usage (Protocol 1).

The plasmid pUAST is often used for the Gal-4–inducible expression of transgenes in flies (Brand and Perrimon, 1993). The multiple cloning site in pUAST contains a range of convenient restriction sites (including EcoRI, BglII, NotI, SacII, XhoI, KpnI, and XbaI) that should be included in the 5′ end of PCR primers to allow directional cloning of the various constructs. Transgenic flies are generated by microinjection of plasmid DNA into the early syncytial Drosophila embryo (Protocol 2 based on the work by Rubin and Spradling [Rubin and Spradling, 1982; Spradling and Rubin, 1982]).

Tissue-specific expression of the $A\beta$ peptides is achieved by crossing the transgenic flies with driver lines that express Gal4 in time- or tissue-specific patterns. A popular driver is the eye-specific GMR-Gal4 line that strongly drives transgene expression, particularly during development. The advantage of expressing only in the eye is that flies producing a highly toxic protein may still be viable. The methods for documenting the resulting rough eye phenotype and the quantification of the progressive degeneration of photoreceptors are covered later in this review. Another approach is to express the $A\beta$ peptides throughout the nervous system. $Elav^{c155}$-Gal-4 is a strong pan-neuronal driver that allows expression of the transgene throughout the brain and retina of the fly and allows pathological processes to be observed under the microscope and also by behavioral assays.

Modeling the Generation of Aβ from Human Amyloid Precursor Protein

A faithful model of the processing of human APP has been created by Greeve and colleagues (2004). They reconstructed the APP-processing machinery by coexpressing human APP, BACE (β-secretase) and fly

Protocol 1:   Generating the Secreted-Aβ_{1-42} Transgene

A PCR reaction was prepared with Taq polymerase and oligonucleotides 1 and 2, annealed at 60° (complementary sequence in lowercase letters) for 30 sec, extended at 72° for 30 sec, and melted at 95° for 30 sec for 30 cycles. The product was a DNA fragment containing the insect-optimized coding sequence for Aβ_{1-42} (underlined, stop codon in bold in oligonucleotide 2). Five percent of a PCR reaction was used as the template for a second reaction using oligonucleotide 2 and oligonucleotide 3 that codes for the insect-derived secretion signal peptide (sequence complementary to template in uppercase letters, start codon in bold) using the same cycling conditions.

The product of the second round PCR was cloned using the Invitrogen topo-TA kit, and 10 clones were sequenced. The use of Taq polymerase gives good yields of PCR products but the absence of a proofreading activity results in mutations in the clones.

Clones with the correct sequence were cut out of the topo-TA kit plasmid (pCR2.1) using the restriction enzymes specific for the sites at the 5' and 3' ends of oligonucleotides 2 and 3, respectively (sites in italics).

1. Aβ forward oligonucleotide:
    5'-CTCAGACCTTCGCCCAG<u>GATGCGGAATTTCGCCATGA</u>
    <u>CAGCGGCTACGAAGTGCATCATCAAAAATTGGT</u>gtttttt
    gcggaagacgtgg-3'
2. Aβ reverse oligonucleotide:
    5'-GATC*CTCGAG***TTA**CGCAATCACCACGCCGCCCACCAT
    CAAGCCAATAATCGCGCCTTTGTTCGAGCccacgtcttccgca
    aaaaac-3'
3. Secretion signal peptide oligonucleotide:
    5'–*gaattc***atg**gcgagcaaagtctcgatccttctcctgctaaccgtccatcttctggc
    tgCTCAGACCTTCGCCCAG-3'

The resulting DNA construct had the following features: *Eco*RI site—start codon—secretion signal peptide—Aβ_{1-42} coding sequence—stop codon—*Xho*I site.

---

presenilin (an essential component of γ-secretase). The APP and BACE sequences were wild type and derived from cloned human cDNA. In contrast, the endogenous presenilin activity in the fly was supplemented by overexpressing either wild-type or mutant *Drosophila* presenilin. The mutants were chosen because they were homologous to familial AD presenilin-1 mutations that increase the generation of Aβ_{1-42} in humans. Although endogenous proteases were able to generate Aβ-like peptides

Protocol 2:   Microinjection of *Drosophila* Embryos

The pUAST and pπ25.7 Δ2-3 wc plasmids were purified using silica columns as found in kits supplied by Qiagen and Promega. Twenty micrograms of the pUAST construct was mixed with 2 μg of pπ25.7 Δ2-3 wc ("helper DNA" that transiently drives transposase at the injection site) and centrifuged at top speed on a bench top centrifuge for 10 min to remove any particulate matter that might block the injection needle. The supernatant was removed, and the DNA was ethanol precipitated, washed with 70% ethanol, and redissolved in Spradling buffer (5 mM KCl, 0.1 mM phosphate buffer, pH 7.8 [Rubin and Spradling, 1982]). The DNA solution was again centrifuged at top speed for 10 min just before loading the microinjection needle. The concentration of the construct DNA was approximately 1 μg/μl.

Injections were performed using Femtotips II needles and microloaders with a transjector 5246 and an micromanipulator 5171 (all from Eppendorf). *Drosophila* embryos were harvested from colonies of white-eyed flies (w^{1118}) comprising 2000 males and females that were approximately 2–7 days old. The flies were transferred from their usual cornmeal media into large collection cages on plates containing grape-juice agar with some fresh yeast at 25°. This was done at least 1 day before collection. During the 2 h before embryo collection, the plates were changed half hourly to remove older embryos and to synchronize egg laying. During the collection period, the flies were allowed to lay for 30 min on freshly yeasted grape juice agar plates. This ensured that all embryos could be injected during the syncytial blastoderm stage, facilitating spread of the injected DNA into the future germ cells. Working in a room maintained at 18° (to slow down embryo development and increase survival), the embryos were washed off the agar plates with tap water using a paintbrush and collected into a fine meshed sieve. The sieve was placed in 50% bleach (sodium hypochlorite) for 2 min to dechlorinate the embryos. The sieve was then rinsed well in tap water to remove the bleach and blotted on tissue to remove excess water to enable easy handling of the embryos.

Working quickly for no longer than 20 min, approximately 180 embryos were lined up on a cover slip (18 × 18 mm) and immobilized on a thin strip of nontoxic double-sided adhesive tape (e.g., Scotch #665) with a gap of one embryo width between each. The embryos were placed on their sides, with the posterior pole of each embryo (the end without the micropyle) orientated toward the mounted needle (Fig. 1). The coverslips were mounted on slides, and the embryos were left to desiccate for approximately 5–15 min by placing the slides in a sealed chamber containing silica gel. Drying the embryos allows maximum uptake of DNA while reducing leakage and bursting of embryos after injection. The embryo was optimally desiccated when it lost its shiny appearance and became slightly wrinkled. After desiccation, the embryos were covered in a dense halocarbon oil (e.g., Volatalef 10S) to enable healing of the injection wound and to prevent further desiccation.

The loaded needle was mounted on the micromanipulator and visualized under the microinjection microscope. The tip of the needle was broken by pushing it against a glass coverslip to give a fine, slightly angled end. Mild positive pressure was applied using a syringe and tubing attached to the needle such that when the needle was in the oil, a small bubble of DNA solution leaks out. The DNA solution made a bubble within the oil that should be a quarter of the width of the embryo; the size of the bubble was adjusted by varying the pressure applied by the syringe. A slide of embryos was placed on the microscope stage, and the embryos were injected one-by-one by moving the microscope stage toward the tip of the needle until it inserts into the posterior pole of the embryo (20% embryo length). Successful injection was visualized as a small clear patch inside the embryo. Once all embryos were injected, those that were too old (cellularized or older embryos) or unfertilized or those that were leaking excessively after injection were discarded. The coverslip carrying the remaining embryos was then transferred to a humid grape juice agar plate with some fresh yeast and left at 18° until the larvae hatch. Care was taken to ensure the plates were left level to prevent the halocarbon oil from draining off the embryos. The wandering larvae were counted and transferred to a tube of normal fly food with yeast and cultured at 25° until adults emerge.

The hatched adults (F_0 generation) all have white eyes. Single F_0 males were backcrossed in separate vials to four virgin w^{1118} (white-eyed) females. Similarly, single F_0 virgin females were crossed to four w^{1118} males. In the subsequent F_1 offspring, flies with any shade of red, orange, or yellow eyes carry the transgene construct and were retained to allow mapping of the insert to a chromosome and for setting up individual fly lines. Ten to 20% of F_0 flies were found to be sterile, and hence more than 100 of the F_1 offspring were tested for red eyes before a cross was rejected.

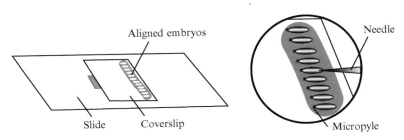

FIG. 1. One hundred and eighty fresh *Drosophila* embryos were aligned on double-sided adhesive tape on a coverslip attached to a microscope slide. The embryos were aligned such that their posterior poles were nearest the edge of the coverslip and were covered with halocarbon oil to prevent excessive dehydration. The antero-posterior orientation of the embryo was determined by observing a small protrusion (the micropyle) at the anterior pole. By moving the microscope stage, the posterior pole of each embryo was pierced by the microinjection needle. The DNA solution was seen as a patch of clear fluid within the posterior pole region of the embryo.

(by presenilin cleavage of APP with a "δ-secretase" cleavage 13 amino acids N-terminal to the normal β-secretase site), the triple transgenic flies that over-expressed APP, BACE, and *Drosophila* presenilin showed the most convincing phenotypes.

Modeling the Tau Pathology of Alzheimer's Disease

The hyperphosphorylation of tau and its subsequent aggregation to form insoluble aggregates is a pathological process common to several neurodegenerative diseases including AD (Lee *et al.*, 2001). The widespread relevance of tau pathology has prompted the generation of several fly models of tau toxicity and has prompted interest in modeling the connection between Aβ generation and subsequent tau phosphorylation and aggregation. Models have been generated in which clear morphological defects (using *C161-Gal4* and *el6E²-Gal4* [Williams *et al.*, 2000]) and subsequently axonal transport and behavioral defects have been detected by overexpressing wild-type human tau in the motor axons of *Drosophila* larvae using *D42-Gal4* (Mudher *et al.*, 2004). Clear phenotypes in the adult fly, including a rough eye, tau phosphorylation, and neurodegeneration, can be detected when a tauopathy-associated mutant of human tau (R406W) is expressed throughout the nervous system using *elav-Gal4* (Wittmann *et al.*, 2001). By coexpression of fly homologs of potentially important kinases (GSK-3β), Shulman and Feany (2003) showed that tau became abnormally phosphorylated and aggregated into neurofibrillary tangles.

Phenotype Assessment in Fly Models of AD

Although the creation of a transgenic fruit fly can be a quick process, in contrast, the detection, optimization, and quantification of the consequent phenotypes require much time and effort. We observe two broad classes of phenotypes in fly models of AD, the first are observations in the fly that are directly equivalent to changes in the human patient; in current models these would include histopathological observations and olfactory memory deficits. The second class of phenotypes consists of surrogate markers of disease. For example, in the fly model, reduced longevity is a useful measure of disease even though AD does not directly cause early death in patients. We have found that the combination of histopathology with longevity and locomotor assessment provides a good spectrum of data.

Histological Assessment of a Drosophila Model of AD

Much can be learned by following the time course of the histological changes that occur in the central nervous system of AD flies. Flies at various stages in the disease process, usually from 0–21 days of adult life (Crowther *et al.*, 2005), but in some studies up to day 60 (Greeve *et al.*, 2004), are decapitated, fixed, embedded, and sectioned. To achieve more aggressive phenotypes the flies may be cultured at 29° rather than the usual 25°; conversely, to achieve a more slowly progressive phenotype the flies may be cultured at 18°. For most applications, wax embedding and sectioning are sufficient to obtain good material for immunohistochemical analysis (Protocol 3). There is a range of monoclonal antibodies that are available for staining both Aβ peptides and Tau (Table I).

Longevity Assays

The longevity of a population of flies provides a robust estimate of their general health and as such it is a surrogate marker of AD. The advantage of a dead-versus-alive phenotype is that it is quick and unambiguous to assess; this is particularly important when considering how to implement the various screening techniques that are discussed in the following. In a screen that may require hundreds of thousands of observations, the convenience of longevity makes it a strong candidate. Furthermore, by use of Kaplan–Meier statistics, the survival of different fly populations may be compared quantitatively (Protocol 4).

Climbing and Other Locomotor Assays

Another surrogate marker of disease that may reflect neuronal dysfunction is the progressive decline in locomotor function that is observed in flies expressing Aβ peptides. The climbing assay described here allows the

---

Protocol 3: Immunohistochemical Analysis of *Drosophila* Brains

At various time points after eclosion, flies expressing Aβ peptides or tau were decapitated and the head placed in 4% w/v paraformaldehyde, 0.1 M phosphate buffer, pH 7.2–7.6, at 4° for 1–5 days before the tissue was embedded in paraffin wax and sectioned at 6 μm. Nonspecific protein binding sites were blocked by a 60-min incubation in PBS containing 10% w/v bovine serum albumin and 0.1% Triton X-100. The sections were washed with PBS and then incubated with the primary antibody for 90 min; in the case of the 4G8 monoclonal antibody, we used a concentration of 20 ng/μl. The sections were washed three times with PBS and then incubated with biotinylated anti-mouse antibody (Vector Laboratories Ltd, Peterborough, UK, 1:200 in PBS) as the secondary antibody and subsequently with avidin-HRP, developing with nickel chloride enhanced DAB as described in the kit.

TABLE I
ANTIBODIES AVAILABLE FOR DETECTING AD-RELATED TRANSGENE PRODUCTS

Antibody	Epitope	Source
19H11	N-term Aβ	Biotrend
1E11	Aβ_{1-8}	Signet Laboratories
2C8	Aβ_{1-16}	AbCam
6E10	Aβ_{1-17}	AbCam
4G8	Aβ_{17-24}	Signet Laboratories
5C3	C-terminus of Aβ_{1-40}	Biotrend
8G4	C-terminus of Aβ_{1-42}	Biotrend
12F4	C-terminus of Aβ_{1-42}	Signet Laboratories
anti-oligomer antiserum (A11)	soluble aggregates of many proteins (e.g., Aβ)	Chemicon International (Kayed *et al.*, 2003)
AT8	Tau phosphorylated at Ser 199 and Ser 202	Endogen
AT10 & AT100	Tau phosphorylated at Thr 212 and Thr 214	Innogenetics
AT180	Tau phosphorylated at Thr 231	Innogenetics
PHF-1	Tau phosphorylated at Ser 396	Albert Einstein College of Medicine, NY
pS396	Tau phosphorylated at Ser 396	Biosource International
pS404	Tau phosphorylated at Ser 404	Biosource International
Tau-1	Tau dephosphorylated at Ser 199 and Ser 202	Chemicon International
DAKO	Total Tau	Dako

Suppliers' details: Biotrend: BIOTREND Chemikalien GmbH, Germany; Signet Laboratories: Signet Laboratories Inc., MA, USA; AbCam: Abcam Ltd., UK; Endogen: Perbio Science UK Ltd., UK; Innogenetics: Autogen Bioclear UK Ltd., UK; Chemicon: CHEMICON International Inc., CA, USA; Dako: Dako Ltd., UK.

simultaneous observation of 20 flies, with each assay taking about 30 sec to perform (Crowther *et al.*, 2005). However, in most people's hands, the data are relatively noisy and, compared with longevity assays, they are time-consuming, making it less appropriate for large-scale screening of drug or genetic interventions (Protocol 5).

Locomotor deficits may also be assessed in the *Drosophila* larva, looking in particular at the velocity of crawling and the time taken for larvae to right themselves when turned over. These assays have been used primarily for determining the effects of tau expression in motor neurones in the larva (Mudher *et al.*, 2004).

Protocol 4:   Longevity Assays

Ten flies expressing the transgene of interest were incubated in 4-inch glass vials containing standard fly food and yeast at an appropriate temperature (range 18–29°). Control flies, the offspring of crossing w^{1118} flies with the Gal4 driver line, were incubated in an identical manner. The number of flies surviving was documented with a frequency that was appropriate to the lifespan of the particular flies. For short-lived flies, daily observations were made; however, for most strains, the flies were counted on days 1, 3, and 5 of a 7-day cycle. At each time point, the number of flies that were observed to die and the number of flies that were lost to follow-up (for example flies that escaped or were accidentally killed) were noted. A computer database may be required to maintain large amounts of data. When all the test and control flies were either dead or lost, the data were analyzed using Kaplan–Meier survival statistics (SPSS 11 statistics package using the log rank calculation). This analysis provides median and mean survival times for a population and determines the significance of any difference in survival times. Typically, reliable data were derived from the assessment of at least 50 flies from three or more independent crosses.

Protocol 5:   Climbing Assay

To assess climbing behavior, 20 flies were placed at the bottom of a clean 3-inch glass vial and a second identical vial was placed above. After 20 sec under red light (Kodak, GBX-2, Safelight Filter [Iijima et al., 2004]), the two vials were separated, and the number of flies at the top and bottom were counted. The proportion of Alzheimer's flies that climbed into the top vial was compared over time with the proportion of control flies (that expressed only the Gal4-elav^{c155} driver). The climbing behavior of a cohort of flies was followed periodically until all flies failed to climb on two consecutive observations.

Rough Eye Phenotype Assessment and the Pseudopupil Assay

The use of the Gal4–GMR line to drive expression of transgenes, primarily in the developing eye, provides phenotypes for two convenient assays of neurotoxicity. The first phenotype is the rough eye, a nonprogressive, developmental abnormality of the ommatidia that make up the compound eye of the fly. Toxic transgene products may produce a range of irregularities in the eye, from subtle misalignments of ommatidia through fusion of ommatidia to the most severe phenotypes. In these cases, the eye is small, there is gross irregularity of ommatidia, and failure of segments of the eye to develop or, most dramatically, the development of necrotic patches (Crowther *et al.*, 2005). The assessment of rough eye phenotypes is qualitative, and so it is important for experiments to be carried out by a single-blinded observer. The data are tabulated as the proportion of flies expressing a particular transgene that have a rough eye. When there is a range in the severity of the rough eye phenotype, it is possible to grade the phenotype and so demonstrate shifts in the spectrum of severity. The inclusion of good control flies in the experimental design is vital, because various *Gal4* driver lines, depending on the temperature of the fly culture, may exhibit a background mild rough eye that probably represents non-specific *Gal4* toxicity. For example, many eye-specific *Gal4* driver lines, when cultured at 29°, develop a rough eye phenotype that is not present when they are cultured at 18°.

The pseudopupil assay (Franceschini, 1972) provides a sensitive and quantitative method for assessing neurodegeneration, because it can show progressive neuronal loss in the eye during adult life (Jackson *et al.*, 1998). Each ommatidium of the compound eye is composed of a regular trapezoidal arrangement of seven photoreceptors, called rhabdomeres. In the presence of toxic transgene products, the regular packing of the rhabdomeres may be disturbed or there may be loss of rhabdomeres. The rhabdomeres are observed using back-illumination of the eyes on a light microscope (Protocol 6). The comparison of the average number of rhabdomeres per ommatidium across a time course for control and AD flies may allow the quantitative demonstration of neuronal loss. Again, it is important to include good control flies that account for the mild toxic effects of overexpressing *Gal4* in the retina in the absence of a transgene.

Pavlovian Olfactory Learning Assays

The Pavlovian olfactory associative learning abilities of flies can be assessed by training flies to associate an electric shock with a particular odor. After training, the flies are allowed to choose between two arms in a T maze, one that has the training odor and the other that has a neutral odor. Normal flies rapidly become averse to the training odor and move toward the control odor (Protocol 7). The loss of this associative learning, or the acceleration of forgetting, may be considered a good analog of the memory impairment seen in AD patients. Iijima and colleagues have shown that Pavlovian olfactory associative learning is impaired in flies

Protocol 6:   Pseudopupil Assay

The pseudopupil assay was performed using a light microscope with oil immersion optics. The test fly was anesthetized and decapitated, and the head was placed on a drop of nail varnish on a microscope slide. Using oil immersion, the head was back-illuminated with the aperture of the light source diaphragm set to a minimum such that the head was viewed in a dark field. The deep layers of the retina were observed by focusing down from the corneal lenses into the eye. The rhabdomeres appeared as bright spots that were counted for each ommatidium observed. Ten to 15 ommatidia were observed from each of 10 eyes, and the average rhabdomere count per ommatidium was calculated. Any particular fly head was observed for a maximum of 20 min to avoid artifactual neurodegeneration.

expressing $A\beta$ peptide that are older than 6 days. Moreover, the degree of impairment is greater in flies expressing $A\beta_{1-42}$ than $A\beta1_{1-40}$ (Iijima *et al.*, 2004).

What Can We Do with a *Drosophila* Model of AD?

Genetic Screens

The most powerful use of a *Drosophila* model of AD is the discovery of interacting genes that rescue or enhance the AD phenotypes. A genetic screen asks a simple question: what random changes to the genome of the fly can change a particular AD phenotype? We have found that the most conveniently quantifiable and statistically sound phenotype is longevity; however, the rough eye phenotype has also been used. A library of fly lines containing unique mutations can be generated in two ways: the first uses a chemical mutagen and the second makes use of mobile genetic elements (P-elements) to disrupt the *Drosophila* genome as insertional mutagens.

Most genetic screens detect dominant effects in the F1 generation caused either by loss of function of one copy of the gene or dominant gain of function. Loss of function phenotypes are seen in chemical mutagenesis, deletion kit screens, and in simple P-element screens. Gain of function phenotypes are seen more commonly in enhancer-P-element screens (EP and GS elements). Often, in chemical mutagenesis screens, the effect of homozygous mutation must be investigated (an F2 screen).

Protocol 7:   Pavlovian Olfactory Associative Learning Assays

Two training odors (0.01% v/v octanol and 0.01% v/v methylcyclohexanol) were used for the associative learning, one to be the aversive odor associated with electric shocks and the other to be the neutral odor. Flies were exposed to electric shocks (12 pulses at 60 V, duration 1.5 sec, interval 12 sec, delivered by electrifiable grids in the fly cages) in the presence of one of the odors and then were exposed to the second odor without electric shocks. Immediately after the training period, the flies were tested for associative learning by placing them in a T-maze in which one of the arms had the aversive odor and the other had the neutral odor. After 120 sec, the number of flies in the aversive and neutral arms was counted. The performance index was calculated by subtracting the number of flies in the aversive arm from the number in the neutral arm and dividing by the total number of flies and multiplying by 100 (Tully and Quinn, 1985). Scores of 70–80% were achieved for control flies throughout life; however, AD flies with learning deficits scored 30–40% (Iijima *et al.*, 2004).

To control for any innate preference for one odor over another, naive flies were placed in the T-maze, and the preference for one odor over the other was quantified by calculating the performance index. Any innate preference was used to correct the data from the associative learning paradigm. A further control was performed to verify that the flies could sense and avoid electric shocks. The flies were placed in the T-maze and electrifiable grids were placed in both arms. Only the grid in one arm of the maze delivered shocks (12 pulses at 60 V, duration 1.5 sec, interval 12 sec) after which the performance index was calculated as described previously.

Chemical Mutagenesis Screens. Random point mutations in genomic DNA can be induced by alkylating agents, the most commonly used is ethylmethane sulfate (EMS) because of its high efficacy paired with its relatively low toxicity (Greenspan, 1997). Single mutagenized males are crossed with balancer females to generate a library of stable stocks that carry mutant chromosomes. Each chromosome, there are essentially three in the fly, is tested for the ability to shorten or prolong the lifespan of the AD fly. Higher resolution mapping of mutations is facilitated by single nucleotide polymorphism (SNP) maps that allow the rapid meiotic mapping of mutations to regions of 50 kb or less (Berger *et al.*, 2001). The laborious nature of these screens can be reduced by exploiting a library of 12,000 lines of flies that carry homozygous viable EMS-induced mutations on the second and third chromosomes (Koundakjian *et al.*, 2004).

Deletion Kit Screens. A deletion kit is a library of fly lines each one of which carries a documented deletion in its DNA (Parks *et al.*, 2004; Ryder *et al.*, 2004). With modern techniques, the generation of "designer" deletions is straightforward; for example, the DrosDel collection facilitates the construction of more than 12,000 deletions all in an isogenic background (Ryder *et al.*, 2004). Crossing a chromosome carrying a deletion into the AD fly allows detection of any enhancer or suppressor activity on AD phenotypes. Initially, large deletions may be screened, and as modifiers are detected, the resolution of the screen may be increased by using progressively smaller deletions to detect candidate genes.

P-Element–Based Screens. P-elements are mobile genetic elements that, in the presence of the enzyme transposase, are able to excise themselves and re-insert, pseudo-randomly, elsewhere in the genome. To perform a screen, a library of fly lines each with a unique P-element insertion is generated by crossing a transposase-negative stock carrying a P-element, inserted on say the X chromosome, with a line that expresses transposase. Single male offspring are then crossed with w^- (white eyed) females to give male offspring that can be shown to have unique autosomal insertions of the P-element by following the red eye of the w^+ marker on the P-element. Libraries of P-element insertions are available, the most exhaustive being the BDGP gene disruption project, which has single P-element insertions associated with 40% of known genes (Bellen *et al.*, 2004). The generation of a novel library containing approximately 3000 unique P-element insertions is manageable and has the power to detect on the order of 30 modifying insertions, most of which will be loss-of-function modifiers.

A second generation P-element, or EP-element (Enhancer Promoter), contains a *Gal4-UAS* enhancer element that is responsive to the yeast transcription factor *Gal4* and can unidirectionally upregulate neighboring gene transcription (Mata *et al.*, 2000; Rorth, 1996). The GS-element (Gene Search) is a further development that possesses bidirectional *Gal4-UAS* elements that can activate transcription of neighboring genes on both sides of an insertion (Aigaki *et al.*, 2001; Toba *et al.*, 1999). The genes detected by EP- and GS-element screens will be a mixture of loss-of-function and gain-of-function modifiers.

Once modifying insertions have been detected, the location of the P-element is determined by inverse PCR (Protocol 8). The identity of the modifier gene is likely to be close to the insertion site; however, this needs to be formally demonstrated by a combination of approaches. The experimental approaches include the overexpression of candidate genes as transgenes, modifying expression levels of candidate genes using fly stocks carrying mutant alleles, making the fly hemizygous for a candidate gene

Protocol 8: Detection of the Site of P-Element Insertion Using Inverse PCR

For each GS insertion line, genomic DNA was extracted from 30 flies using the Qiagen Tissue DNeasy kit. A centrifugation step (13,000 rpm for 5 min) was incorporated into the manufacturers' protocol after the lysis step and before the first ethanol precipitation to remove fly debris. One microgram of DNA was digested with the restriction enzyme Sau3A I at 37° for 2 h (Table II) with digestion of genomic DNA being confirmed on a 1% agarose gel. Sau3A I cut in the middle of the GS element and again regularly throughout the *Drosophila* genome. The restriction enzyme was inactivated by incubation at 65° for 20 min. Digested DNA in dilute solution was then circularized using T4 DNA ligase at 14° overnight (Table III) to provide a template for the inverse PCR reaction.

Circularized DNA was ethanol precipitated by adding 40 μl 3 M sodium acetate, pH 5.4, and 1 ml of 100% ethanol and incubating on ice for 20 min. The DNA was pelleted by centrifugation at 13,000 rpm for 15 min at 4° and the pellet rinsed with 1 ml of ice-cold 70% ethanol. The DNA was then pelleted by centrifugation at 13,000 rpm for 10 min at 4° before air-drying and resuspension in 10 μl dH$_2$O.

This protocol generated two circularized DNA fragments that contain portions of the GS element. The 5' and 3' ends of the GS element and the contiguous fragment of genomic DNA were amplified using inverse PCR. The primer sequences used to amplify each end were as follows:

5p out: 5'-CTGAATAGGGAATTGGGAATTCG-3'
5p in: 5'-CTCCGTAGACGAAGCGCCTCTATTT-3'
3p out: 5'-TAATTCAAACCCCACGGACATGC-3'
3p in: 5'-GCAAAGCTTGGCTGCAGGTCGAGC-3'

Thermo-start PCR Master Mix (AbGene, UK) PCR reaction was prepared (Table IV) and performed on a PTC-200 (BioRad, UK) thermocycler (Table V). When the PCR was complete, 25 μl was visualized on a 0.8% agarose gel and the remaining 25 μl was used to generate DNA for sequencing using the Qiagen PCR Purification Kit. The resulting sequences were BLASTed on Flybase to view the site of integration (http://flybase.bio.indiana.edu/blast/ accessed September 2005).

using chromosome deletion kits or by knocking down expressing using RNAi transgenes (Reichhart *et al.*, 2002).

Drug Screens

The fly model of AD may be used for screening candidate therapeutic compounds. The most compelling advantages of a fly system are the short time taken for the appearance of disease markers, the small quantities of

TABLE II
Sau3A I Reaction

10 × BSA (supplied by New England Biolabs, UK)	2 μl
Distilled H_2O	5 μl
10× Sau3A I enzyme buffer	2 μl
Sau3A I enzyme (20 U)	1 μl
1 μg genomic DNA	10 μl

TABLE III
Circularization of the Sau3A I Fragments

Sau3A I restriction reaction	20 μl
Distilled H_2O	340 μl
10× T4 DNA ligase buffer	40 μl
T4 DNA ligase (600 U)	1.5 μl

TABLE IV
Thermo-Start PCR Reaction Mix

Thermo-start PCR Master Mix: DNA polymerase, reaction buffer, $MgCl_2$ and dNTPs	45 μl
Circularized template DNA	3 μl
Forward primer	1 μl
Reverse primer	1 μl

TABLE V
Inverse PCR Parameters

Step	Temperature and duration
1	95° for 15 min (to activate Thermo-start enzyme)
2	95° for 30 sec
3	59° for 1 min
4	72° for 2 min
5	Cycle to step 2 34–39 times
6	72° for 7 min
7	Store at 4°

test compound required, and the low cost of maintaining fly cultures. There is, however, very limited knowledge about *Drosophila* pharmacokinetics and about how drug absorption differs between the larva and the adult. It is our impression that absorption of some compounds, such as Congo red, is better in the larva. As a rule of thumb, we take the effective concentration of a test compound in cell culture experiments and then administer the drug at 5× and 50× this concentration in fly food. Instant fly food is reconstituted with a drug solution of the required concentration, and the parent flies lay their eggs onto this food; in this way, the developing embryos and larvae are exposed (Protocol 9). When the adult AD flies hatch, they are fed on standard fly food with a streak of yeast paste that contains the required concentration of drug. Cultures yielding control flies without the AD transgene are performed under the same conditions to ensure there is no nonspecific effect of the drug on fly survival. Flies treated with the drugs are compared with control AD flies that are cultured under identical conditions without the drug, but with any solvents that might be present in the treatment group. For example when drugs must be dissolved

---

Protocol 9: Assay of Drug Efficacy

Quadruplicate cultures of flies were set up for each control or treatment condition. For each vial, 20 *Elav*^c155-Gal-4 virgin females were mated with 5 males carrying the transgene of interest for 24 h at 25° on standard fly food. Fly food containing the test compound or control solvent was prepared by thoroughly mixing a 12-ml solution of the compound with 12 g of instant fly food (Philip Harris Education, UK). The reconstituted food was divided evenly and packed down into four 3-inch glass culture tubes. Four hundred microliters of distilled water and a few grains of dried yeast were added to the surface of the food. The flies were allowed to lay for 24 h at 25° before moving them to 29°. On the fourth day the parent flies were removed and 100 μl of fresh yeast solution (1 g yeast in 1 ml of distilled water) was added to each tube. The tubes, now containing developing larvae, were returned to the 29° incubator until they enclosed on days 9–12.

The progeny were collected, and the flies carrying the transgene and the driver chromosome, but not any balancer or marker chromosomes, were retained and assessed in groups of 10 for their longevity (Protocol 4). Adult flies were cultured on unyeasted standard fly food supplemented with a 50-mg streak of yeast paste containing the drug or control solvent. The yeast paste was prepared by thoroughly mixing 1 ml of drug solution, or control solvent, with 1 g dry yeast and stored at −20°.

in DMSO, it is important to include DMSO in the control food, because DMSO is lethal at 0.1% v/v, but not at 0.01% v/v.

Conclusions

The current fly models of AD, based on the expression of $A\beta$ peptides, provide a platform for discovering how toxic aggregates of these peptides result in neuronal dysfunction and death. The power of the genetic tools in *Drosophila* will allow us to identify new and unexpected pathological processes and provide clear molecular targets for drug design. The clarity and celerity of the markers of AD in the fly make both genetic and drug screens both rapid and cheap. The use of the fly for drug screening will require basic research to understand their pharmacodynamics and pharmacokinetics; however, even at this early stage, promising compounds can be identified for further investigation.

Acknowledgments

We thank Drs. Sara Imarisio, Cambridge Institute for Medical Research, University of Cambridge, and Steven Russell, Department of Genetics, University of Cambridge for their critical reading of the manuscript.

References

Aigaki, T., Ohsako, T., Toba, G., Seong, K., and Matsuo, T. (2001). The gene search system: Its application to functional genomics in Drosophila melanogaster. *J. Neurogenet.* **15,** 169–178.

Bellen, H. J., Levis, R. W., Liao, G., He, Y., Carlson, J. W., Tsang, G., Evans-Holm, M., Hiesinger, P. R., Schulze, K. L., Rubin, G. M., Hoskins, R. A., and Spradling, A. C. (2004). The BDGP gene disruption project: Single transposon insertions associated with 40% of *Drosophila* genes. *Genetics* **167,** 761–781.

Berger, J., Suzuki, T., Senti, K. A., Stubbs, J., Schaffner, G., and Dickson, B. J. (2001). Genetic mapping with SNP markers in *Drosophila*. *Nat. Genet.* **29,** 475–481.

Braak, H., and Braak, E. (1994). Morphological criteria for the recognition of Alzheimer's disease and the distribution pattern of cortical changes related to this disorder. *Neurobiol. Aging* **15,** 355–356; discussion 379–380.

Brand, A. H., and Perrimon, N. (1993). Targeted gene expression as a means of altering cell fates and generating dominant phenotypes. *Development* **118,** 401–415.

Chyung, A. S., Greenberg, B. D., Cook, D. G., Doms, R. W., and Lee, V. M. (1997). Novel beta-secretase cleavage of beta-amyloid precursor protein in the endoplasmic reticulum/intermediate compartment of NT2N cells. *J. Cell. Biol.* **138,** 671–680.

Chyung, J. H., Raper, D. M., and Selkoe, D. J. (2005). Gamma-secretase exists on the plasma membrane as an intact complex that accepts substrates and effects intramembrane cleavage. *J. Biol. Chem.* **280,** 4383–4392.

Cook, D. G., Forman, M. S., Sung, J. C., Leight, S., Kolson, D. L., Iwatsubo, T., Lee, V. M., and Doms, R. W. (1997). Alzheimer's A beta(1-42) is generated in the endoplasmic reticulum/intermediate compartment of NT2N cells. *Nat. Med.* **3,** 1021–1023.

Crowther, D. C. (2002). Familial conformational diseases and dementias. *Hum. Mutat.* **20,** 1–14.

Crowther, D. C., Kinghorn, K. J., Miranda, E., Page, R., Curry, J. A., Duthie, F. A., Gubb, D. C., and Lomas, D. A. (2005). Intraneuronal Ab, non-amyloid aggregates and neurodegeneration in a *Drosophila* model of Alzheimer's disease. *Neuroscience* **132,** 123–135.

Dickson, D. W. (1997). The pathogenesis of senile plaques. *J. Neuropathol. Exp. Neurol.* **56,** 321–339.

Doerflinger, H., Benton, R., Shulman, J. M., and St Johnston, D. (2003). The role of PAR-1 in regulating the polarised microtubule cytoskeleton in the *Drosophila* follicular epithelium. *Development* **130,** 3965–3975.

Finelli, A., Kelkar, A., Song, H. J., Yang, H., and Konsolaki, M. (2004). A model for studying Alzheimer's Abeta42-induced toxicity in *Drosophila* melanogaster. *Mol. Cell Neurosci.* **26,** 365–375.

Fossgreen, A., Brückner, B., Czech, C., Masters, C. L., Beyreuther, K., and Paro, R. (1998). Transgenic *Drosophila* expressing human amyloid precursor protein show g-secretase activity and blistered wing phenotype. *Proc. Natl. Acad. Sci. USA* **95,** 13703–13708.

Franceschini, N. (1972). Pupil and pseudopupil in the compound eye of *Drosophila*. *In* "Information Processing in the Visual System of *Drosophila*" (R. Wehner, ed.), pp. 75–82. Springer, Berlin.

German, D. C., and Eisch, A. J. (2004). Mouse models of Alzheimer's disease: Insight into treatment. *Rev. Neurosci.* **15,** 353–369.

Glenner, G. G., and Wong, C. W. (1984a). Alzheimer's disease and Down's syndrome: Sharing of a unique cerebrovascular amyloid fibril protein. *Biochem. Biophys. Res. Commun.* **122,** 1131–1135.

Glenner, G. G., and Wong, C. W. (1984b). Alzheimer's disease: Initial report of the purification and characterization of a novel cerebrovascular amyloid protein. *Biochem. Biophys. Res. Commun.* **120,** 885–890.

Goedert, M., Wischik, C. M., Crowther, R. A., Walker, J. E., and Klug, A. (1988). Cloning and sequencing of the cDNA encoding a core protein of the paired helical filament of Alzheimer disease: Identification as the microtubule-associated protein tau. *Proc. Natl. Acad. Sci. USA* **85,** 4051–4055.

Greenspan, R. (1997). "Fly Pushing: The Theory and Practice of *Drosophila* Genetics." Cold Spring Harbor Laboratory Press, New York.

Greeve, I., Kretzschmar, D., Tschape, J. A., Beyn, A., Brellinger, C., Schweizer, M., Nitsch, R. M., and Reifegerste, R. (2004). Age-dependent neurodegeneration and Alzheimer-amyloid plaque formation in transgenic *Drosophila*. *J. Neurosci.* **24,** 3899–3906.

Hoyert, D. L., Kung, H. C., and Smith, B. L. (2005). Deaths: Preliminary data for 2003. *Natl. Vital Stat. Rep.* **53,** 1–48.

Iijima, K., Liu, H. P., Chiang, A. S., Hearn, S. A., Konsolaki, M., and Zhong, Y. (2004). Dissecting the pathological effects of human A{beta}40 and A{beta}42 in *Drosophila*: A potential model for Alzheimer's disease. *Proc. Natl. Acad. Sci. USA* **101,** 6623–6628.

Jackson, G. R., Salecker, I., Dong, X., Yao, X., Arnheim, N., Faber, P. W., MacDonald, M. E., and Zipursky, S. L. (1998). Polyglutamine-expanded human huntingtin transgenes induce degeneration of *Drosophila* photoreceptor neurons. *Neuron* **21,** 633–642.

Kayed, R., Head, E., Thompson, J. L., McIntire, T. M., Milton, S. C., Cotman, C. W., and Glabe, C. G. (2003). Common structure of soluble amyloid oligomers implies common mechanism of pathogenesis. *Science* **300,** 486–489.

Koo, E. H., and Squazzo, S. L. (1994). Evidence that production and release of amyloid beta-protein involves the endocytic pathway. *J. Biol. Chem.* **269,** 17386–17389.

Koundakjian, E. J., Cowan, D. M., Hardy, R. W., and Becker, A. H. (2004). The Zuker collection: A resource for the analysis of autosomal gene function in *Drosophila melanogaster*. *Genetics* **167,** 203–206.

Lambert, M. P., Barlow, A. K., Chromy, B. A., Edwards, C., Freed, R., Liosatos, M., Morgan, T. E., Rozovsky, I., Trommer, B., Viola, K. L., Wals, P., Zhang, C., Finch, C. E., Krafft, G. A., and Klein, W. L. (1998). Diffusible, nonfibrillar ligands derived from Abeta1-42 are potent central nervous system neurotoxins. *Proc. Natl. Acad. Sci. USA* **95,** 6448–6453.

Lee, V. M., Goedert, M., and Trojanowski, J. Q. (2001). Neurodegenerative tauopathies. *Annu. Rev. Neurosci.* **24,** 1121–1159.

Link, C. D. (2005). Invertebrate models of Alzheimer's disease. *Genes Brain Behav.* **4,** 147–156.

Mata, J., Curado, S., Ephrussi, A., and Rorth, P. (2000). Tribbles coordinates mitosis and morphogenesis in *Drosophila* by regulating string/CDC25 proteolysis. *Cell* **101,** 511–522.

Mudher, A., Shepherd, D., Newman, T. A., Mildren, P., Jukes, J. P., Squire, A., Mears, A., Berg, S., MacKay, D., Asuni, A. A., Bhat, R., and Lovestone, S. (2004). GSK-3beta inhibition reverses axonal transport defects and behavioural phenotypes in *Drosophila*. *Mol. Psychiatry* **9,** 522–530.

Parks, A. L., Cook, K. R., Belvin, M., Dompe, N. A., Fawcett, R., Huppert, K., Tan, L. R., Winter, C. G., Bogart, K. P., Deal, J. E., Deal-Herr, M. E., Grant, D., Marcinko, M., Miyazaki, W. Y., Robertson, S., Shaw, K. J., Tabios, M., Vysotskaia, V., Zhao, L., Andrade, R. S., Edgar, K. A., Howie, E., Killpack, K., Milash, B., Norton, A., Thao, D., Whittaker, K., Winner, M. A., Friedman, L., Margolis, J., Singer, M. A., Kopczynski, C., Curtis, D., Kaufman, T. C., Plowman, G. D., Duyk, G., and Francis-Lang, H. L. (2004). Systematic generation of high-resolution deletion coverage of the *Drosophila melanogaster* genome. *Nat. Genet.* **36,** 288–292.

Price, D. L., Tanzi, R. E., Borchelt, D. R., and Sisodia, S. S. (1998). Alzheimer's disease: Genetic studies and transgenic models. *Annu. Rev. Genet.* **32,** 461–493.

Refolo, L. M., Sambamurti, K., Efthimiopoulos, S., Pappolla, M. A., and Robakis, N. K. (1995). Evidence that secretase cleavage of cell surface Alzheimer amyloid precursor occurs after normal endocytic internalization. *J. Neurosci. Res.* **40,** 694–706.

Reichhart, J. M., Ligoxygakis, P., Naitza, S., Woerfel, G., Imler, J. L., and Gubb, D. (2002). Splice-activated UAS hairpin vector gives complete RNAi knockout of single or double target transcripts in *Drosophila melanogaster*. *Genesis* **34,** 160–164.

Rorth, P. (1996). A modular misexpression screen in *Drosophila* detecting tissue-specific phenotypes. *Proc. Natl. Acad. Sci. USA* **93,** 12418–12422.

Rubin, G. M., and Spradling, A. C. (1982). Genetic transformation of *Drosophila* with transposable element vectors. *Science* **218,** 348–353.

Ryder, E., Blows, F., Ashburner, M., Bautista-Llacer, R., Coulson, D., Drummond, J., Webster, J., Gubb, D., Gunton, N., Johnson, G., O'Kane, C. J., Huen, D., Sharma, P., Asztalos, Z., Baisch, H., Schulze, J., Kube, M., Kittlaus, K., Reuter, G., Maroy, P., Szidonya, J., Rasmuson-Lestander, A., Ekstrom, K., Dickson, B., Hugentobler, C., Stocker, H., Hafen, E., Lepesant, J. A., Pflugfelder, G., Heisenberg, M., Mechler, B., Serras, F., Corominas, M., Schneuwly, S., Preat, T., Roote, J., and Russell, S. (2004). The DrosDel collection: A set of P-element insertions for generating custom chromosomal aberrations in *Drosophila melanogaster*. *Genetics* **167,** 797–813.

Selkoe, D. J. (2000). Toward a comprehensive theory for Alzheimer's disease. Hypothesis: Alzheimer's disease is caused by the cerebral accumulation and cytotoxicity of amyloid beta-protein. *Ann. N. Y. Acad. Sci.* **924,** 17–25.

Shulman, J. M., and Feany, M. B. (2003). Genetic modifiers of tauopathy in *Drosophila*. *Genetics* **165**, 1233–1242.

Spradling, A. C., and Rubin, G. M. (1982). Transposition of cloned P elements into *Drosophila* germ line chromosomes. *Science* **218**, 341–347.

St Johnston, D. (2002). The art and design of genetic screens: *Drosophila melanogaster*. *Nat. Rev. Genet.* **3**, 176–188.

Struhl, G., and Greenwald, I. (1999). Presenilin is required for activity and nuclear access of Notch in *Drosophila*. *Nature* **398**, 522–525.

Thinakaran, G., Teplow, D. B., Siman, R., Greenberg, B., and Sisodia, S. S. (1996). Metabolism of the "Swedish" amyloid precursor protein variant in neuro2a (N2a) cells. Evidence that cleavage at the "beta-secretase" site occurs in the Golgi apparatus. *J. Biol. Chem.* **271**, 9390–9397.

Toba, G., Ohsako, T., Miyata, N., Ohtsuka, T., Seong, K. H., and Aigaki, T. (1999). The gene search system. A method for efficient detection and rapid molecular identification of genes in *Drosophila melanogaster*. *Genetics* **151**, 725–737.

Tully, T., and Quinn, W. G. (1985). Classical conditioning and retention in normal and mutant *Drosophila melanogaster*. *J. Comp. Physiol. [A]* **157**, 263–277.

Walsh, D. M., Klyubin, I., Fadeeva, J. V., Cullen, W. K., Anwyl, R., Wolfe, M. S., Rowan, M. J., and Selkoe, D. J. (2002). Naturally secreted oligomers of amyloid beta protein potently inhibit hippocampal long-term potentiation *in vivo*. *Nature* **416**, 535–539.

Wilcock, G. K., and Esiri, M. M. (1982). Plaques, tangles and dementia. A quantitative study. *J. Neurol. Sci.* **56**, 343–356.

Wild-Bode, C., Yamazaki, T., Capell, A., Leimer, U., Steiner, H., Ihara, Y., and Haass, C. (1997). Intracellular generation and accumulation of amyloid beta-peptide terminating at amino acid 42. *J. Biol. Chem.* **272**, 16085–16088.

Williams, D. W., Tyrer, M., and Shepherd, D. (2000). Tau and tau reporters disrupt central projections of sensory neurons in *Drosophila*. *J. Comp. Neurol.* **428**, 630–640.

Wittmann, C. W., Wszolek, M. F., Shulman, J. M., Salvaterra, P. M., Lewis, J., Hutton, M., and Feany, M. B. (2001). Tauopathy in *Drosophila*: Neurodegeneration without neurofibrillary tangles. *Science* **293**, 711–714.

Xu, H., Sweeney, D., Wang, R., Thinakaran, G., Lo, A. C., Sisodia, S. S., Greengard, P., and Gandy, S. (1997). Generation of Alzheimer beta-amyloid protein in the trans-Golgi network in the apparent absence of vesicle formation. *Proc. Natl. Acad. Sci. USA* **94**, 3748–3752.

Ye, Y., Lukinova, N., and Fortini, M. E. (1999). Neurogenic phenotypes and altered Notch processing in *Drosophila* presenilin mutants. *Nature* **398**, 525–529.

[16]   Modeling Polyglutamine Pathogenesis in *C. elegans*

By HEATHER R. BRIGNULL*, JAMES F. MORLEY*,
SUSANA M. GARCIA, and RICHARD I. MORIMOTO

Abstract

A growing number of human neurodegenerative diseases are associated with disruption of cellular protein folding homeostasis, leading to the appearance of misfolded proteins and deposition of protein aggregates and inclusions. Recent years have been witness to widespread development of invertebrate systems (specifically *Drosophila* and *Caenorhabditis elegans*) to model these disorders, bringing the many advantages of such systems, particularly the power of genetic analysis in a metazoan, to bear on these problems. In this chapter, we describe our studies using the nematode, *C. elegans*, as a model to study polyglutamine expansions as occur in Huntington's disease and related ataxias. Using fluorescently tagged polyglutamine repeats of different lengths, we have examined the dynamics of aggregate formation both within individual cells and over time throughout the lifetime of individual organisms, identifying aging as an important physiological determinant of aggregation and toxicity. Expanding on these observations, we demonstrate that a genetic pathway regulating longevity can alter the time course of aging-related polyglutamine-mediated phenotypes. To identify novel targets and better understand how cells sense and respond to the appearance of misfolded and aggregation-prone proteins, we use a genome-wide RNA interference-based genetic screen to identify modifiers of age-dependent polyglutamine aggregation. Throughout these studies, we used fluorescence-based, live-cell biological and biophysical methods to study the behavior of these proteins in a complex multicellular environment.

Introduction

Misfolded proteins, aggregates, and inclusion bodies are hallmarks of a range of neurodegenerative disorders including Alzheimer's disease (AD), Parkinson's disease (PD), prion disorders, amyotrophic lateral sclerosis (ALS), and polyglutamine diseases, which include Huntington's disease

---

* These authors contributed equally to this work.

METHODS IN ENZYMOLOGY, VOL. 412
Copyright 2006, Elsevier Inc. All rights reserved.

0076-6879/06 $35.00
DOI: 10.1016/S0076-6879(06)12016-9

(HD) and related ataxias (Kakizuka, 1998; Kopito and Ron, 2000; Stefani and Dobson, 2003). Each of these disorders exhibits aging-dependent onset, selective neuronal vulnerability despite widespread expression of the related proteins, and a progressive, usually fatal, clinical course. Deposition of intracellular or extracellular protein aggregates is a well-conserved pathological feature and has been the focus of extensive investigation, as described later (Stefani and Dobson, 2003). Despite differences in the underlying genes involved, inheritance, and clinical presentation, the similarities observed have led to the idea of shared pathogenic mechanisms and the hope that insights into one process may be generalized to others.

In support of this premise is growing evidence that the cellular protein quality control system seems to be an underlying common denominator of these diseases (Dobson, 2001). For example, genes involved in protein folding and degradation, including molecular chaperones and components of the proteasome, have been shown to modulate onset, development, and progression in models of multiple neurodegenerative diseases (Bonini, 2002; Chan *et al.*, 2002; Cummings *et al.*, 1998). Furthermore, it has been suggested that despite the absence of sequence homology, different disease-related proteins share a common ability to adopt similar proteotoxic conformations (Kayed *et al.*, 2003; O'Nuallain and Wetzel, 2002) and that these might be used as therapeutic targets.

Invertebrate Models of Neurodegenerative Disease

Some of these disorders, including the polyglutamine diseases, exhibit clear familial inheritance, allowing the use of genetic studies and positional cloning to identify single gene alterations underlying the disorders (Kawaguchi *et al.*, 1994; Koide *et al.*, 1994; La Spada *et al.*, 1991; Orr *et al.*, 1993). Other diseases are more commonly sporadic, but rare familial forms have allowed the identification of candidate genes that could reveal insights into pathology. These include mutations of amyloid precursor protein in AD, parkin and α-synuclein in PD, and superoxide dismutase in ALS (Kamino *et al.*, 1992; Laing and Siddique, 1997; Lucking *et al.*, 2000; Mizuno *et al.*, 2001; Polymeropoulos *et al.*, 1997; Rosen *et al.*, 1993). Identification of these genes has led to an explosion of models to investigate the underlying pathology, identify factors and pathways that modify the disease process, and test potential therapeutic interventions.

Mouse and cell culture models continue to be mainstays to complement clinical study of these disorders for their complexity and simplicity, respectively. Recently, invertebrate models, in particular *Drosophila* and *C. elegans,* have been used to great advantage in the study of neurodegenerative disease (Driscoll and Gerstbrein, 2003; Link, 2001; Thompson and Marsh, 2003;

Westlund *et al.*, 2004). As described in more detail later for *C. elegans*, these systems represent an attractive intermediate by combining sufficient complexity to allow investigation of both cellular and behavioral phenotypes with simplicity that facilitates rapid, high-throughput testing of hypotheses.

Pathogenic mechanisms in neurodegeneration often involve a gain of function toxicity allowing these disorders to be modeled by transgenic over-expression of human disease–related proteins regardless of the presence or absence of a clear ortholog. For example, expression of polyglutamine-containing proteins is neurotoxic in both *Drosophila* retinal neurons and *C. elegans* chemosensory or mechanosensory neurons despite the absence of clear disease gene orthologs (Faber *et al.*, 1999, 2002; Parker *et al.*, 2001; Warrick *et al.*, 1998). Similar strategies have been used to examine the toxicity of APP, SOD, and α-synuclein in flies and worms (Feany and Bender, 2000; Lee *et al.*, 2005; Link, 1995, 2001; Oeda *et al.*, 2001; Shulman *et al.*, 2003; Wexler *et al.*, 2004). However, it is clear that no single model recapitulates all features of a given disease. A discrepancy common with many models is that disease-related proteins are toxic in most cell types where they have been expressed without appropriate tissue specificity. Despite the idiosyncrasies of different models, each provides unique insights that clearly validate the general approach. Presumably, features observed across disparate models, with different inherent assumptions and sources of variability, are most likely to represent key events underlying pathology.

C. elegans Models of Polyglutamine Disease

We have focused our studies largely on polyglutamine expansions as occur in Huntington's disease and related movement disorders including several spinocerebellar ataxias and Kennedy's disease (Orr, 2001; Ross, 2002; Trottier *et al.*, 1995; Zoghbi and Orr, 2000). Expression of expanded polyglutamine, with or without flanking sequences from the endogenous proteins, or when inserted into an unrelated protein, is sufficient to recapitulate pathological features of the diseases in multiple model systems (Davies *et al.*, 1997; Mangiarini *et al.*, 1996; Ordway *et al.*, 1997). This suggests that polyglutamine expansions play a central role in these disorders and supports an approach in which expression of isolated polyglutamine expansions without flanking sequences could lend insight into shared features of these disorders.

A novel feature of the polyglutamine diseases is polymorphism of the repeat length among individuals, with an apparent pathogenic threshold of 35–40 residues. Molecular genetic studies have established that Huntingtin alleles from normal chromosomes contain fewer than 30–34 CAG repeats, whereas those from affected chromosomes contain greater than 35–40 repeats (Andrew *et al.*, 1993). Analysis of patient databases has established a strong inverse correlation between repeat length and age of onset (Andrew *et al.*, 1993; Brinkman *et al.*, 1997). Similar breakpoints and length dependence are seen for the other polyglutamine repeat diseases suggesting a 35–40 residue threshold at which the disease gene products are converted to a proteotoxic state.

Here, we describe our experience using the nematode, *C. elegans*, as a model to study polyglutamine expansions. *C. elegans* is a roundworm that in its free-living form can be found in a variety of soil habitats throughout the world. In the laboratory, the animals can be readily cultured in large numbers on agar plates seeded with a lawn of *E. coli*, as wild-type adult animals reach a maximum length of approximately 1–1.5 mm. The adult stage is preceded by progression through embryonic development and four larval stages. This life cycle is completed in approximately 3 days under typical growth conditions (20–25°). Taken together with an adult lifespan of approximately 2 weeks under normal conditions, this allows experiments to be designed and carried out quite rapidly compared with other metazoan model systems.

A self-fertilizing hermaphrodite form comprises greater than 99% of *C. elegans* in normal populations, although males arise at a low frequency of ~0.01% (owing to meiotic nondisjunction) allowing construction and maintenance of genetic stocks. The hermaphrodite body plan is relatively simple, composed of only 959 somatic cells. However, this small number of cells is sufficient for the formation of multiple complex tissues types including intestine, muscle, hypodermis, and a fully differentiated nervous system. Thus, despite its simplicity and ease of handling in the laboratory, studies in *C. elegans* can offer insight into processes unique to complex multicellular organisms. In addition, the *C. elegans* system has well-established techniques for forward and reverse genetics, a wide range of mutants, and a fully sequenced genome, all of which enhance the usefulness of this model. We highly recommend two monographs published by Cold Spring Harbor Laboratory Press titled "The Nematode *C. elegans*" and "*C. elegans* II" for further information on *C. elegans*. Additional information and updates on *C. elegans* are available online as "WormBook" (http://www.wormbook.org).

In the studies described here, we have examined the fate of aggregation-prone polyglutamine proteins in a living organism and, in particular, the potential influence of physiological and genetic modifiers on these processes. For example, we wished to understand the relationship between aging and the ability of organisms to sense and respond to the appearance of misfolded and aggregation-prone proteins. In addition, we studied the expression of polyglutamine proteins in different tissue types, namely muscle cells and neurons, to better understand the effect of cell-type specific factors.

Using genome-wide RNA interference (RNAi) analysis in *C. elegans*, we sought to identify the comprehensive set of proteins or complexes that modulate the cellular protein folding environment, which we refer to as the "protein folding buffer." Throughout these studies, we used powerful fluorescence-based cell biological and biophysical methods (described here in a series of detailed protocols) to examine the behavior of these proteins in a complex multicellular environment.

Visualization of Protein in a Live, Multicellular Organism

Previous studies *in vitro* had demonstrated that polyglutamine peptides exhibited nucleation-dependent aggregation kinetics proportional to repeat length (Chen *et al.*, 2001; Scherzinger *et al.*, 1997), reminiscent of the inverse correlation between polyglutamine length and age-at-onset observed in the human diseases (Chen *et al.*, 2002). However, studies on polyglutamine-mediated toxicity in animal models had examined only the extreme ends of the spectrum comparing the effects of very short (<30Q) or very long (>60Q) repeats. We chose to examine the distinctive properties of polyglutamine proteins at the apparent pathogenic threshold (Q30–Q40) in the crowded macromolecular environment of the cell.

We generated a series of transgenic animals expressing polyglutamine repeats of different lengths in body-wall muscle cells, with greater resolution near the apparent threshold (Q0, Q19, Q29, Q33, Q35, Q40, Q44, Q64, and Q82) fused to YFP. In young adult animals (3–4 days old), we found a clear length-dependent shift in the subcellular localization of the proteins, because repeat lengths of Q35 or less were localized diffusely as soluble proteins (Fig. 1A), but repeat lengths of Q40 and greater were localized in aggregates (Fig. 1A).

To determine whether the properties of polyglutamine proteins observed in muscle cells were tissue specific, we expressed a range of polyglutamine proteins in the nervous system of *C. elegans*. Polyglutamine proteins in *C. elegans* neurons displayed Q-length–dependent changes in protein localization similar to that observed in muscle cells. Expression of YFP alone or YFP fused to Q19, Q35, or Q40 resulted in a soluble distribution pattern (Fig. 1B). The distribution of Q67::YFP or Q86::YFP in the nervous system was distinctly different from smaller glutamine tracts, with the protein localized to discreet foci (Fig. 1B). Although the trend of length-dependent changes in the localization of polyglutamine proteins is consistent between muscle and neurons in *C. elegans*, the aggregation threshold seems to be lower in muscle cells at this level of resolution. The use of additional assays described here, including Fluorescence Recovery After Photobleaching (FRAP), reveals that Q40 is also the threshold for changes

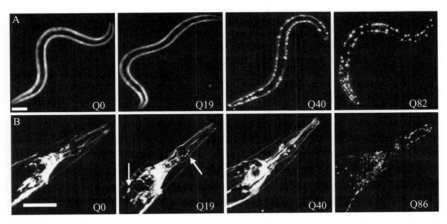

FIG. 1. Length-dependent aggregation of polyglutamine::YFP fusion proteins in *C. elegans*. (A) Fluorescent images of 3–4-day-old *C. elegans* expressing different lengths of polyglutamine::YFP in body-wall muscle cells. Bar = 0.1 mm. (B) Fluorescent images of *C. elegans* expressing different lengths of polyglutamine::YFP throughout the nervous system of 4-day-old animals. Images are flattened, confocal z-stacks of *C. elegans* head, scale bar = 50 μm. Neuronal cell bodies are significantly smaller than body-wall muscle cells and foci, such as those indicated by the arrows in Q19 (B) are cell bodies. In both tissue types, expressing YFP or Q19::YFP results in diffuse fluorescence in all expressing cells, whereas animals expressing >80 glutamine repeats exhibited distribution into fluorescent foci. Animals expressing Q40::YFP in muscle cells showed a polymorphic phenotype with both diffuse and punctate fluorescence distribution (A), whereas no changes in Q40::YFP distribution are visible in the neuronal model at this level of microscopic resolution (B).

in the properties of polyglutamine proteins in neurons as well as in muscles. Together, these models establish a system allowing the simultaneous visualization of aggregation and toxicity in multiple tissue types in a single organism, allowing further studies to identify physiological and genetic modifiers.

Generating Transgenic *C. elegans–Expressing Polyglutamine Proteins*

Plasmid Construction. Construction of polyglutamine vectors for expression in muscle cells was as described previously (Morley *et al.*, 2002). Pan-neuronal polyglutamine expression was achieved by cloning 3.5 kb of DNA upstream of the *C. elegans* gene, F25B3.3, along with the first five codons with a point mutation of the start codon to ATC (gift of Dr. David Pilgrim). This fragment was cloned into pPD95.79 (gift of Dr. Andrew Fire) between the *Pst1* and *BamH1* sites generating $P_{F25B3.3}$. Glutamine tracts were established by sequencing and inserted into a second vector containing CFP (pECFP-N1, Clonetech) between the *KpnI* and *BamH1* sites. Polyglutamine::CFP fragments were generated by addition of a 3′ *Bsm1* site following Q(n)CFP in pECFP-N1 by PCR amplification. The resulting Q(n)CFP fragments and $P_{F25B3.3}$ were digested with *Kpn1/Bsm1* and ligated. Constructs containing polyglutamine expansions labeled with YFP were generated by removing CFP from $P_{F25B3.3}Q(n)$::CFP and inserting YFP from pEYFP-N1 (Clonetech) between *Age1* and *Not1* sites.

Generation of Transgenic Strains. Nematodes were raised and handled using standard methods unless otherwise noted (Brenner, 1974). Generation of transgenic animals expressing polyglutamine proteins in muscle cells was as described previously (Morley *et al.*, 2002). For generation of transgenic animals expressing pan-neuronal polyglutamine, DNA encoding $P_{F25B3.3}Q(0, 19, 29, 40, 67, 86)$::CFP or YFP was injected into wild-type animals (N2) at 50 ng/μl. Of note, in *C. elegans*, heritable transgenes result from the assembly of injected DNAs into large extrachromosomal arrays (Mello *et al.*, 1991). Depending on copy number and other factors, expression of the gene of interest may vary somewhat in independent lines; therefore, it is important to isolate and analyze multiple lines for a given transgene to ensure homogeneity of phenotype. This is also important to ensure that any phenotypes are not the result solely of transgene expression levels, although we suggest using a biochemical approach, such as Western blots, to compare protein expression levels between transgenic lines. Finally, it is of critical importance that transgenic lines are frozen as soon as possible after they have been generated. We have observed that animals kept in continuous culture, especially those expressing large, toxic glutamine tracts, eventually adapt to the transgene. This has been more evident in various behavioral assays used to establish toxicity rather than aggregation. The time required for animals to adapt varies widely depending on the transgene and various stressors, so we cannot recommend how long lines should be kept in culture. However, we do suggest thawing lines before any new or critical assay.

The Use of Fluorescence Imaging in C. elegans

High-Resolution Imaging in C. elegans. Animals used for live imaging are mounted on a glass slide with a pad of 3% agarose. The agarose pad provides a "cushion" for the animals to sink into when the coverslip is added and prevents them from being crushed, allowing individual animals to be repeatedly imaged over time. Animals are immobilized by 1 mM levamisole, an acetylcholine agonist that causes permanent contraction of muscle cells. It is important to note that although levamisole paralyzes the body-wall muscle cells of *C. elegans*, animals should continue to display erratic pharyngeal pumping, indicating that they have survived the mounting process.

Imaging and fluorescence-based biophysical experiments presented here were performed on a Zeiss LSM, 510 Meta confocal microscope. For *C. elegans*, a 63× water objective with a 1.4 numerical aperture was used. Because animals are mounted in agarose, a water objective is preferable to an oil immersion objective. For optimal imaging, animals should be well spaced on the slide and the coverslip carefully sealed to prevent the agarose pad from drying out and desiccating the animal during imaging.

Aggregate Quantification. Animals expressing polyglutamine aggregates in muscle cells were viewed at 100× magnification using a stereomicroscope equipped for fluorescence, and the number of polyglutamine aggregates was counted. Aggregates were defined as discrete structures with boundaries distinguishable from surrounding fluorescence on all sides. Aggregate size in muscle cells, when measured using confocal microscopy, typically ranged from 1×5 μm. At 100× magnification, we were able to detect >80% of aggregates observable at higher magnifications. Repeated aggregate counts by the same observer and independent observers in blinded analyses varied by less than 10%. Polyglutamine aggregate size varies depending on tissue type, and aggregates in neurons were smaller (typically less than 2 μm in diameter), because their growth is limited by the small size of neurons relative to body-wall muscle cells.

Biophysical Analysis of Polyglutamine Proteins in C. elegans

A clear advantage provided by this system was our ability to readily visualize aggregates in an intact organism. This allowed us to apply fluorescence-based cell biological techniques to examine the properties of polyglutamine proteins in the living animal. For example, we used Fluorescence Recovery After Photobleaching (FRAP) to determine whether the shift in polyglutmine::YFP cellular distribution corresponded to a transition of these proteins from a soluble to an aggregated state. FRAP and other biophysical techniques can be adapted for *C. elegans* with very few changes to the processes established for studying cultured cells.

Fluorescence Recovery After Photobleaching. One of the most critical requirements for FRAP is a good sample; in this case a well-mounted animal exhibits minimal movement and provides access to the appropriate cells. The process of mounting *C. elegans* for microscopy was described previously; however it is often difficult to find animals that are sufficiently still and in the correct orientation. Therefore, we recommend preparing at least two slides with ~5–10 animals mounted on each slide. To minimize movement, animals should not be mounted immediately before FRAP experiments. We typically allow at least 30 min before experiments for maximal effectiveness of levamisole treatment.

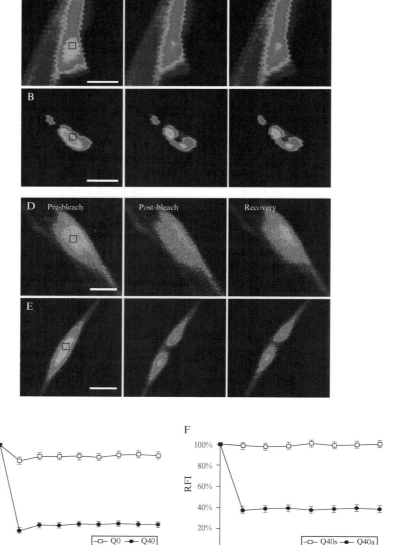

FIG. 2. FRAP analysis of polyglutamine::YFP solubility in living animals. FRAP distinguishes between Q0::YFP (A) and a Q40::YFP foci (B) in muscle cells. Images presented are representative and include an image before photobleaching (pre-bleach) of the boxed area (black box), immediately after photobleaching (post-bleach) and after a recovery period (time

The data presented here were generated by FRAP analysis on animals expressing various lengths of polyglutamine proteins fused to a C-terminal YFP. Therefore, both imaging and photobleaching used a laser with a wavelength of 514 nm, specific for YFP. Excitation for imaging purposes was performed at 0.1% laser power transmission. This low percentage of transmission was selected for imaging to minimize photobleaching while still providing sufficient image quality for data analysis. Photobleaching used 100% transmission of the 514 nm laser over a region of interest (ROI) with iterations sufficient to photobleach the sample. The size of the ROI, number of photobleaching iterations and the imaging transmission should be adjusted on the basis of the specific properties of the sample. For example, *C. elegans* muscle cells are much larger than neurons, and, consequently, polyglutamine aggregates become larger in muscle cells than in neurons, as indicated previously. Therefore, in muscle cells, the ROI was fixed at an area of 3.6 μm^2 (25 × 25 pixels), whereas in neurons, the ROI was 10 × 10 pixels. Once these parameters are determined, they must remain consistent between experiments for data analysis.

Images for analysis should be collected before and after photobleaching. The duration of the recovery period is dependent on the protein of interest but should continue until fluorescent recovery has plateaued. In the data presented in Fig. 2, the ROI was bleached for 4 sec (five iterations at 100% laser power), after which an image was collected every 30 sec for up to 5 min.

FRAP Data Analysis. Relative fluorescence intensity (RFI) was determined using the equation: $RFI = (T_t/C_t)/(T_0/C_0)$. T_0 indicates the intensity of fluorescence in the "test" ROI before photobleaching, and T_t is intensity in the same area at a given time after bleaching. Data should be normalized against an unbleached area in the same cell as a control for photobleaching and autofluorescence. Therefore, C_0 is the fluorescence intensity of a control area before bleaching and C_t the same area at a given time after bleaching (Phair and Misteli, 2000). In choosing the control area, it is better to pick an area within the same cell being bleached. This is especially important for

---

indicated in accompanying graph). As expected, Q0 recovers rapidly from photobleaching (A), whereas Q40 protein in foci does not recover from photobleaching (B), consistent with an insoluble protein. Bar = 2 μm. (C) Results from FRAP are quantified by determining the relative fluorescence intensity (RFI) for each time point the graph represents the average of analysis of a minimum of five independent measurements. Error bars indicate SEM. In neurons, FRAP was able to distinguish between a soluble (40s, D) and insoluble species (40a, E) of the same Q40::YFP protein, quantified in (F). Scale bars indicate 2 μm. Quantification of neuronal data is the mean of five neurons for 40s and 10 neurons for 40a. Error bars indicate SEM. (See color insert.)

FRAP experiments in *C. elegans,* because even when anesthetized, animals may move unexpectedly, changing the intensity in the ROI without affecting the background or other cells. Any movement of the cell of interest will skew data automatically generated by the ROI function in the AIM software associated with a Zeiss LSM confocal microscope. The intensity data generated by the ROI function quantifies intensity in the same location for each time point so if there is any movement, results will only be accurate until that point. Therefore, when intensity of an ROI is being determined, each time frame *must* be examined individually for any shift in the ROI. This is a significant difference from FRAP analysis in cells that remain quite still by comparison. Data resulting from analysis using the preceding equation will generate an RFI for every time point at which an image is collected and is used to generate the graphs shown in Fig. 2.

FRAP Analysis of Polyglutamine Proteins in a Multicellular Organism. We used FRAP to determine whether the shift in polyglutmine::YFP cellular distribution corresponded to a transition of these proteins from a soluble to an aggregated state. Furthermore, FRAP can be performed on individual cells in a live animal; significant advantages to traditional biochemical approaches to identifying aggregates. We reasoned that soluble YFP-tagged proteins in the cytoplasm would diffuse freely and recover rapidly after photobleaching. In contrast, fluorescent proteins with limited mobility within an aggregate should exhibit little or very slow recovery after photobleaching. Photobleaching of diffuse Q0::YFP from young animals led to an immediate 100% fluorescence recovery (Fig. 2A, C), consistent with the biophysical properties of soluble proteins; this result has also been demonstrated biochemically by SDS-PAGE and immunoblotting of bulk samples (Nollen *et al.,* 2004; Satyal *et al.,* 2000). In contrast, the fluorescent signal associated with foci in muscle cells of Q40 animals does not recover after photobleaching consistent with restriction within aggregates (Fig. 2B, C). These results are consistent with the polyglutamine length-dependent changes in localization observed visually and provide a quantitative measure of solubility that can be used to define aggregation in *C. elegans.*

When FRAP is applied to polyglutamine proteins in *C. elegans* neurons, a similar trend is observed: as the length of glutamine tracts increases, protein solubility decreases. However, FRAP experiments on multiple neurons distributed throughout a single animal revealed that Q40::YFP recovery from photobleaching varies widely. Q40 protein solubility ranges from rapid (Fig. 3D, F), similar to that observed for soluble Q19, to completely immobile Q40::YFP (Fig 3E, F), similar to the Q86 aggregates despite the absence of overt visual foci in Q40 animals. The small size and complex structures of neurons in *C. elegans* prevented visual detection of

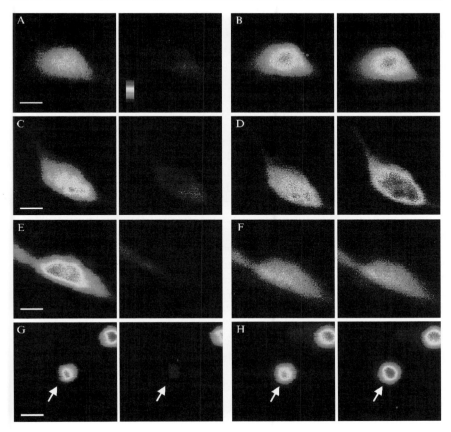

Fig. 3. Expanded polyglutamine proteins FRET *in vivo*. Q86 protein in neuronal aggregates exhibits FRET, indicating close and roughly ordered interactions at the molecular level. YFP photobleaching is seen on the YFP channel (A, C, E, and G) and its effect on the CFP donor in the same cell (B, D, F, H). CFP and YFP coexpression (A, B) do not FRET, $E^c = 0.08$ (± standard deviation of 0.16), $n = 5$. Animals expressing CFP::YFP do FRET (C, D), $E^c = 0.25$ (± 0.08), $n = 20$. Neurons coexpressing Q19::CFP and Q19::YFP (E, F) do not FRET, $E^c = -0.08$ (± 0.05), $n = 15$, whereas coexpression of Q86::CFP and Q86::YFP (G, H) does produce FRET, $E^c = 0.22$ (± 0.08), $n = 17$. Cells shown are representative of FRET experiments. Intensity is by a color scale (G), where blue is least intense and red is most intense. Scale bar = 2 μm. (See color insert.)

changes in protein solubility at low magnification, as were observed in muscle cells. These data provide an example of the usefulness of FRAP as a technique for establishing subtle changes in protein solubility in specific cells of a live, metazoan model.

Fluorescence Resonance Energy Transfer (FRET). FRET is a technique developed principally for *in vitro* biochemical studies and applied to cell culture to address *in vivo* questions of macromolecular interactions. The resolution of this technique is in the nanometer range; FRET is maximal at 50 Å and will not occur if proteins are more than 100 Å apart when using CFP and YFP fluorophores. The second condition required for FRET is that proteins are appropriately oriented to enable dipole–dipole interactions between fluorophores; proteins must be roughly parallel for energy transfer between fluorophores (Miyawaki, 2003; Miyawaki and Tsien, 2000). This technique has been widely used to show protein interactions *in vitro* and in cell culture (Kim *et al.*, 2002; Tsien, 1998), and we were interested in establishing whether FRET could be performed in *C. elegans* to determine whether polyglutamine proteins are sufficiently ordered and in molecular proximity for energy transfer to occur between CFP (donor) and YFP (acceptor).

FRET experiments in *C. elegans* were performed by an indirect method; combined donor and acceptor photobleaching (Berney and Danuser, 2003; Wouters *et al.*, 1998). When fluorophores are in sufficient proximity and the appropriate orientation, energy is transferred from the donor to the acceptor. This results in decreased donor intensity and increased acceptor intensity. In conditions where FRET is occurring, photobleaching of the acceptor blocks energy transfer causing donor intensity to increase. Thus, measuring donor intensity before and after acceptor photobleaching is an indirect measurement of FRET, whereas determining acceptor emission in the presence or absence of donor excitation would be a direct measure of FRET.

We chose to measure FRET indirectly, using the acceptor photobleaching technique, to minimize the potential for excitation bleed-through. YFP is also excited by the 458 nm laser used to excite CFP in *C. elegans*. Therefore, we measured FRET indirectly by quantifying CFP emission before and after YFP photobleaching and only needed to control for emission bleed-through. There are two approaches to controlling emission bleed-through. First is careful selection of emission filters that do not allow any overlapping emissions, although this may result in some loss of signal intensity. The data presented here used this approach, and the setup included an HFT 458/514 primary dichroic beam splitter, a 480–520 nm bandpass filter on the CFP detection channel, and a bandpass filter from 524–546 on the YFP detection channel.

The second option for setting up the microscope to minimize bleed-through is to capture both donor and acceptor emission simultaneously using a filter such as the Zeiss "Meta Detector." This detector, when used in the "Lambda Scan Mode" will scan across all emission wavelengths and bin all emissions into 10.7 nm wavelengths. The resulting spectra can be unmixed using the linear unmixing algorithm to maximize the available signal while minimizing spectral overlap.

For imaging, excitation of CFP and YFP was at 5% transmission of 458 nm laser line. YFP photobleaching was done with 514 nm laser at 100% transmission. To photobleach YFP in neurons of *C. elegans*, an ROI of 25 × 25 pixels was bleached for 50 iterations (~10 sec) while using a 63× water objective, numerical aperture 1.4, at 9× scan zoom. Conditions were optimized to photobleach YFP throughout the entire *C. elegans'* neuron so the ROI selected was larger than that used in FRAP (10 × 10) in which only a portion of each neuron was photobleached. Both the duration and area of photobleaching should be adjusted for the sample, cell, and fluorophore of interest. Images should be obtained before and after photobleaching.

FRET Data Analysis. Normally, FRET data are presented as ratio images to provide a visual representation in the change in donor intensity after acceptor photobleaching. However, even immobilized *C. elegans* twist slightly during imaging causing three-dimensional misalignment of sequential images that hinders alignments for ratio images. Therefore, representative images from FRET experiments display individual images before and after YFP bleaching for both acceptor and donor intensity (Fig. 3). Calculation of E^c included a control area, outside of the photobleached region, to correct for any movement of the animal. Because the entire cell of interest is photobleached, the control was an adjacent cell.

There are several methods to determine molecular interactions indicated by FRET, and these are reviewed extensively in Berney and Danuser (2003). Of particular interest is their observation that when using the acceptor photobleaching technique, incomplete photobleaching increases the error in calculations of FRET efficiency. Therefore, we excluded from analysis cells with more than 10% acceptor fluorescence intensity remaining after photobleaching to minimize error. Calculating the FRET coefficient provides an indication of how efficient FRET is in the sample of interest. We calculated the FRET coefficient using the equation $(E^c) = 1 - (C_{ab}/C)(T/T_{ab})$ where ab = after bleaching, C = unbleached control cell, T = experimental cell, where YFP is bleached (Berney and Danuser, 2003).

FRET Reveals Intermolecular Interactions of Polyglutamine Proteins in a Multicellular Organism. To establish whether FRET could be observed in single cells of a live animal, we generated as a positive control *C. elegans* expressing a CFP::YFP chimera with a flexible linker separating the two fluorophores (Kim *et al.*, 2002). FRET experiments performed on the neuronal CFP::YFP positive control by photobleaching the YFP acceptor (Fig. 3C) resulted in increased donor (CFP) intensity as expected, $E^c = 0.25$

($\pm$ 0.08) (Fig. 3D). As a negative control, animals coexpressing CFP and YFP from separate constructs were tested. After YFP photobleaching (Fig. 3A), there was no visible increase in CFP intensity (Fig. 3B) demonstrating, as expected, that FRET was not occurring, $E^c = 0.08$ ($\pm$ 0.16). With FRET efficiencies from control animals as reference points, *C. elegans* expressing both Q19::CFP and Q19::YFP were tested. Q19::YFP photobleaching (Fig. 3E) had no visible effect on CFP intensity, $E^c = -0.08$ ($\pm$ 0.05) (Fig. 3F), similar to negative control animals (Fig. 3B). In contrast, animals coexpressing Q86::CFP and Q86::YFP showed an increase in CFP intensity (Fig. 3H) after YFP photobleaching, $E^c = 0.22$ ($\pm$ 0.08), (Fig. 3G). FRET positive aggregates were detected in a wide range of neurons and visible redistribution of Q86 into foci, combined with SDS resistance, FRAP, and FRET data, shows that large polyglutamine expansions form insoluble, ordered aggregates in neurons throughout the nervous system of *C. elegans*.

Combining Visualization of Polyglutamine Proteins with other C. elegans *Techniques*

Behavioral Assays Determine Polyglutamine-Mediated Cellular Dysfunction. Motility in *C. elegans* can be assayed as described in Morley *et al.* (2002). Individual animals at the indicated ages were picked to fresh plates and their tracks recorded (Fig. 4A) at indicated time intervals using a CCD camera and Leica dissection stereomicroscope (8× magnification), and distance traveled was determined using a ruler calibration macro in the Openlab software program (Improvision). Dividing this distance by the time interval gave the motility index for each animal. Statistical significance of the results was determined by a chi-squared test. Because the polyglutamine transgenes were carried on extrachromosomal arrays, some animals in the population were nontransgenic and consequently provided internal controls. Motility values for wild-type (N2) and nontransgenic control groups were indistinguishable from one another.

Young adult animals expressing polyglutamine expansions of ≤ 35 exhibited motility similar to wild type (Fig. 4B). In contrast, we observed a ~10-fold reduction in motility of young adult Q82 animals (Fig. 4B). Q40 animals, which had aggregates in some cells but not in others, exhibited an intermediate motility defect with a high degree of variation in the intensity of loss of motility across a population (Fig. 4B) that corresponded directly with the degree of aggregate formation in any given animal (Morley *et al.*, 2002).

Aging Influences the Threshold for Polyglutamine Aggregation and Toxicity. Although aggregation-prone proteins are expressed throughout the lifetime of patients, pathology associated with diseases such as

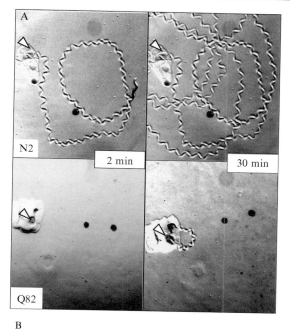

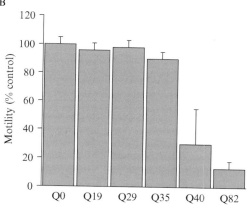

FIG. 4. Behavioral analysis of polyglutamine-mediated cellular dysfunction. (A) Time-lapse micrographs show tracks left by 5-day-old wild type (N2) and Q82 animals 2 and 30 min after being placed at the position marked by the arrowhead. (B) Quantification of motility index for 4- to 5-day-old Q0, Q19, Q29, Q35, Q40, and Q82 animals reveal polyglutamine length-dependent toxicity. Data are mean ± SD for at least 50 animals of each type as percentage of N2 motility. (Morley *et al.*, 2002. Copyright 2002, National Academy of Sciences, U.S.A.)

Huntington's, Alzheimer's, and Parkinson's are typically not manifest until middle or old age. This motivated us to study the behavior of polyglutamine proteins during aging. Indeed, a major advantage of the *C. elegans* system was our ability to follow the behavior of fluorescent polyglutamine proteins in individual animals over their entire lifetimes. We found that the threshold for aggregate formation described previously for young animals was dynamic as the animals aged. At 3 days of age or less, only animals expressing Q40 or greater exhibit aggregates (Fig. 5A). However, at 4–5 days of age, the threshold shifts as aggregates appear in Q33 and Q35 animals (Fig. 5A). The threshold again shifts to Q29 with appearance of aggregates in aged animals (>9–10 days) (Fig. 5A). In all cases, we observed a coordinated age-dependent loss of motility relative to controls that paralleled the accumulation of polyglutamine aggregates (Fig. 5B). Thus, the threshold for polyglutamine aggregation and toxicity is not static or strictly repeat length-dependent, likely reflecting a balance of different factors including repeat length and changes in the cellular protein folding environment over time.

On the basis of these results, we wondered whether this behavior resulted from the intrinsic properties of a protein motif, or whether changes over time reflect the influence of aging-related alterations in the cellular physiology. The idea that the molecular determinants of longevity might influence polyglutamine-mediated toxicity is supported by observations that the time until polyglutamine-mediated pathology develops—days in *C. elegans*, weeks in *Drosophila*, months in mice, and years in humans—correlates approximately with the lifespan of the organism. Here, again, the *C. elegans* model was a tremendous advantage, because the availability of mutants with extended lifespans allowed us to test these ideas directly.

To accomplish this, we generated transgenic animals expressing Q82:: YFP in the background of animals with extended longevity. In *C. elegans*, *age-1* encodes a phosphoinositide-3 kinase that functions in an insulin-like signaling pathway (ILS), and mutations in this gene can extend lifespan by 1.5–2 fold (Guarente and Kenyon, 2000; Morris *et al.*, 1996). When expressed in long-lived animals, both aggregation (Fig. 5C) and toxicity (Fig. 5D) of polyglutamine proteins were substantially delayed (Morley *et al.*, 2002). Furthermore, the suppression of polyglutamine aggregation and toxicity was entirely dependent on the activity of *daf-16*, a forkhead transcription factor that functions downstream of *age-1* in the ILS pathway and is required for extended lifespan (Guarente and Kenyon, 2000; Lin *et al.*, 1997; Ogg *et al.*, 1997). Thus, the dual effects of *age-1* on longevity and polyglutamine-mediated toxicity share a common genetic pathway.

Genome-Wide RNAi Screening Defines Novel Regulators of Polyglutamine Aggregation and Toxicity. The dynamic threshold for polyglutamine

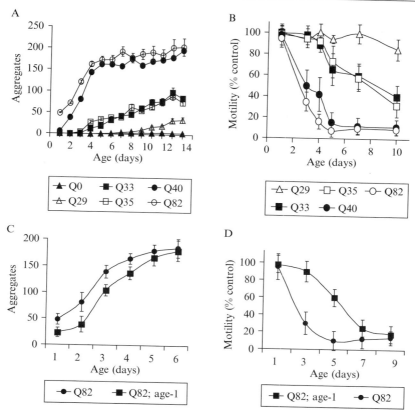

FIG. 5. Influence of aging on polyglutamine aggregation and toxicity. (A) Repeat length-dependent accumulation of polyglutamine aggregates during aging. Data are mean ± SEM. Twenty-four animals of each type are represented at day 1. Cohort sizes decreased as animals died during the experiment, but each data point represents at least five animals. (B) Motility index as a function of age for the same cohorts of animals described in A. Data are mean ± SD as a percentage of age-matched Q0 animals. Mutations in the *C. elegans* gene, *age-1* extend lifespan and alter polyglutamine pathogenesis. (C) When polyglutamine expressing animals are crossed into an *age-1* background we observe that extending lifespan in *C. elegans* delays the onset of aggregation. Data are mean ±SEM. (D) Increased lifespan also delays the onset of polyglutamine-mediated toxicity as measured by motility. Data are mean ± SD for 30 animals of each type. Motility of nontransgenic wild-type and *age-1* animals was similar to that of wildtype (N2). (Morley *et al.*, 2002. Copyright 2002, National Academy of Sciences, U.S.A.).

aggregation and toxicity demonstrates that the fate of these proteins is not an all-or-none outcome solely on the basis of repeat length and implies the presence of a cellular buffering system to prevent proteotoxicity. Thus, this model provided a substrate to which we could apply the power of *C. elegans* forward and reverse genetics to define components of the cellular buffer and uncover novel modifiers of polyglutamine-induced pathology.

The results previously described identify ILS as a genetic pathway that can influence the course of polyglutamine-mediated phenotypes. What other pathways might exert similar effects? Numerous overexpression and genetic studies in mammalian (Cummings *et al.*, 2001) and *Drosophila* (Fernandez-Funez *et al.*, 2000; Warrick *et al.*, 1999) models have identified various enhancers or suppressors of polyglutamine-mediated aggregation and toxicity. This is well-illustrated by the large number of approaches in which various molecular chaperones either alone or in combination have been shown to influence polyglutamine-mediated phenotypes (Bonini, 2002; Carmichael *et al.*, 2000; Chai *et al.*, 1999; Cummings *et al.*, 1998, 2001; Warrick *et al.*, 1999). Although one could argue which of these modifiers is the "key" to determining the fate of aggregation-prone proteins, we interpret these results to suggest that the transition of polyglutamine proteins from a soluble to an aggregated state is the result of a delicate balance in which multiple pathways are involved. To identify the complete protein-folding buffer involved in polyglutamine transition from a soluble to an aggregated state, we used a genome-wide RNAi approach.

RNAi is a commonly used reverse-genetics approach in *C. elegans* (Fire *et al.*, 1998; Wang and Barr, 2005) allowing the targeted down-regulation of specific genes by introduction of small fragments of cognate double-stranded RNAs. Although initially used to test the function of single or small groups of candidate genes, this technique has been adapted to genome-wide screens with the Ahringer laboratory's construction of a library consisting of 16,757 bacterial clones covering 86% of the predicted *C. elegans* genome (Fraser *et al.*, 2000; Kamath *et al.*, 2003). RNAi has the additional advantages of allowing the detection of lethal positives and the immediate identification of the target genes. However, genome-wide RNAi screens may miss certain genes because of variation in mRNA depletion or relative inefficiency of RNAi in specific tissues, of which neurons are an example. Nevertheless, this approach offers an extremely powerful and rapid tool to identify the set of genes that modify a given phenotype.

To identify genes that prevent polyglutamine aggregate formation we used *C. elegans* strains expressing polyglutamine lengths close to the aggregation threshold, Q33 and Q35 strains, in an RNAi genetic screen.

Genome-wide RNAi Screening in C. elegans. The full-genome RNAi screen was performed in a 96-well format by feeding RNAi bacteria to animals in liquid culture. Construction of the RNAi library used in the genetic screen was described originally by Ahringer's group, and our initial use was described in Nollen *et al.* (Ashrafi *et al.*, 2003; Fraser *et al.*, 2000;

Nollen *et al.*, 2004). The RNAi library used is now commercially available thru "MRC geneservice" at http://www.hgmp.mrc.ac.uk/geneservice/index. shtml.

Animals were synchronized by NaOCl bleaching and overnight hatching in M9. In each well, 10–15 L1 larval-stage animals were suspended in 50 μl of M9 plus (M9, 10 μg/ml cholesterol, 50 μg/ml ampicillin, 12 μg/ml tetracycline, 200 μg/ml isopropyl β-D-thiogalactoside [IPTG] and 0.1 μg/ml fungizone) and added to 80 μl of an overnight culture of RNAi bacteria induced by IPTG for 4 h. The animals were grown at 23° with continuous shaking at 150 rpm (New Brunswick Scientific Incubation shaker). Q35 animals were scored for foci formation by visual inspection after 72 h. RNAi-producing bacteria that induced more than five visible foci in >30% of the animals in a well were scored as aggregation enhancers. All positive RNAi clones were confirmed in an independent experiment and scored for aggregate formation in Q0, Q24, and Q33 animals. A gene was scored as a positive when it induced early onset of aggregation in both Q33- and Q35-expressing animals but not in Q0 and Q24 animals. This criteria ensured that the genes identified where polyglutamine expansion specific. The gene targets of the positive RNAi clones were verified by sequencing of the insert of the RNAi plasmids.

A total of 186 modifiers were identified, comprising five major classes: genes involved in RNA metabolism, protein synthesis, protein folding, protein trafficking, and protein degradation (Nollen *et al.*, 2004). Examples of some of these genes are RNA helicases, splicing factors, and transcription factors for RNA metabolism; initiation and elongation factors and ribosomal subunits for protein synthesis; chaperonins and Hsp70 family members for protein folding; nuclear import and cytoskeletal genes for protein trafficking; and proteasomal genes for protein degradation. Although numerous distinct biochemical activities are represented by the different modifiers, a common feature is an expected imbalance between protein synthesis, folding, and degradation triggered by their disruption. In this context, these five classes can then be further grouped in two major categories: genes whose disruption leads to an increase in misfolded protein production and genes that when disrupted lead to a decreased clearance of misfolded proteins and proper protein turnover. Together, these results reveal that the transition between soluble and aggregated states of polyglutamine proteins is regulated by a much more complex integration of events, extending beyond the immediate involvement of chaperone-mediated folding and proteasomal degradation. In addition these findings validate an approach allowing our group and others to screen RNAi libraries against model systems for multiple neurodegenerative diseases to identify similarities and differences in the set of modifiers observed to better understand the conserved or unique aspects of their pathophysiology.

Discussion

C. elegans *Models of Polyglutamine-Mediated Toxicity and Aggregation*

We describe here two *C. elegans* models expressing isolated polyglutamine repeats in neurons or in muscle cells to investigate not only the pathogenesis of HD and related disorders but also to uncover the genes and associated pathways that regulate the protein folding environment. Our model differs from the human diseases in that we expressed isolated polyglutamine motifs rather than those in the context of disease-related proteins. Although protein context and posttranslational modifications of flanking sequences are increasingly recognized to play a role in pathology (Chai *et al.*, 2001; Chen *et al.*, 2003), we reasoned that studying isolated polyglutamine expansions would lend insight into the conserved features underlying the different pathologies. In support of this idea, our genome wide RNAi screen identified multiple molecular chaperones that have been implicated in polyglutamine diseases, AD and PD. Furthermore, expression of isolated polyglutamine tracts in both neurons and body-wall muscle cells of *C. elegans* recapitulated numerous features of the human diseases, including polyglutamine length-dependent aggregation and toxicity. More subtle features were also observed, specifically variability in aggregate formation and severity of motility defect observed in Q40 animals (Figs. 1 and 4). These observations relied on unique characteristics of *C. elegans*, in particular its' transparency, which enabled us to apply fluorescence-based biophysical assays such as FRAP and FRET to a live, multicellular animal. These studies represent the initial stages of applying biophysical approaches to *C. elegans,* and future studies may extend the use of these techniques to generate more precise, quantitative data on the polyglutamine proteins *in vivo*. These might include examining the relative proportions of soluble versus insoluble polyglutamine protein in specific cells or determining how solubility and/or intermolecular interactions of polyglutamine proteins compare with other aggregation prone proteins.

Using *C. elegans to Identify Modifiers of Polyglutamine Pathogenesis*

Another important characteristic of *C. elegans* in our studies was the animals' short lifespan and the availability of genetic mutants effecting lifespan. The ability to combine visualization of polyglutamine aggregates in live animals with mutations conferring longevity enabled us to demonstrate that increasing lifespan delays polyglutamine aggregation and toxicity. These results suggest a novel link between the genetic regulation of aging and aging-related disease and once again use unique characteristics of *C. elegans*.

In subsequent studies, we and others have demonstrated that the molecular link between these pathways is, in part, maintained by factors that detect and respond to misfolded proteins, heat shock transcription factor (HSF) and molecular chaperones/heat shock proteins. For example, it has been shown that inhibition of HSF-1 function leads to decreased lifespan and an accelerated aging phenotype in *C. elegans* (Garigan *et al.*, 2002; Hsu *et al.*, 2003; Morley and Morimoto, 2004). Conversely, overexpression of HSF-1 in *C. elegans* extends lifespan (Hsu *et al.*, 2003; Morley and Morimoto, 2004). Research on the molecular mechanisms in both areas, aging and molecular chaperones, is active and continuously identifies new players in these pathways. Future work using *C. elegans* will take advantage of new mutants generated from ongoing research to apply the approaches described here.

In addition to our targeted approach to specific genetic pathways for their role as potential modifiers, we used an unbiased genome-wide RNAi screen to identify a set of proteins and pathways that influence polyglutamine aggregation. The identification of five classes of polyglutamine suppressors provides a better understanding of the breadth of pathways and cellular complexes involved in sensing protein damage. In particular, our findings suggest a model in which each step in the birth, life, and death of a protein influences the capacity of a cell to maintain the conditions necessary for protein folding. For example, we found that perturbation of the RNA-processing machinery was associated with accelerated aggregation of Q35::YFP. Although a direct effect is possible, it seems more likely that this results from an increased burden of abnormal proteins requiring the activity of the protein folding buffer. Under such conditions, aggregation-prone proteins normally degraded escape quality control, leading to aggregation and toxicity. Uncovering the role of these genes in the disease process additionally provides new potential therapeutic targets.

Together, these studies have revealed a common set of factors that link the genetic regulation of protein homeostasis, stress responsiveness, and longevity. Using *C. elegans* as a model system, we have combined a variety of approaches, including RNAi, genetics, and behavioral assays with the visualization of polyglutamine proteins in a live, aging, multicellular model to examine polyglutamine pathogenesis. Extending the visualization of polyglutamine proteins to include FRAP and FRET has made it possible to apply quantitative criteria of solubility and intermolecular interactions to visible changes in protein localization and further our understanding of polyglutamine pathogenesis in a highly sensitive, cell-specific manner. In the future, the continued application of this combination of approaches will provide additional insight on the mechanisms underlying protein aggregation and how it contributes to pathology in human disease.

Acknowledgments

We thank members of the Morimoto laboratory past and present, who contributed to this work both intellectually and technically. H. R. B. was supported by the Cellular and Molecular Biology of Disease Training Grant T32 GM08061 from the National Institute of General Medical Sciences (NIGMS) to Northwestern University. J. F. M. was supported by a Medical Scientist Training Grant from NIGMS to Northwestern University and an individual NRSA from the National Institute of Neurological Disease and Stroke. S. M. G. was supported by a PhD fellowship Praxis XXI BD/21451/99 from Fundação para a Ciência e Tecnologia. These studies were also supported by grants from NIGMS (GM38109), the National Institutes of Aging, the Huntington Disease Society of America Coalition for the Cure, and the Daniel F. and Ada L. Rice Foundation.

References

Andrew, S. E., Goldberg, Y. P., Kremer, B., Telenius, H., Theilmann, J., Adam, S., Starr, E., Squitieri, F., Lin, B., Kalchman, M. A., Graham, R. K., and Hayden, M. (1993). The relationship between trinucleotide (CAG) repeat length and clinical features of Huntington's disease. *Nat. Genet.* **4**, 398–403.

Ashrafi, K., Chang, F. Y., Watts, J. L., Fraser, A. G., Kamath, R. S., Ahringer, J., and Ruvkun, G. (2003). Genome-wide RNAi analysis of *Caenorhabditis elegans* fat regulatory genes. *Nature* **421**, 268–272.

Berney, C., and Danuser, G. (2003). FRET or no FRET: A quantitative comparison. *Biophys. J.* **84**, 3992–4010.

Bonini, N. M. (2002). Chaperoning brain degeneration. *Proc. Natl. Acad. Sci. USA* **99**(Suppl. 4), 16407–16411.

Brenner, S. (1974). The genetics of *Caenorhabditis elegans*. *Genetics* **77**, 71–94.

Brinkman, R. R., Mezei, M. M., Theilmann, J., Almqvist, E., and Hayden, M. R. (1997). The likelihood of being affected with Huntington disease by a particular age, for a specific CAG size. *Am. J. Hum. Genet.* **60**, 1202–1210.

Carmichael, J., Chatellier, J., Woolfson, A., Milstein, C., Fersht, A. R., and Rubinsztein, D. C. (2000). Bacterial and yeast chaperones reduce both aggregate formation and cell death in mammalian cell models of Huntington's disease. *Proc. Natl. Acad. Sci. USA* **97**, 9701–9705.

Chai, Y., Koppenhafer, S. L., Bonini, N. M., and Paulson, H. L. (1999). Analysis of the role of heat shock protein (Hsp) molecular chaperones in polyglutamine disease. *J. Neurosci.* **19**, 10338–10347.

Chai, Y., Wu, L., Griffin, J. D., and Paulson, H. L. (2001). The role of protein composition in specifying nuclear inclusion formation in polyglutamine disease. *J. Biol. Chem.* **276**, 44889–44897.

Chan, H. Y., Warrick, J. M., Andriola, I., Merry, D., and Bonini, N. M. (2002). Genetic modulation of polyglutamine toxicity by protein conjugation pathways in *Drosophila*. *Hum. Mol. Genet.* **11**, 2895–2904.

Chen, H. K., Fernandez-Funez, P., Acevedo, S. F., Lam, Y. C., Kaytor, M. D., Fernandez, M. H., Aitken, A., Skoulakis, E. M., Orr, H. T., Botas, J., and Zoghbi, H. Y. (2003). Interaction of Akt-phosphorylated ataxin-1 with 14-3-3 mediates neurodegeneration in spinocerebellar ataxia type 1. *Cell* **113**, 457–468.

Chen, S., Berthelier, V., Yang, W., and Wetzel, R. (2001). Polyglutamine aggregation behavior *in vitro* supports a recruitment mechanism of cytotoxicity. *J. Mol. Biol.* **311**, 173–182.

Chen, S., Ferrone, F. A., and Wetzel, R. (2002). Huntington's disease age-of-onset linked to polyglutamine aggregation nucleation. *Proc. Natl. Acad. Sci. USA* **99**, 11884–11889.

Cummings, C. J., Mancini, M. A., Antalffy, B., DeFranco, D. B., Orr, H. T., and Zoghbi, H. Y. (1998). Chaperone suppression of aggregation and altered subcellular proteasome localization imply protein misfolding in SCA1. *Nat. Genet.* **19**, 148–154.

Cummings, C. J., Sun, Y., Opal, P., Antalffy, B., Mestril, R., Orr, H. T., Dillmann, W. H., and Zoghbi, H. Y. (2001). Over-expression of inducible HSP70 chaperone suppresses neuropathology and improves motor function in SCA1 mice. *Hum. Mol. Genet.* **10**, 1511–1518.

Davies, S. W., Turmaine, M., Cozens, B. A., DiFiglia, M., Sharp, A. H., Ross, C. A., Scherzinger, E., Wanker, E. E., Mangiarini, L., and Bates, G. P. (1997). Formation of neuronal intranuclear inclusions underlies the neurological dysfunction in mice transgenic for the HD mutation. *Cell* **90**, 537–548.

Dobson, C. M. (2001). Protein folding and its links with human disease. *Biochem. Soc. Symp.* **68**, 1–26.

Driscoll, M., and Gerstbrein, B. (2003). Dying for a cause: Invertebrate genetics takes on human neurodegeneration. *Nat. Rev. Genet.* **4**, 181–194.

Faber, P. W., Alter, J. R., MacDonald, M. E., and Hart, A. C. (1999). Polyglutamine-mediated dysfunction and apoptotic death of a *Caenorhabditis elegans* sensory neuron. *Proc. Natl. Acad. Sci. USA* **96**, 179–184.

Faber, P. W., Voisine, C., King, D. C., Bates, E. A., and Hart, A. C. (2002). Glutamine/proline-rich PQE-1 proteins protect *Caenorhabditis elegans* neurons from huntingtin polyglutamine neurotoxicity. *Proc. Natl. Acad. Sci. USA* **99**, 17131–17136.

Feany, M. B., and Bender, W. W. (2000). A Drosophila model of Parkinson's disease. *Nature* **404**, 394–398.

Fernandez-Funez, P., Nino-Rosales, M. L., de Gouyon, B., She, W. C., Luchak, J. M., Martinez, P., Turiegano, E., Benito, J., Capovilla, M., Skinner, P. J., McCall, A., Canal, I., Orr, H. T., Zoghbi, H. Y., and Botas, J. (2000). Identification of genes that modify ataxin-1-induced neurodegeneration. *Nature* **408**, 101–106.

Fire, A., Xu, S., Montgomery, M. K., Kostas, S. A., Driver, S. E., and Mello, C. C. (1998). Potent and specific genetic interference by double-stranded RNA in *Caenorhabditis elegans*. *Nature* **391**, 806–811.

Fraser, A. G., Kamath, R. S., Zipperlen, P., Martinez-Campos, M., Sohrmann, M., and Ahringer, J. (2000). Functional genomic analysis of *C. elegans* chromosome I by systematic RNA interference. *Nature* **408**, 325–330.

Garigan, D., Hsu, A. L., Fraser, A. G., Kamath, R. S., Ahringer, J., and Kenyon, C. (2002). Genetic analysis of tissue aging in *Caenorhabditis elegans*: A role for heat-shock factor and bacterial proliferation. *Genetics* **161**, 1101–1112.

Guarente, L., and Kenyon, C. (2000). Genetic pathways that regulate ageing in model organisms. *Nature* **408**, 255–262.

Hsu, A. L., Murphy, C. T., and Kenyon, C. (2003). Regulation of aging and age-related disease by DAF-16 and heat-shock factor. *Science* **300**, 1142–1145.

Kakizuka, A. (1998). Protein precipitation: A common etiology in neurodegenerative disorders? *Trends Genet.* **14**, 396–402.

Kamath, R. S., Fraser, A. G., Dong, Y., Poulin, G., Durbin, R., Gotta, M., Kanapin, A., Le Bot, N., Moreno, S., Sohrmann, M., Welchman, D. P., Zipperlen, P., and Ahringer, J. (2003). Systematic functional analysis of the *Caenorhabditis elegans* genome using RNAi. *Nature* **421**, 231–237.

Kamino, K., Orr, H. T., Payami, H., Wijsman, E. M., Alonso, M. E., Pulst, S. M., Anderson, L., O'Dahl, S., Nemens, E., White, E., *et al.* (1992). Linkage and mutational analysis of familial Alzheimer disease kindreds for the APP gene region. *Am. J. Hum. Genet.* **51**, 998–1014.

Kawaguchi, Y., Okamoto, T., Taniwaki, M., Aizawa, M., Inoue, M., Katayama, S., Kawakami, H., Nakamura, S., Nishimura, M., Akiguchi, M., Kimura, J., Narumiya, S., and Kakizuka, A. (1994). CAG expansions in a novel gene for Machado–Joseph disease at chromosome 14q32.1. *Nat. Genet.* **8,** 221–228.

Kayed, R., Head, E., Thompson, J. L., McIntire, T. M., Milton, S. C., Cotman, C. W., and Glabe, C. G. (2003). Common structure of soluble amyloid oligomers implies common mechanism of pathogenesis. *Science* **300,** 486–489.

Kim, S., Nollen, E. A., Kitagawa, K., Bindokas, V. P., and Morimoto, R. I. (2002). Polyglutamine protein aggregates are dynamic. *Nat. Cell. Biol.* **4,** 826–831.

Koide, R., Ikeuchi, T., Onodera, O., Tanaka, H., Igarashi, S., Endo, K., Takahashi, H., Kondo, R., Ishikawa, A., Hayashi, A., Saito, M., Tomoda, A., Miike, T., Naito, H., Ikuta, F., and Tsuji, S. (1994). Unstable expansion of CAG repeat in hereditary dentatorubral-pallidoluysian atrophy (DRPLA). *Nat. Genet.* **6,** 9–13.

Kopito, R. R., and Ron, D. (2000). Conformational disease. *Nat. Cell Biol.* **2,** E207–E209.

La Spada, A. R., Wilson, E. M., Lubahn, D. B., Harding, A. E., and Fischbeck, K. H. (1991). Androgen receptor gene mutations in X-linked spinal and bulbar muscular atrophy. *Nature* **352,** 77–79.

Laing, N. G., and Siddique, T. (1997). Cu/Zn superoxide dismutase gene mutations in amyotrophic lateral sclerosis: Correlation between genotype and clinical features. *J. Neurol. Neurosurg. Psychiatry* **63,** 815.

Lee, V. M., Kenyon, T. K., and Trojanowski, J. Q. (2005). Transgenic animal models of tauopathies. *Biochim. Biophys. Acta* **1739,** 251–259.

Lin, K., Dorman, J. B., Rodan, A., and Kenyon, C. (1997). daf-16: An HNF-3/forkhead family member that can function to double the life-span of *Caenorhabditis elegans. Science* **278,** 1319–1322.

Link, C. D. (1995). Expression of human beta-amyloid peptide in transgenic *Caenorhabditis elegans. Proc. Natl. Acad. Sci. USA* **92,** 9368–9372.

Link, C. D. (2001). Transgenic invertebrate models of age-associated neurodegenerative diseases. *Mech. Ageing Dev.* **122,** 1639–1649.

Lucking, C. B., Durr, A., Bonifati, V., Vaughan, J., De Michele, G., Gasser, T., Harhangi, B. S., Meco, G., Denefle, P., Wood, N. W., Agid, Y., and Brice, A. (2000). Association between early-onset Parkinson's disease and mutations in the parkin gene. French Parkinson's Disease Genetics Study Group. *N. Engl. J. Med.* **342,** 1560–1567.

Mangiarini, L., Sathasivam, K., Seller, M., Cozens, B., Harper, A., Hetherington, C., Lawton, M., Trottier, Y., Lehrach, H., Davies, S. W., and Bates, G. P. (1996). Exon 1 of the HD gene with an expanded CAG repeat is sufficient to cause a progressive neurological phenotype in transgenic mice. *Cell* **87,** 493–506.

Mello, C. C., Kramer, J. M., Stinchcomb, D., and Ambros, V. (1991). Efficient gene transfer in *C. elegans*: Extrachromosomal maintenance and integration of transforming sequences. *EMBO J.* **10,** 3959–3970.

Miyawaki, A. (2003). Visualization of the spatial and temporal dynamics of intracellular signaling. *Dev. Cell* **4,** 295–305.

Miyawaki, A., and Tsien, R. Y. (2000). Monitoring protein conformations and interactions by fluorescence resonance energy transfer between mutants of green fluorescent protein. *Methods Enzymol.* **327,** 472–500.

Mizuno, Y., Hattori, N., Kitada, T., Matsumine, H., Mori, H., Shimura, H., Kubo, S., Kobayashi, H., Asakawa, S., Minoshima, S., and Shimizu, N. (2001). Familial Parkinson's disease. Alpha-synuclein and parkin. *Adv. Neurol.* **86,** 13–21.

Morley, J. F., Brignull, H. R., Weyers, J. J., and Morimoto, R. I. (2002). The threshold for polyglutamine-expansion protein aggregation and cellular toxicity is dynamic and influenced by aging in *Caenorhabditis elegans*. *Proc. Natl. Acad. Sci. USA* **99,** 10417–10422.

Morley, J. F., and Morimoto, R. I. (2004). Regulation of longevity in *Caenorhabditis elegans* by heat shock factor and molecular chaperones. *Mol. Biol. Cell* **15,** 657–664.

Morris, J. Z., Tissenbaum, H. A., and Ruvkun, G. (1996). A phosphatidylinositol-3-OH kinase family member regulating longevity and diapause in *Caenorhabditis elegans*. *Nature* **382,** 536–539.

Nollen, E. A., Garcia, S. M., van Haaften, G., Kim, S., Chavez, A., Morimoto, R. I., and Plasterk, R. H. (2004). Genome-wide RNA interference screen identifies previously undescribed regulators of polyglutamine aggregation. *Proc. Natl. Acad. Sci. USA* **101,** 6403–6408.

O'Nuallain, B., and Wetzel, R. (2002). Conformational Abs recognizing a generic amyloid fibril epitope. *Proc. Natl. Acad. Sci. USA* **99,** 1485–1490.

Oeda, T., Shimohama, S., Kitagawa, N., Kohno, R., Imura, T., Shibasaki, H., and Ishii, N. (2001). Oxidative stress causes abnormal accumulation of familial amyotrophic lateral sclerosis-related mutant SOD1 in transgenic *Caenorhabditis elegans*. *Hum. Mol. Genet.* **10,** 2013–2023.

Ogg, S., Paradis, S., Gottlieb, S., Patterson, G. I., Lee, L., Tissenbaum, H. A., and Ruvkun, G. (1997). The Fork head transcription factor DAF-16 transduces insulin-like metabolic and longevity signals in *C. elegans*. *Nature* **389,** 994–999.

Ordway, J. M., Tallaksen-Greene, S., Gutekunst, C. A., Bernstein, E. M., Cearley, J. A., Wiener, H. W., Dure, L. S.t., Lindsey, R., Hersch, S. M., Jope, R. S., Albin, R. L., and Detloff, P. J. (1997). Ectopically expressed CAG repeats cause intranuclear inclusions and a progressive late onset neurological phenotype in the mouse. *Cell* **91,** 753–763.

Orr, H. T. (2001). Beyond the Qs in the polyglutamine diseases. *Genes Dev.* **15,** 925–932.

Orr, H. T., Chung, M. Y., Banfi, S., Kwiatkowski, T. J., Jr., Servadio, A., Beaudet, A. L., McCall, A. E., Duvick, L. A., Ranum, L. P., and Zoghbi, H. Y. (1993). Expansion of an unstable trinucleotide CAG repeat in spinocerebellar ataxia type 1. *Nat. Genet.* **4,** 221–226.

Parker, J. A., Connolly, J. B., Wellington, C., Hayden, M., Dausset, J., and Neri, C. (2001). Expanded polyglutamines in *Caenorhabditis elegans* cause axonal abnormalities and severe dysfunction of PLM mechanosensory neurons without cell death. *Proc. Natl. Acad. Sci. USA* **98,** 13318–13323.

Phair, R. D., and Misteli, T. (2000). High mobility of proteins in the mammalian cell nucleus. *Nature* **404,** 604–609.

Polymeropoulos, M. H., Lavedan, C., Leroy, E., Ide, S. E., Dehejia, A., Dutra, A., Pike, B., Root, H., Rubenstein, J., Boyer, R., Stenroos, E. S., Chandrasekharappa, S., Athanassiadou, A., Papapetropoulos, T., Johnson, W. G., Lazzarini, A. M., Duvoisin, R. C., Di Iorio, G., Golbe, L. I., and Nussbaum, R. L. (1997). Mutation in the alpha-synuclein gene identified in families with Parkinson's disease. *Science* **276,** 2045–2047.

Rosen, D. R., Siddique, T., Patterson, D., Figlewicz, D. A., Sapp, P., Hentati, A., Donaldson, D., Goto, J., O'Regan, J. P., Deng, J. P., *et al.* (1993). Mutations in Cu/Zn superoxide dismutase gene are associated with familial amyotrophic lateral sclerosis. *Nature* **362,** 59–62.

Ross, C. A. (2002). Polyglutamine pathogenesis: Emergence of unifying mechanisms for Huntington's disease and related disorders. *Neuron* **35,** 819–822.

Satyal, S. H., Schmidt, E., Kitagawa, K., Sondheimer, N., Lindquist, S., Kramer, J. M., and Morimoto, R. I. (2000). Polyglutamine aggregates alter protein folding homeostasis in *Caenorhabditis elegans*. *Proc. Natl. Acad. Sci. USA* **97,** 5750–5755.

Scherzinger, E., Lurz, R., Turmaine, M., Mangiarini, L., Hollenbach, B., Hasenbank, R., Bates, G. P., Davies, S. W., Lehrach, H., and Wanker, E. E. (1997). Huntingtin-encoded polyglutamine expansions form amyloid-like protein aggregates *in vitro* and *in vivo*. *Cell* **90,** 549–558.

Shulman, J. M., Shulman, L. M., Weiner, W. J., and Feany, M. B. (2003). From fruit fly to bedside: Translating lessons from *Drosophila* models of neurodegenerative disease. *Curr. Opin. Neurol.* **16,** 443–449.

Stefani, M., and Dobson, C. M. (2003). Protein aggregation and aggregate toxicity: New insights into protein folding, misfolding diseases and biological evolution. *J. Mol. Med.* **81,** 678–699.

Thompson, L. M., and Marsh, J. L. (2003). Invertebrate models of neurologic disease: Insights into pathogenesis and therapy. *Curr. Neurol. Neurosci. Rep.* **3,** 442–448.

Trottier, Y., Lutz, Y., Stevanin, G., Imbert, G., Devys, D., Cancel, G., Saudou, F., Weber, C., David, G., Tora, L., Agid, Y., Brice, A., and Mandel, J. L. (1995). Polyglutamine expansion as a pathological epitope in Huntington's disease and four dominant cerebellar ataxias. *Nature* **378,** 403–406.

Tsien, R. Y. (1998). The green fluorescent protein. *Annu. Rev. Biochem.* **67,** 509–544.

Wang, J., and Barr, M. M. (2005). RNA interference in *Caenorhabditis elegans*. *Methods Enzymol.* **392,** 36–55.

Warrick, J. M., Chan, H. Y., Gray-Board, G. L., Chai, Y., Paulson, H. L., and Bonini, N. M. (1999). Suppression of polyglutamine-mediated neurodegeneration in *Drosophila* by the molecular chaperone HSP70. *Nat. Genet.* **23,** 425–428.

Warrick, J. M., Paulson, H. L., Gray-Board, G. L., Bui, Q. T., Fischbeck, K. H., Pittman, R. N., and Bonini, N. M. (1998). Expanded polyglutamine protein forms nuclear inclusions and causes neural degeneration in *Drosophila*. *Cell* **93,** 939–949.

Westlund, B., Stilwell, G., and Sluder, A. (2004). Invertebrate disease models in neurotherapeutic discovery. *Curr. Opin. Drug Discov. Dev.* **7,** 169–178.

Wexler, N. S., Lorimer, J., Porter, J., Gomez, F., Moskowitz, C., Shackell, E., Marder, K., Penchaszadeh, G., Roberts, S. A., Gayan, J., Brocklebank, D., Cherny, S. S., Cardon, L. R., Gray, J., Dlouhy, S. R., Wiktorski, S., Hodes, M. E., Conneally, P. M., Penney, J. B., Gusella, J., Cha, J. H., Irizarry, M., Rosas, D., Hersch, S., Hollingsworth, Z., MacDonald, M., Young, A. B., Andresen, J. M., Housman, D. E., De Young, M. M., Bonilla, E., Stillings, T., Negrette, A., Snodgrass, S. R., Martinez-Jaurrieta, M. D., Ramos-Arroyo, M. A., Bickham, J., Ramos, J. S., Marshall, F., Shoulson, I., Rey, G. J., Feigin, A., Arnheim, N., Acevedo-Cruz, A., Acosta, L., Alvir, J., Fischbeck, K., Thompson, L. M., Young, A., Dure, L., O'Brien, C. J., Paulsen, J., Brickman, A., Krch, D., Peery, S., Hogarth, P., Higgins, D. S., Jr., and Landwehrmeyer, B. (2004). Venezuelan kindreds reveal that genetic and environmental factors modulate Huntington's disease age of onset. *Proc. Natl. Acad. Sci. USA* **101,** 3498–3503.

Wouters, F. S., Bastiaens, P. I., Wirtz, K. W., and Jovin, T. M. (1998). FRET microscopy demonstrates molecular association of non-specific lipid transfer protein (nsL-TP) with fatty acid oxidation enzymes in peroxisomes. *EMBO J.* **17,** 7179–7189.

Zoghbi, H. Y., and Orr, H. T. (2000). Glutamine repeats and neurodegeneration. *Annu. Rev. Neurosci.* **23,** 217–247.

Section III

Computational Approaches in Theory

[17] Nucleation: The Connections Between Equilibrium and Kinetic Behavior

By Frank A. Ferrone

Abstract

This chapter describes the thermodynamics that govern the formation of nuclei, the least stable species in the reaction path of large, linear aggregates. In the approach described here, parameters are used that have direct molecular interpretations, such as contact energies of the molecular species. The extensive work on sickle hemoglobin is used as a model. An important result is that the nucleus size is expected to vary with initial conditions, such as the initial monomer concentration. Another unexpected result of some generality is that motion of the molecules within the nucleus recovers significant amounts of entropy that would be lost on complete immobilization.

Introduction

In a nucleation-controlled aggregation process, the rate of aggregate formation is directly related to the concentration of nuclei. The purpose of this chapter is to explore the thermodynamics that govern the formation of the nuclei. If the concentration of the nuclei is denoted $c_i{}^*$, then equilibrium nucleation theory considers them to be in equilibrium with monomers (i.e., $c_i{}^* = K_{i*}c^{i*}$), where i^* is the number of molecules in the nucleus, and, of course, K_{i*} is the equilibrium constant for that process. (It will be assumed here that the conditions are dilute enough not to require activity coefficients. A review of crowding and polymerization can be found elsewhere [Ferrone and Rotter, 2004]). Therefore, what concerns us in this chapter is understanding K_{i*}, and i^* itself. It is precisely here that structure–function relationships will be manifest. This presupposes that nucleation control has been established, and the reader is referred elsewhere for determination of the appropriateness of nucleation control as the governing paradigm (Ferrone, 1999).

What is presently understood about describing the stability of nuclei in biological aggregation has been learned largely from work in sickle hemoglobin. This exploration of the underlying basis for the various nucleation parameters is the logical next step once a kinetic mechanism is described and, in turn, allows insights into the details of the interactions. Particularly useful in revealing such details is the contrast between the stability of

METHODS IN ENZYMOLOGY, VOL. 412
0076-6879/06 $35.00
DOI: 10.1016/S0076-6879(06)12017-0

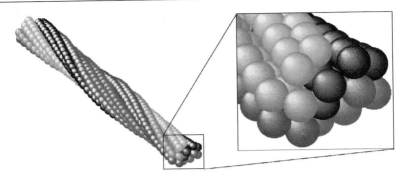

FIG. 1. The hemoglobin polymer. Each sphere is a molecule of sickle hemoglobin. The structure is a helical winding of seven pairs of double strands, which in turn are half-staggered structures. The thickness of the structure permits three-dimensional close-packed clusters to provide nuclei without invoking the details of the transformation from three-dimensional clustering to effectively one-dimensional growth.

nuclei, which are typically small aggregates, and the stability of the final structure.

Sickle hemoglobin differs from many aggregation diseases in that a quaternary structure change empowers the assembly, rather than a change in secondary–tertiary structure (i.e., a folding change). This makes sickle hemoglobin a molecule whose behavior is remarkably like a large sticky sphere and thus amenable to simplifications in analysis. The structure formed in sickle hemoglobin polymerization is a thick, ropelike arrangement composed of seven pairs of double-strands (Dykes *et al.,* 1979), as shown in Fig. 1. Although clearly not addressing some of the features fundamental to misfolding diseases, the analysis of sickle hemoglobin has provided a solid understanding of the assembly process per se. The generalization to folding diseases can thus be made from a stable platform. In what follows, we will first consider the association of fixed structure molecules and then describe some generalizations required to incorporate folding changes.

Equilibrium

Although assembly takes place in solution, the assembly process itself is well described by a model whose treatment is equivalent to the deposition of a gas to a solid. The dispersion of the proteins in solution and lack of long-range interaction gives them the properties of a gas. The protein interactions with its surrounding solvent can be described by an average interaction that is the same for every protein in the solution, and thus effectively recalibrates the energy zero point. It is useful to use the chemical potential for analysis, because it represents an average free energy and also has the desirable simplicity of being additive for the most part. In the solution, the chemical potential of a monomeric protein is:

$$\mu_{soln} = \mu_{ST} + \mu_{SR} + RT \ln c \qquad (1)$$

The subscript S here represents solution, and the second subscript represents translational and rotational motion. If the solution molecules are dense enough, it is necessary to introduce a term for molecular crowding. This activity coefficient γ multiplies c to give an "effective concentration" (i.e., the activity). The translational and rotational terms describe the motional freedom a monomer possesses and quantitatively represent the entropy they must give up in an association process. These terms depend on the size of the molecules in question and are derived in statistical mechanics texts (see, e.g., Hill's exemplary text [1986]). They are:

$$\mu_{ST} = -RT \left| \ln \left(\frac{2\pi m k T}{h^2} \right)^{3/2} + 1 - \ln \frac{N_o}{V_o} \right| \qquad (2)$$

and

$$\mu_{SR} = -RT \ln \frac{\sqrt{\pi}}{\sigma} \left(\frac{8\pi^2 I k T}{h^2} \right)^{3/2} \qquad (3)$$

Here R is the gas constant, T the absolute temperature, m the mass of the molecule, k, Boltzman's constant, h, Planck's constant, and N_o, Avogadro's constant. I is the moment of inertia of the molecule, which for a sphere of size r is $2/5\, mr^2$. σ is a symmetry number. V_o is the reference volume, related to the reference state. For hemoglobin, it is customary to use a 1 mM reference state, and thus N_o/V_o is 6.023×10^{20} molecules per liter. It is important that this number coordinate with the units for concentration c used in Eq. (1). For hemoglobin, the sum of these terms ($\mu_{ST} + \mu_{SR}$) is -35 kcal/mol, which is thus the measure of the free energy (because of motional entropy) that a 1 mM solution possesses. Smaller molecules will have a number closer to zero. For example, for a protein of the size of Aβ40, $\mu_{ST} + \mu_{SR}$ would be approximately -29 kcal/mol.

The aggregate is viewed in a crystalline approximation, that is, it has a chemical potential given by:

$$\mu_{agg} = \mu_{PC} + \mu_{PV} \qquad (4)$$

The leading subscript stands for "polymer," whereas the second subscript, respectively, describes contacts and internal vibrations. There is no $\ln c$ term here. The contact energies μ_{PC} are the free energies arising from the

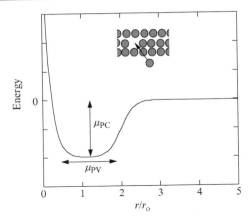

FIG. 2. Schematic potential–energy well for a molecule within a polymeric aggregate. The depth of the well is what gives the contact energy μ_{PC}, whereas the width of the energy well is related to the vibrational chemical potential, μ_{PV}. The x-axis schematically shows the displacement of the molecule from r_o, its average position in the aggregate. In an aggregate, each of six degrees of freedom would have a well, describing the various linear and rotational motions.

various contact sites within the polymer. If the sites are well enough isolated, then individual energies can be estimated from standard energies that are determined elsewhere. Most notably, hydrophobic energies are amenable to such an approach, and if the structure of the aggregate is known, even better approximations of these can be obtained.

μ_{PV} is an unusual term. It is the contribution made by the movement of the molecules around their equilibrium positions in the aggregate, as illustrated in Fig. 2. These vibrations are *not* vibrations within the molecules themselves but are the movements of the molecules as if they were hard spheres jiggling about their mean locations. Such an idea is not new and, for example, has been in use in the field of water droplet condensation for a considerable time (Abraham, 1974). What is novel is that the size of this term, relative to the others, is more substantial than is seen in water droplet condensation, as will be discussed in the following.

At equilibrium, the solution and aggregate chemical potentials are equal:

$$\mu_{\text{soln}} = \mu_{\text{agg}} \tag{5}$$

and this gives a simple equation for solubility, or critical concentration, viz.:

$$\text{RT} \ln c_s = \mu_{PC} + \mu_{PV} - \mu_{ST} + \mu_{SR} \tag{6}$$

(and, of course, if crowded, the concentration would be multiplied by an activity coefficient γ_s). This equation specifically excludes any internal differences in the molecules in the solution or aggregated forms. Note that low solubility (O'Nuallain *et al.*, 2005; Sengupta *et al.*, 2003) creates an experimental problem of distinguishing between equilibrium and a system that has simply run to completion by exhaustion of the monomers. Measurement of solubility is a critical feature in describing any equilibrium properties.

Clusters

Whereas the aggregated phase is viewed as infinitely large, nucleation is a phenomenon that comes from finite size aggregates, which we will call clusters. As a whole entity, the cluster can translate and rotate, yet internally it has contact and vibrational energy. Thus, a cluster of size i has chemical potential:

$$\mu_i = \mu_{iT} + \mu_{iR} + \mu_{iC} + \mu_{iV} + RT \ln c_i \tag{7}$$

The cluster with the smallest concentration of all clusters on the reaction path is the nucleus. Thus, the strategy is to equate the chemical potential of a cluster of size i with that of i monomers, solve for c_i, and then find what value of i (called i^*) minimizes c_i. The reasons for going through the details here is to allow easier adaptation if a particular assumption is to be modified for a future problem.

Translation and rotation of a cluster can be easily related to the corresponding term for the monomer because Eq. (2) and (3) must be modified transparently for a larger object. Thus, in Eq. (2), the cluster mass is simply the mass of i molecules, and:

$$\mu_{iT} = -3/2 \, RT \ln i + \mu_{ST} \tag{8}$$

In Eq. (3), the moment of inertia goes up by the increased mass *and* the increased size of the cluster. If the cluster can be approximated as spherical, the moment of inertia I increases as $i^{5/3}$. Therefore:

$$\mu_{iR} = -3/2 \, RT \ln (I_i/I) = -5/2 \, RT \ln i + RT \ln v\rho + \mu_{SR} \tag{9}$$

The added term of $\ln v\rho$ is to account for the different density of the nucleus relative to the monomer, and ρ is the relative density, whereas v is the specific volume of the monomer. Thus,

$$\mu_{iR} + \mu_{iT} = -4 \, RT \ln i + RT \ln v\rho + \mu_{ST} + \mu_{SR} \tag{10}$$

The contact energy is not proportional to the size of the cluster, but it arises from the number of contacts, which depends on the specific geometry. Let $\delta(i)$ represent the fraction of contacts present in a cluster of size i, relative to an infinitely large aggregate. Therefore, $i\delta(i)$ is the number of contacts present in a cluster of size i. Thus, the contact energy in a cluster of size i is given by:

$$\mu_{iC} = i\delta(i)\mu_{PC} \tag{11}$$

The cluster vibrational term, μ_{iV}, is the result of the normal modes of vibration of a cluster. There will be six normal modes added for each molecule added, viz. three for linear vibrations and three for torsions and bending, *except* that the cluster, being allowed to rotate and translate, reserves 6 degrees of freedom for that motion. Thus there are $6(i-1)$ degrees of vibrational freedom and, therefore, that many vibrational modes. The assumption that has been successfully used in sickle hemoglobin is that only the number of modes increases as the cluster gets larger, rather than the frequency of the vibrations themselves. Hence,

$$\mu_{iV} = (i - 1)\mu_{PV} \tag{12}$$

Linear Structures

For illustration, consider a simple linear chain, with one contact region between elements (Fig. 3). The sequence of numbers of contacts for successively large clusters is given by $i\delta(i) = i - 1$ as shown in the figure. For this case the contact energy is:

$$\mu_{iC} = (i - 1)\mu_{PC} \tag{13}$$

where μ_{PC} is the contact energy of a single bond, as is found in the infinite polymer.

Therefore, equating the chemical potential of a cluster to that of its i constituents,

$$i\mu_{soln} = \mu_i \tag{14}$$

# of molecules	1	2	3	4	5	6 ...
# of bonds	0	1	2	3	4	5 ...

FIG. 3. A simple linear polymerizing system. Every molecule added contributes exactly one bond or contact. This system does not nucleate without other features such as internal energy changes.

$$i(\mu_{ST} + \mu_{SR} + RT \ln c_o) = -4RT \ln i + RT \ln v\rho + \mu_{ST} + \mu_{SR}$$
$$+(i-1)\mu_{PC} + (i-1)\mu_{PV} + RT \ln c_i \quad (15)$$

Using the solubility equation, we can simplify to get:

$$i\, RT \ln (c_o/c_s) = -4\, RT \ln i + RT \ln v\rho - \mu_{PC} - \mu_{PV} + RT \ln c_i \quad (16)$$

or

$$\ln c_i = i \ln S + 4 \ln i - \ln v\rho + (\mu_{PC} + \mu_{PV})/RT \quad (17)$$

S is defined as the supersaturation of the solution, here simply given by the ratio of concentration to solubility, c/c_s. (In crowded solutions, S would be the ratio of activities rather than concentrations.) To find the nucleus, c_i is minimized by taking the derivative relative to i and setting to zero. This gives $i^* = -4/\ln S$ (i^* is used because now this is the value of the minimum c_i). Because S must be >1, this gives a *negative* value for the nucleus size, a decidedly unphysical outcome. Linear chains alone cannot have a nucleus: if the solution is supersaturated (meaning $S > 1$), any size aggregate is more stable than monomers.

A similar result applies to a double-stranded chain as drawn in Fig. 4. Let all contacts be equal in strength. The first molecule has no contacts. The second molecule has 1, for an average of 1/2 per molecule. The third molecule adds two contacts: now there is an average of 1 per molecule. The fourth molecule adds two contacts again, adding one per molecule. The net number of contacts is thus as shown, leading to $\mu_{iC} = (2i - 3)\,\mu_{PC}$ ($i > 1$). (In the infinite double strand, on the average, each molecule contacts four neighbors, and so gets assigned four "half-bonds" for a total of two bonds per molecule.) Because once again (for $i > 1$) μ_{iC} is proportional to i, there is again no nucleus beyond dimers. There is a discontinuity between the dimer and trimer, and this permits a dimeric nucleus. (Note that a nucleus of one is also possible here.) Finite size nuclei depending on the bond

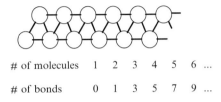

# of molecules	1	2	3	4	5	6	...
# of bonds	0	1	3	5	7	9	...

FIG. 4. Two-strand linear polymerizing system. After the second molecule, every added molecule contributes two contacts. The monomer–dimer transition, however, has only one added contact, and in contrast to Fig. 3 could formally provide a nucleus at that step.

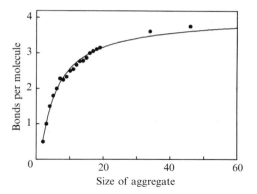

FIG. 5. Contacts vs size for close packed spheres. The points are taken from Ginnel (1961). The line is a fit of $N\delta(i)$ where $\delta(i)$ is defined in Eq. (18), and N is the asymptotic number of such contacts as the cluster size becomes infinite. $\delta_1 = 1.29 \pm 0.04$ and $\delta_2 = 0.84 \pm 0.06$. The value of N is 4.14 for the polymer structure shown in Fig. 1 and was determined to be 4.18 ± 0.08 in an unconstrained fit to the points.

strengths in detail and not dominated by the geometry are possible when μ_{iC} depends on i in a nonlinear way, i.e., the function $i\delta(i)$ must be nonlinear. Two- and three-dimensional geometries have this property.

Three-Dimensional Clusters

The simplest conceptual three-dimensional arrangement, and the one that applies to sickle hemoglobin (recall Fig. 1) is that of close-packed spheres. The number of contacts made between close packed spheres in small clusters has been determined (Ginnel, 1961). The function $i\delta(i)$ will be most useful in generating an analytic solution if it has the form (Ferrone *et al.*, 1980):

$$i\delta(i) = i + \delta_1 \ln i + \delta_2 \qquad (18)$$

This form does an excellent job of describing the empirically counted contacts (Ginnel, 1961), as shown in Fig. 5. When this is used, (Eq.) 17 becomes:

$$\ln c_i = i \ln S + (4 + \delta_1\mu_{PC}/RT) \ln i - \ln v\rho + ((1 - \delta_2)\mu_{PC} + \mu_{PV})/RT \qquad (19)$$

Figure 6 shows the effective free energy barriers for this three-dimensional model. For strong enough μ_{PC}, this model exhibits a real nucleus. Taking the derivative of Eq. (19) generates the nucleus size in analytic form:

$$i* = -\frac{4 + \delta_1 \mu_{PC}/RT}{\ln S} \tag{20}$$

Despite the negative sign, $i*$ is positive because the energy μ_{PC} is a stabilizing one, and thus negative itself. Consequently, if $\delta_1 \mu_{PC}/RT$ is negative enough (i.e., smaller than -4) Eq. (20) will yield a positive value of $i*$. The form of this nucleus expression is quite general, because many possible geometries can be parameterized approximately by Eq. (18), and the geometric factors will appear as different values for the constant δ_1. It might seem inconsistent to consider spherical clusters that form a fibrous ropelike structure. When the transformation occurs beyond the nucleus size, the problem does not create inconsistencies; in the case of sickle hemoglobin polymers, the diameter is large enough to include rather large nuclei (recall Fig. 1). This equation can also generate a nucleus size of approximately 1 for a limited range of supersaturation, although a large enough variation of S will cause $i*$ to vary (cf. following).

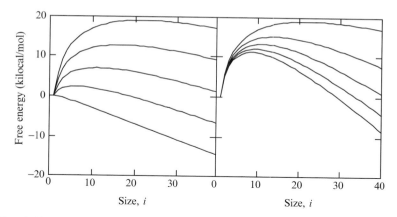

FIG. 6. Free energy barriers. The energy difference $\mu_i - i\mu$ ($= RT \ln c_i$) is shown as a function of size, on the basis of Eq. (19). This energy difference represents the energy of an aggregate relative to its constituents; monomers have been set to zero. Small aggregates are less stable than the monomers (which have been set to a common zero), causing the initial steps to be unfavorable. The size at which each curves peaks is the nucleus, which clearly varies with the conditions. Note that ultimately the aggregates are more stable than monomers. (A) S is fixed at 2, and μ_{PC} ranges from 2–6 kcal/mol (highest curve) in steps of 1 kcal/mol. Because supersaturation is fixed in the graph while μ_{PC} varies, each curve would require lower concentration to retain the same supersaturation (because the change in μ_{PC} would be changing the solubility). (B) μ_{PC} is fixed at 6 kcal/mol, and S varies from 2–7 (lowest curve).

The most remarkable feature of this equation is that the nucleus size depends on the supersaturation S, which in turn depends on the initial concentration. The physical reason for this is quite intuitive. In equilibrating monomers with the cluster, the monomers must surrender some entropy they had by virtue of their ability to wander in the solution. That entropy is the origin of the ln c term in the chemical potential of Eq. (1). The more concentrated the solution, the less space each molecule has been "assigned" as its own, and thus the less onerous is the prospect of joining a less mobile cluster.

The experimenter controls the degree to which i^* varies. Varying initial concentration 10-fold will cause ln S to change by 2.3. For $S = 1.3$, a modest supersaturation, the nucleus size would change by almost 10 for a change of 10 in S. Changing S by a factor of two would increase the nucleus size a factor of 3.6. Note the importance of the solubility here. It is not so much the change in concentration per se, but the change relative to solubility that changes nucleus size. This inevitable variation in nucleus size, when confronted with experimental error or differences in conditions between experiments (especially in different laboratories), can easily make for a confusing situation.

Internal Changes

What if there are internal changes (e.g., a refolding)? Energies effectively internal to the monomer now must be considered. Let μ_{SI} be the solution phase internal chemical potential and μ_{PI} be the polymeric (aggregated) chemical potential. The solubility equation (Eq. [6]) becomes:

$$RT \ln c_s = \mu_{PC} + \mu_{PV} - \mu_{ST} + \mu_{SR} + (\mu_{PI} - \mu_{SI}) \qquad (21)$$

The parenthetical term is likely to be positive (destabilizing). (If the refolded structure were more stable, the solution molecules would spontaneously convert without aggregation.) The contacts in μ_{PC} are those that are found in the aggregate and may well occur between regions that would not be exposed in the initial solution structure. Thus, these are the contacts that stabilize the refolded form in the presence of the other molecules in the aggregate.

For the creation of the cluster in the presence of internal rearrangements, there is also an internal cluster term μ_{iI}. If it is assumed that the refolding is all-or-none, then the internal energy appears for every molecule in the cluster, and Eq. (19) remains the same, with the same Eq. (20) for the nucleus size. The nucleus size itself does depend on the internal changes in structure, because those internal changes alter solubility and thus the supersaturation.

A Nucleus of 1?

Recent analysis of polyGln aggregation led to an unusual, but model, independent conclusion (viz. that the nucleus size was unity for this assembly). Although the analysis that led to the conclusion is straightforward, its implications are not, and it is worth a brief discussion of what does or does not constitute a nucleus of 1. For nucleated polymerization, the central and necessary idea is that the monomers present at time zero face an uphill climb over a barrier to generate stable polymers. If a monomer must undergo a conformational change before it can then polymerize, even if all added monomers make a more stable system, then the system formally nucleates. It is also possible for a nucleus with a thermodynamic form given by Eq. (20) to simply have i^* close to 1. So long as i^* is larger than 1, even if less than 2, the system has a small barrier over which aggregates must climb. If a monomer simply must find another monomer, but faces no unfavored states along the way, the system climbs no barrier, does not nucleate, and the growth curves will be different from nucleated assembly.

The Concentration of Nuclei

Given a nucleus i^*, the next task is to determine the concentration of nuclei, c_{i*}. This is done by straightforward substitution. To make the expressions somewhat more compact, we introduce the variable:

$$\xi = -(4 + \delta_1 \mu_{PC}/RT) \qquad (22)$$

Now the nucleus expression is particularly concise, $i^* = \xi/\ln S$, and it follows that:

$$\ln c_{i*} = \xi \ln(\ln S) + \xi - \xi \ln \xi + - \ln v\rho + ((1 - \delta_2)\mu_{PC} + \mu_{PV})/RT \qquad (23)$$

Only the first term varies with concentration; the remaining terms are constants (some of which will change inevitably if solubility changes).

How does this relate to analysis of kinetics? The key point is that the homogeneous nucleation rate f will be proportional to the concentration of nuclei c_{i*}. Thus, homogeneous nucleation provides insights into the equilibrium behavior and energetics. In turn, $c_i^* = K_{i*}c^{i^*}$ and thus it might have been expected that $\ln c_i^*$ would have a slope linear in $\ln c$. For a sufficiently large range of date, this does not occur, because i^* is not a constant. To illustrate the curvature of concentration dependence, Fig. 7 shows a plot of the log of the homogeneous nucleation rate, which is proportional to $\ln(\ln S)$, as a function of supersaturation, S. Taken over a small region, the graph might appear linear, and with any noise in the data,

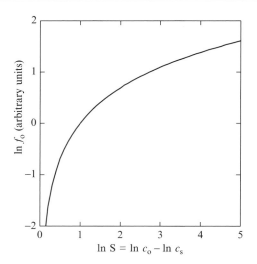

FIG. 7. Homogeneous nucleation rate f as a function of supersaturation S. Although small segments may appear linear, it is evident that over a wide range, the log of nucleation rate is not proportional to log c. Although the x-axis is log S, it is equivalent to log c minus an offset, and thus the nonlinear behavior would be evident even if solubility is not known (so S could not be determined).

the nonlinearities would be further masked. Nonetheless, the full range is clearly nonlinear. Note that the x-axis is $\ln S$ and therefore is equal to log c_o − log c_s. Hence, even if S could not be determined because the solubility was not known, a log–log plot of nucleation rate vs initial concentration would resemble this figure; however, in that case, the x-axis would be shifted by an unknown amount (viz., $\ln c_s$).

The Outcome

What does this analysis yield? Analysis of the rate of homogeneous nucleation, as the preceding makes clear, generates a value for the variable ξ, which in turn gives μ_{PC}, the average contact strength. This assumes that the geometric factor δ_1 is known from some structural information. This energy μ_{PC} is a fundamental quantity of the interaction and by construction is related directly to the molecular contacts. As Eq. (6) makes clear, the solubility alone cannot provide such information, because that must include the vibrational chemical potential, as well as any internal refolding energies.

TABLE I
ENERGIES INVOLVED IN SICKLE HEMOGLOBIN POLYMERIZATION (KCAL/MOL) (25°)

Overall free energy (solubility)	−1.5
Solution rotational and translational energy	−35.5
Contact energy	−7.5
Vibrational energy in polymer	−26.5

It is worth considering the magnitudes of these quantities, shown in Table I for sickle hemoglobin (Cao and Ferrone, 1997). It is remarkable how well vibrational motions compensate for the loss of motion possessed by the free molecules in solution. The vibrational motions in question have a macroscopic manifestation in the bending and twisting of the polymers. This represents a coherent (i.e., long range) form of the bending and torsional vibrations. The calculation of the modes of motion is a complex problem even for structures that are known at high resolution, and so this information must remain somewhat empirical. But it provides useful insight in that structures that are less flexible will also be less stable if the contact energy remains constant.

This illustrates a result that at first seems paradoxical. At fixed supersaturation, stronger intermolecular contacts lead to *larger* nuclei (according to Eq. [20]) and thus slower nucleation rates (because it is harder to make the larger nuclei). One might have thought that stronger bonds would lead to faster, not slower, nucleation rates. The source of this paradox is that solubility was held constant, for to fix solubility as μ_{PC} changes means that μ_{PV} is changing as well. Physically, keeping S fixed means that the strengthening of the contacts is necessarily being offset by a presumed decrease in mobility of the molecules within the clusters. The reason such a tradeoff slows nucleation despite constant supersaturation is that, as clusters form, the vibrational modes are recovered in full for all clusters, in contrast to the contact energies that increase more slowly (viz., as the number of intermolecular contacts rises). This very behavior is seen in a sickle hemoglobin double mutant, HbC-Harlem, which forms crystals instead of flexible fibers, and with an immensely slowed rate, as a result of a mutation in a primary contact area (Ivanova *et al.*, 2001).

Other Approaches

This approach differs somewhat from traditional nucleation theories, where it is common to use a surface tension and a bulk energy. The same features are present, however. The difference between surface and bulk has been handled in the preceding treatment explicitly by including molecular bonds and the $\delta(i)$ function. A surface-tension approach is logical for describing water vapor clusters, for example, because the surface tension is an independently measurable quantity. In contrast, the surface tension of a cluster of aggregating molecules is virtually unmeasurable by any known technique. On the other hand, contacts between molecules are the stock-in-trade of modern protein chemistry, and the vibrational entropy contribution, as mentioned, should be accessible (at least in terms of systematic changes) through observation of the flexibility of the fibrillar assemblies.

It is also worth mentioning the well-known approach used by Oosawa (Oosawa and Asakura, 1975) that involved a linear trimer that converted to a triangular geometry as the nucleation event. This is not a case of equilibrium nucleation, however, because equilibrium mandates that the best configurations are constantly produced. If a triangular trimer had the best stability, it, and not the linear trimer, is the correct three-molecule species to consider.

Finally, it is worth noting that there are approaches such as helix-closure in which nucleation is the consequence of an abrupt event. A simple version of such a discontinuity was seen in Fig. 4, when the formation of a trimer caused a different change in stability than seen in the dimer or in subsequent i-mers. In constructing any such scenario, if equilibrium nucleation is to be used, it is important that the best-case structure (viz., with most bonds) be used. For example, consider a hypothetical model in which a 12-mer helically closes on itself like a lock-washer in such a way that the 13-th member of the string has an added contact beyond that which simply lengths the string. Then the 12-mer would have 11 lateral bonds, and the 13-mer would have 12 lateral bonds plus one axial bond of closure. (Thereafter, every molecule would add one lateral and one axial bond). Such a model assumes that the competing structure of a 12-mer with two lines of six molecules is less stable than a structure with the molecules in a simple line. The preceding two-line configuration has 10 lateral bonds plus six axials as opposed to the 11 lateral bonds plus 0 axials. Thus, for such a lock-washer nucleus model to be internally consistent, six axial bonds could not be stronger than one of the lateral bonds, because otherwise the structure would prefer to grow as a patch of the surface rather than a string that closes.

Heterogeneous Nucleation

Sickle hemoglobin also possesses a secondary nucleation pathway, in which nuclei can form preferentially on the surface of other polymers. This produces an autocatalysis coupled with high reaction order. The description of heterogeneous nucleation has been described elsewhere (Ferrone et al., 1985, 2002) and will not be given in detail here, because it is a relatively simple extension of the previous ideas. The critical conceptual trick is to recognize that nucleation barriers are computed as an equilibrium process and, as such, do not depend on the pathway by which the state is reached. Hence, even though heterogeneous nuclei grow on the surface from monomers, their equilibrium constant is the same as that computed by considering the process of forming a cluster in solution and attaching it whole to the polymer. The only new terms are those that involve connection of the polymer and cluster. The reader is referred to the literature (Ferrone *et al.*, 1985, 2002) for further details.

Acknowledgment

The author thanks W. A. Eaton and J. Hofrichter with whom most of these ideas were originally developed.

References

Abraham, F. F. (1974). "Homogeneous Nucleation Theory." Academic Press, New York.

Cao, Z., and Ferrone, F. A. (1997). Homogeneous nucleation in sickle hemoglobin. Stochastic measurements with a parallel method. *Biophys. J.* **72,** 343–372.

Dykes, G. W., Crepeau, R. H., and Edelstein, S. J. (1979). Three dimensional reconstruction of 14-filament fibers of hemoglobin S. *J. Mol. Biol.* **130,** 451–472.

Ferrone, F. (1999). Analysis of protein aggregation kinetics. *Methods Enzymol.* **309,** 256–274.

Ferrone, F. A., Hofrichter, J., and Eaton, W. A. (1985). Kinetics of sickle hemoglobin polymerization II: A double nucleation mechanism. *J. Mol. Biol.* **183,** 611–631.

Ferrone, F. A., Hofrichter, J., Sunshine, H., and Eaton, W. A. (1980). Kinetic studies on photolysis-induced gelation of sickle cell hemoglobin suggest a new mechanism. *Biophys. J.* **32,** 361–377.

Ferrone, F. A., Ivanova, M., and Jasuja, R. (2002). Heterogeneous nucleation and crowding in sickle hemoglobin: An analytic approach. *Biophys. J.* **82,** 399–406.

Ferrone, F. A., and Rotter, M. A. (2004). Crowding and the polymerization of sickle hemoglobin. *J. Mol. Recognition* **17,** 497–504.

Ginnel, R. (1961). Geometric Basis of Phase Change. *J. Chem. Phys.* **34,** 992–998.

Hill, T. L. (1986). "An Introduction to Statistical Thermodynamics." Dover Publications, New York.

Ivanova, M., Jasuja, R., Krasnosselskaia, L., Josephs, R., Wang, Z., Ding, M., Horiuchi, K., Adachi, K., and Ferrone, F. A. (2001). Flexibility and nucleation in sickle hemoglobin. *J. Mol. Biol.* **314,** 851–861.

O'Nuallain, B., Shivaprasad, S., Kheterpal, I., and Wetzel, R. (2005). Thermodynamics of Ab (1-40) amyloid fibril elongation. *Biochemistry* **44,** 12709–12718.

Oosawa, F., and Asakura, S. (1975). "Thermodynamics of the Polymerization of Protein." Academic Press, New York.

Sengupta, P., Garai, K., Sahoo, B., Shi, Y., Callaway, D. J., and Maiti, S. (2003). The amyloid beta peptide (Abeta(1-40)) is thermodynamically soluble at physiological concentrations. *Biochemistry* **42,** 10506–10513.

[18] Amyloid Fibril Structure Modeling Using Protein Threading and Molecular Dynamics Simulations

By Jun-tao Guo *and* Ying Xu

Abstract

The elucidation of the structure of amyloid fibrils is an important step toward understanding the mechanism of amyloid formation and developing new reagents that could inhibit fibril formation. Here we describe an approach to modeling amyloid fibril structures using computational techniques, including protein threading and molecular dynamics simulations. Specifically, we introduce these methods using Aβ amyloid fibril modeling as an example. First, the amyloid protein sequence is threaded against a set of structural templates. Structural models are generated on the basis of threading alignments and are then subjected to molecular dynamic simulations to assess the stabilities of the model.

Introduction

The amyloidoses are disorders characterized by the extracellular accumulation of amyloid fibrils (Sipe, 1992). Although amyloid precursor proteins do not share any homology with respect to their amino acid sequences, the structural characteristics of the amyloid fibrils derived from these peptides and proteins are remarkably similar. For example, electron microscopy (EM) reveals that amyloid fibrils are straight, unbranched, with a diameter of 70–120 Å (Serpell, 2000). X-ray diffraction patterns of oriented amyloid fibrils indicate that they share a common cross-β structural motif, in which the β-strands run perpendicular to the long axis of the fibrils, whereas the hydrogen bonds between β-strands are parallel to the axis (Sunde and Blake, 1997). Because of the importance to understanding the functional roles of amyloid fibrils in amyloidoses and the potential bearing on the rational design of therapeutics, high-resolution structures have been the subject of intense research. As of today, however, high-resolution structural characterization of amyloid fibrils using traditional experimental structure solution methods, such as x-ray crystallography or solution NMR, have not been successful, because of the insolubility and noncrystalline nature of the amyloid fibrils. At the same time, other experimental approaches, such as fiber diffraction (Sunde and Blake, 1998), electron microscopy (Jimenez *et al.*, 2002), hydrogen-deuterium exchange

METHODS IN ENZYMOLOGY, VOL. 412
0076-6879/06 $35.00
DOI: 10.1016/S0076-6879(06)12018-2

(HX) (Kheterpal *et al.*, 2000), solid state NMR (Petkova *et al.*, 2002, 2005), small angle neutron scattering (Lu *et al.*, 2003), limited proteolysis (Kheterpal *et al.*, 2001), and electron paramagnetic resonance spectroscopy (EPR) (Jayasinghe and Langen, 2004; Torok *et al.*, 2002), have yielded valuable low-resolution data.

Under the current circumstances, computational techniques have been considered as a preferred approach to building structural models of the amyloid fibrils, to testing the stabilities of the modeled structures, and to studying the fibril formation process (Guo *et al.*, 2006; Zanuy *et al.*, 2004). There are two classes of approaches to modeling amyloid fibril structures. One class of methods predicts amyloid fibril structures through first generating fibril structural models and testing the model stability using molecular dynamics (MD) simulations. This type of approach bypasses the fibril formation process and studies the chemical interactions that stabilize the fibril structure. The stability of proposed oligomer models can then be assessed using molecular dynamics simulations. The second class of approaches simulates the fibril formation process, which includes modeling of conformational changes from the native globular protein, seed formation, protofilament formation, and protofilaments packing. In this chapter, we mainly focus on the first class of approaches; the second approach is discussed in detail in other chapters (see Chapters 19 and 20 in this volume).

The first step of the approach is to construct monomer and oligomer models. For fibrils formed by short peptides, the models can be generated manually (Ma and Nussinov, 2002a,b). For fibrils formed by longer or full-length amyloid sequences, protein threading (Bowie *et al.*, 1991; Jones *et al.*, 1992) seems to be a natural and useful approach for modeling monomer structures, because protein structures in Protein Data Bank (PDB) (Berman *et al.*, 2000) with cross-β features might hold folding patterns of amyloid fibrils (Jenkins and Pickersgill, 2001; Wetzel, 2002). It has been suggested that in amyloidoses, proteins may have converted into the "primordial" structure rather than remaining in their evolved folded states (Dobson, 2002; Pickersgill, 2003). Protein threading identifies a structural homolog or analog through aligning the query sequence onto template structures and finds the best possible template through evaluating sequence–structure alignment using empirical energy functions. Given recent suggestions that fibrils are stabilized by forces common to all proteins, not by forces particular to a specific sequence (Bucciantini *et al.*, 2002; Kayed *et al.*, 2003), threading should be a good tool for identifying structural motifs in the PDB structures that might possibly share similar structural features to those of amyloid fibrils (Guo *et al.*, 2006).

We use Aβ amyloid as an example to discuss how to apply the protein threading approach to model the fibril structure and to assess the stability of the constructed models using MD simulations. The general procedure should be applicable to other fibril structure modeling as well.

Aβ Amyloid Fibril Structure Studies Using Computational Approaches

The amyloid fibrils are composed of amyloid-β (Aβ) peptides, 40–42 amino acid fragments derived from the amyloid precursor protein (APP) (Yankner, 1996). Computational modeling of fibril structures has been reported on both short and long Aβ peptides (Guo et al., 2004; Li et al., 1999; Ma and Nussinov, 2002b). Although these models differ in many structural details, they generally fall into two main categories: anti-parallel and parallel β-pleated sheet arrangements. Motivated by the compelling evidence from the solid state NMR (Petkova et al., 2002) and liquid suspension EPR (Torok et al., 2002) studies on full-length Aβ fibrils, which suggests that the peptides in the fibril core are in an in-register, parallel arrangement, several parallel Aβ models have been proposed (Guo et al., 2004; Lakdawala et al., 2002; Petkova et al., 2002).

In our modeling of the amyloid fibril core structure composed of Aβ (15–36) (Guo et al., 2004), Aβ(15–36) is used as the query sequence, because a number of studies have suggested that Aβ(15–36) is involved in the core formation of Aβ amyloid fibrils (Guo et al., 2004; Kheterpal et al., 2001; Torok et al., 2002; Williams et al., 2004).

Computational Modeling of Aβ Oligomers

A flowchart for computational modeling of amyloid fibril structure is shown in Fig. 1. First a single-chain model of the peptide sequence is modeled using a threading approach. Then the protofilament or multichain models are constructed on the basis of protein structural templates with either parallel or antiparallel conformations. After the models are built, an evaluation is carried out to make sure that the models have good quality. The molecular dynamics simulations are then performed, and the simulation results are analyzed to check the stabilities of the models and the detailed chemical interactions. Models can be refined on the basis of the analysis from molecular dynamics simulations and new experimental data.

Protein Threading Using PROSPECT

Protein threading is carried out on Aβ(15–36) sequence using PROS-PECT (http://csbl.bmb.uga.edu/downloads/#prospect) (Kim et al., 2003; Xu and Xu, 2000) against a preselected structural template database.

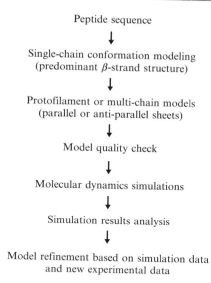

Peptide sequence

↓

Single-chain conformation modeling
(predominant β-strand structure)

↓

Protofilament or multi-chain models
(parallel or anti-parallel sheets)

↓

Model quality check

↓

Molecular dynamics simulations

↓

Simulation results analysis

↓

Model refinement based on simulation data
and new experimental data

FIG. 1. Flowchart for computational modeling of amyloid fibril structures.

PROSPECT uses a knowledge-based energy function, and it guarantees to find the globally optimal sequence–structure alignment under the given energy function (Xu and Xu, 2000).

Threading Energy Functions. PROSPECT uses three knowledge-based energy terms: mutation ($E_{mutation}$), singleton ($E_{singleton}$), and pairwise ($E_{pairwise}$) energies, plus a gap penalty function (E_{gap}) and a secondary structure match function (E_{ss}) (Kim *et al.*, 2003; Xu and Xu, 2000). The overall threading score is measured by the following function:

$$E_{total} = \omega_m E_{mutation} + \omega_s E_{singleton} + \omega_p E_{pairwise} + \omega_g E_{gap} + \omega_{ss} E_{ss} \quad (1)$$

where ω_m, ω_s, ω_p, ω_g, and ω_{ss} are scaling factors for the corresponding energy terms. The default set of these weighting factors in PROSPECT is chosen through optimizing the overall alignment accuracy on a training set (Xu and Xu, 2000). The mutation energy describes the compatibility of substituting one amino acid type by another. The new version of PROS-PECT uses profile–profile alignment to calculate the mutation energy (Kim *et al.*, 2003). The singleton energy measures the fitness of each of the amino acids of the query sequence to its aligned structural environment in the structural template, defined in terms of secondary structure and solvent accessibility. The pairwise energy measures the mutual preference of two amino acid types that are spatially close. PROSPECT provides many different options for running the system. For example, a user can adjust the scaling factors and turn on/off some of the energy terms. The user's manual of PROSPECT is available at http://csbl.bmb.uga.edu/downloads/prospect_manual/.

Threading Templates. Parallel β-helical-type models have been proposed for other types of amyloid fibrils including polyglutamine (Perutz *et al.*, 2002), mammalian prions (Wille *et al.*, 2002), and insulin amyloid fibrils (Jimenez *et al.*, 2002). Thermodynamic analysis of the stabilities of fibrils derived from proline mutants of $A\beta(1-40)$ also points to a well-known folding motif, the parallel β-helical fold, as a possible model for $A\beta$ fibrils (Williams *et al.*, 2004). Therefore, our structural templates include all the left-handed and right-handed parallel β-helical proteins defined by SCOP (Structural Classification of Proteins, http://scop.berkeley.edu/) (Murzin *et al.*, 1995) in the FSSP (Families of Structurally Similar Proteins) database (Holm and Sander, 1996) that do not share significant sequence similarity ($<25\%$ sequence identity). This set consists of 21 β-helical structures, listed by their PDB codes (with the fifth letter indicating the chain name, if any): 1fwya, 1qrea, 1lxa, 3tdt, 1kqaa, 2xat, 1m8nb (left-handed parallel β-helical proteins), 1hf2a, 1rmg, 1bhe, 1czfa, 1h80a, 1dbga, 1tyv, 1qcxa, 1air, 1qjva, 1daba, 1ee6a, 1kq5a, and 1ezga (right-handed parallel β-helical proteins). Representative non-β-helical all-β proteins that contain predominantly parallel and antiparallel β-sheet structures are also selected from the PDB as controls. The template files of the preceding protein chains are then generated using PROSPECT's make_template program. For example, to generate the template file using protein chain 1qjva, a user can simply use the following command:

    make_template –pdbfile 1qjv.pdb –c A

where *–pdbfile* and *–c* specify the pdb name and the chain identifier, respectively. A user can also use structural domains of a protein as threading templates by specifying the domain boundaries when generating template files:

    make_template –pdbfile 1qjv.pdb –c A –d 56 328

This command will generate a threading template file using residues 56–328 of the A chain of protein 1qjv.

Protein Threading and Model Construction. The sequence profile of $A\beta$ (15–36) is generated using PSI-BLAST (Altschul *et al.*, 1997), which is threaded using PROSPECT against all the template structures as described previously. The alignment scores indicate the sequence-structure compatibility. Figure 2 shows an alignment example. The top line is the query sequence $A\beta(15-36)$, and the bottom line is the sequence of the structural

```
Alignment(1m8nb):
QKLVFFAEDVGSNKGAIIGLMV
     |    . .|    .
-TCVNTNSQITANSQCVKSTAT
```

FIG. 2. Threading results of Aβ(136) against template 1m8nb.

template (1m8nb) aligned to the query sequence. Using the highest-scoring alignments from the threading analysis and the strong evidence from solid state NMR and EPR studies that Aβ monomers are in in-register, parallel β-sheet organization in Aβ fibrils, multimer (Aβ15–36) models are generated using MODELLER (http://salilab.org/modeller/) (Sali and Blundell, 1993). To construct a 6-mer structure, six successive rungs of β-helices of template structures are aligned to six Aβ(15–36) sequences. The alignment files for MODELLER are generated as follows: one copy of Aβ(15–36) is aligned with corresponding residues of template structure as shown in the threading alignments; five other Aβ(15–36) peptides are placed in successive rungs of the template, one peptide per rung, in an in-register fashion. A segment of glycine residues is added to connect the N- and C-termini of adjacent Aβ peptides. Figure 3 shows a sample alignment file for MODELLER. Model building using MODELLER is simply done using its default parameters. After the model is constructed, the glycine connectors are removed. The resulted model contains six chains of Aβ(15–36). Figure 4 shows one of the structures based on template 1lxa, a left-handed parallel β-helical fold.

Model Quality Evaluation

The stereochemical quality of the computed models is evaluated using PROCHECK (Laskowski et al., 1993). PROCHECK takes protein structures as inputs and outputs a number of plots in Postscript format, which include the standard Ramachandran plot, Ramachandran plots by residue type, and a few others. The Ramachandran plot shows the phi–psi torsion angles for all residues in the structure. The darkest areas correspond to the most favored regions. As shown in Fig. 5, in a right-handed β-helical model (1bhe as template) with six Aβ(16–35) molecules, approximately 79.4% of the residues are in the most favored regions. One residue is in a disallowed region (white area). The left-handed model has a better quality, with 86% of residues being in the most favored regions and none in the disallowed regions (Guo et al., 2004). For a typical (high-resolution) X-ray protein structure, approximately 90% of its dihedral angles lie within the preferred region of the Ramachandran plot, whereas this number for a typical

```
>P1;abeta
sequence:abeta:1 : :147 : :target sequence: unknown: :
QKLVFFAEDVGSNKGAIIGLMVGGGQKLVFFAEDVGSNKGAIIGLMVGGGQKLVFFAEDV
GSNKGAIIGLMVGGGQKLVFFAEDVGSNKGAIIGLMV--GGG---QKLVFFAEDVGSNKG
AIIGLMVGGGQKLVFFAEDVGSNKGAIIGLMV*
>P1;1lxa
structure:1lxa:1   : :125 :
----MIDKSAFVHPTAIVEE-------GASIGANAHIGPFCIVGP-------HVEIGEGT
VLKSHVVVNG-------HTKIGRDNEIYQFASIGEVNQDLKYAGEPTRVEIGDRNRIRES
VTIHRGTVQGGGLTKVGSDNLLMINAHIAH--*
```

FIG. 3. A sample alignment file prepared for MODELLER for a six-rung Aβ model.

FIG. 4. A structural model with six rungs of Aβ(15–36).

NMR structure in the PDB is approximately 80% (Doreleijers *et al.*, 1998; Laskowski *et al.*, 1993). Therefore, a few deviations are not unusual even in experimentally determined structures. In general, one would hope to have more than 90% of the residues in the most favored regions. But, if several errors cluster in the same region of the model, misalignment errors might have occurred. In addition to PROCHECK, there are several other programs that can be used to evaluate various aspects of the quality of a structural model. For example, an energy profile of the model can be calculated using PROSAII program (Sippl, 1993), which reports the energy for each position of the model. Peaks of positive energy in the profile might indicate errors in the model.

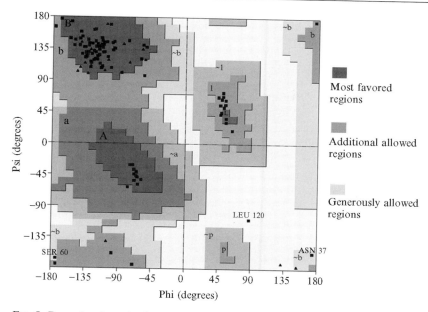

FIG. 5. Ramachandran plot for a model constructed using 1bhe as a template, generated using PROCHECK. The plot shows that 79.4% of the residues are in the most favored regions; 19.6% of the residues are either in the additional allowed or generally allowed regions. Only one residue is in the disallowed regions.

Molecular Dynamics Simulations

Molecular dynamics simulations are used to check the stability of predicted structural models. We have used GROMACS (http://www.gromacs.org/, version 3.1) (Lindahl *et al.*, 2001) for our simulation studies. In addition to the nice feature that MD simulations can be run in parallel, GROMACS also provides a suite of tools for analysis of simulation results. A structure model is first put into a suitably sized box, of which the minimal distance from the peptide to the box wall is 0.8 nm. Then the box is solvated with the simple point charge (SPC) water (Berendsen *et al.*, 1981). After the system is energy minimized to remove any bad van der Waals contacts, counter ions are added to the system to provide a neutral simulation system if the net charge of the system is not zero. The entire system is then energy minimized again. Molecular dynamics simulations using the GROMACS force field are carried out under NPT conditions (constant number of particle, pressure and temperature) at 300 K and under periodic boundary conditions. The Particle Mesh Ewald (PME) method (Darden *et al.*, 1993) and a distance cutoff of 10Å are used for long-range electrostatic interactions and van der Waals interactions, respectively. The time step was 2 femtoseconds (fs). All bonds including hydrogen atoms are constrained using the linear constraint solver (LINCS) algorithm (Hess *et al.*, 1997). A constant pressure of 1 bar in all three directions is used with a coupling constant of $\tau p = 1.0$ picoseconds (ps). The simulation is carried out for a number of nanoseconds (ns) after 30 ps of equilibration.

Below are the detailed steps of using GROMACS to do molecular dynamics simulations. We use 'amy_prot' to represent the starting structure for simulations. Three configuration files, em.mdp, pr.mdp, and full.mdp, are used to specify the key parameters for a simulation such as the simulation time, time step, and simulation method. A user can get the detailed information about all the parameters of GROMACS by checking the user's manual at http://www.gromacs.org/documentation/paper_manuals.php.

1. Generate structure and topology files of the model in GROMACS format using *pdb2gmx*.

   pdb2gmx -f amy_prot.pdb -p amy_prot.top -o amy_prot.gro.

2. Solvate the protein in a cubic box with SPC water using programs *editconf* and *genbox*;

   editconf -bt cubic -f amy_prot.gro -o temp.gro -d 0.8 –c

   genbox -cp temp.gro -cs -p amy_prot.top -o amy_prot_solv.gro.

3. Energy minimization

   a. Preprocessing using program *grompp*

      grompp -f em.mdp -v -c amy_prot_solv.gro -o afterem.tpr -p amy_prot.top

      After this step, the user might get a warning that the net charge is not zero. The user can use genion program of GROMACS and the afterem.tpr file to add counter ions to neutralize any net charge. For example, if the model system has a net charge of +6, one will need to add 6 chloride ions to neutralize the system.

      genion -s afterem.tpr -o amy_prot_ion.gro -nname Cl -nn 6

      The topology file is then modified to reflect the changes in solvent molecules and is used to run *grompp* again.

      grompp -f em.mdp -v -c amy_prot_ion.gro -o afterem.tpr -p amy_prot.top

   b. Energy minimization using *mdrun*

      mdrun -s afterem.tpr -o afterem.trr -c afterem.gro -e afterem.edr

4. Short molecular dynamics simulations with position restraints:

   grompp -f pr.mdp -c afterem.gro –pamy_prot.top -o pr.tpr

   mdrun -s pr.tpr -o pr.trr -c afterpr.gro -e afterpr.edr -g prlog

5. Full molecular dynamics simulations:
    grompp -f full.mdp -c afterpr.gro -p amy_prot.top -o full.tpr
    mdrun -v -s full.tpr -o full.trr -c afterfull.gro -e afterfull.edr
    A file full.trr with simulation trajectories is generated that could
    be analyzed later.

Simulation Result Analysis.

Analysis of simulation result can be done using the tools provided by GROMACS, which include tools for simulation trajectory analysis, for monitoring the root mean square deviation (RMSD) change with respect to the starting structure, the changes of the total number of hydrogen bonds (Guo *et al.*, 2004), and for analyzing energy variations. DSSP (*D*efinition of *S*econdary *S*tructure of *P*roteins) program is used for the secondary structure analysis (Kabsch and Sander, 1983). Figure 6 shows secondary structure changes during a simulation at each position in each of the six strands of

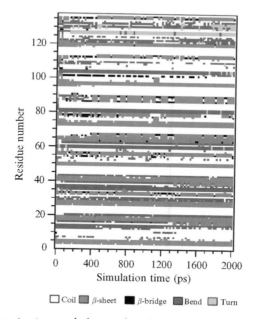

FIG. 6. Secondary structure analysis at each position in each of the six strands during the molecular dynamics simulations of a right-handed Aβ(16–35) 6-mer model. Residue number of each successive rung of β-helix: 1–22, 23–44, 45–66, 67–88, 89–110, 111–132. Different shade represents a different type of secondary structure.

a right-handed $A\beta(16-35)$ 6-mer model. The sequential positions for each successive rung of β-helix are 1–22, 23–44, 45–66, 67–88, 89–110, and 111–132, respectively. Structures can be visualized using Rasmol (http://www.umass. edu/microbio/rasmol/) or Pymol (http://pymol.sourceforge.net). As shown in Fig. 6, some β-sheet conformations (red) at the start of the simulation disappeared during the simulation. In our left-handed model, all the starting β-sheet structures are conserved throughout the simulation; in addition, more sheet conformations are formed during the course of simulation (Guo *et al.*, 2004), suggesting that the sequence fits the left-handed model better than the right-handed model.

Our simulation results show that the left-handed β-helical model generated using the threading approach is in good agreement with much of the experimental data on $A\beta$ amyloid fibrils. The simulation results also show that the degree of stability of $A\beta$ amyloid fibril model may well depend on the overall structural fold, whereas the model based on right-handed β-helical proteins, although also a parallel β-sheet arrangement, does not show the same degree of stability as the left-handed model does. During the simulation, the right-handed model maintains fewer backbone hydrogen bonds than the left-handed model. In the left-handed parallel model, both hydrogen bonds and stabilizing hydrophobic interactions are maximized (Guo *et al.*, 2004).

The β-helical model also provides new insights into the hierarchical structure of amyloid fibrils. A schematic model for the β-strand arrangement and protofilaments packing has been proposed (Guo *et al.*, 2004) in which each fibril consists of six protofilaments. The fibril model is consistent with much of the experimental data including the size of protofilaments (Serpell and Smith, 2000).

Discussion

This chapter describes an approach to modeling amyloid fibril structures using protein threading and molecular dynamics simulations. Our prediction method through combining computational approaches with low-resolution experimental data has shown to be promising in providing atomic structural models. Our model is consistent with many experimental observations (Guo *et al.*, 2004). However, structural modeling should be an evolving process rather than a static model report. New experimental results might become available after a modeling work is finished; new advances in experimental techniques may provide more accurate experimental data in the future. Therefore, a predicted model needs to be refined through incorporating the most recent experimental data. For example, previous computational work focused on antiparallel conformations on the basis of experimental data available at that time (Chaney *et al.*, 1998; Li *et al.*, 1999). Solid state

NMR and site-directed spin labeling analysis, however, suggest that the fibrils are organized with peptides in parallel conformation (Antzutkin et al., 2000; Benzinger et al., 1998; Torok et al., 2002), which leads to the development of several parallel models (Guo et al., 2004; Lakdawala et al., 2002; Ma and Nussinov, 2002b). Recently, cysteine mutation experiments revealed the side-chain orientations of some residues of the peptide in the $A\beta$ fibril (Shivaprasad and Wetzel, 2006). If this information is not reflected in computational models, a revision of the model is clearly needed. Currently, we are working to incorporate this type of information into our model.

A more challenging issue in amyloid fibril structure study is to model the packing of protofilaments. Currently, the detailed interactions between protofilaments are not well understood. For example, it has been shown that the N-terminal part of $A\beta$ peptide is not involved in the fibril core structure formation (Kheterpal et al., 2001; Williams et al., 2004), but how the N-terminal of the full-length $A\beta$ peptide folds and how to position them in the packed protofilaments is still an open question. Although there are many challenging issues in the computational prediction of amyloid conformations, modeling techniques coupled with experimental data seem to be the best approach so far.

Acknowledgments

The work of J. T. G. and Y. X. is, in part, supported by National Institutes of Health (R01 AG18927), National Science Foundation (DBI-0354771/ITR-IIS-0407204), and a "Distinguished Cancer Scholar" grant from the Georgia Cancer Coalition.

References

Altschul, S. F., Madden, T. L., Schaffer, A. A., Zhang, J., Zhang, Z., Miller, W., and Lipman, D. J. (1997). Gapped BLAST and PSI-BLAST: A new generation of protein database search programs. *Nucleic Acids Res.* **25**, 3389–3402.

Antzutkin, O. N., Balbach, J. J., Leapman, R. D., Rizzo, N. W., Reed, J., and Tycko, R. (2000). Multiple quantum solid-state NMR indicates a parallel, not antiparallel, organization of beta-sheets in Alzheimer's beta-amyloid fibrils. *Proc. Natl. Acad. Sci. USA* **97**, 13045–13050.

Benzinger, T. L., Gregory, D. M., Burkoth, T. S., Miller-Auer, H., Lynn, D. G., Botto, R. E., and Meredith, S. C. (1998). Propagating structure of Alzheimer's beta-amyloid(10–35) is parallel beta-sheet with residues in exact register. *Proc. Natl. Acad. Sci. USA* **95**, 13407–13412.

Berendsen, H. J. C., Postma, J. P. M., van Gunsteren, W. F., and Hermans, J. (1981). Interaction models for water in relation to protein hydration. *In* "Intermolecular Forces" (B. Pullman, ed.), pp. 331–342. Reidel Publishing Company, Dordrecht.

Berman, H. M., Westbrook, J., Feng, Z., Gilliland, G., Bhat, T. N., Weissig, H., Shindyalov, I. N., and Bourne, P. E. (2000). The Protein Data Bank. *Nucleic Acids Res.* **28,** 235–242.

Bowie, J. U., Luthy, R., and Eisenberg, D. (1991). A method to identify protein sequences that fold into a known three-dimensional structure. *Science* **253,** 164–170.

Bucciantini, M., Giannoni, E., Chiti, F., Baroni, F., Formigli, L., Zurdo, J., Taddei, N., Ramponi, G., Dobson, C. M., and Stefani, M. (2002). Inherent toxicity of aggregates implies a common mechanism for protein misfolding diseases. *Nature* **416,** 507–511.

Chaney, M. O., Webster, S. D., Kuo, Y. M., and Roher, A. E. (1998). Molecular modeling of the Abeta1-42 peptide from Alzheimer's disease. *Protein Eng.* **11,** 761–767.

Darden, T., York, D., and Pedersen, L. (1993). Particle mesh Ewald: An N.log(N) method for Ewald sums in large systems. *J. Chem. Phys.* **98,** 10089–10090.

Dobson, C. M. (2002). Getting out of shape. *Nature* **418,** 729–730.

Doreleijers, J. F., Rullmann, J. A., and Kaptein, R. (1998). Quality assessment of NMR structures: A statistical survey. *J. Mol. Biol.* **281,** 149–164.

Guo, J.-T., Hall, C. K., Xu, Y., and Wetzel, R. (2006). *In* Modeling protein aggregate assembly and structure (Y. Xu, *et al.*, eds.). "Computational Methods for Protein Structure Prediction and Modeling." Springer. In press.

Guo, J. T., Wetzel, R., and Xu, Y. (2004). Molecular modeling of the core of Abeta amyloid fibrils. *Proteins* **57,** 357–364.

Hess, B., Bekker, H., Berendsen, H. J. C., and Fraaije, J. G. E. M. (1997). LINCS: A linear constraint solver for molecular simulations. *J. Comp. Chem.* **18,** 1463–1472.

Holm, L., and Sander, C. (1996). Mapping the protein universe. *Science* **273,** 595–603.

Jayasinghe, S. A., and Langen, R. (2004). Identifying structural features of fibrillar islet amyloid polypeptide using site-directed spin labeling. *J. Biol. Chem.* **279,** 48420–48425.

Jenkins, J., and Pickersgill, R. (2001). The architecture of parallel beta-helices and related folds. *Prog. Biophys. Mol. Biol.* **77,** 111–175.

Jimenez, J. L., Nettleton, E. J., Bouchard, M., Robinson, C. V., Dobson, C. M., and Saibil, H. R. (2002). The protofilament structure of insulin amyloid fibrils. *Proc. Natl. Acad. Sci. USA* **99,** 9196–9201.

Jones, D. T., Taylor, W. R., and Thornton, J. M. (1992). A new approach to protein fold recognition. *Nature* **358,** 86–89.

Kabsch, W., and Sander, C. (1983). Dictionary of protein secondary structure: Pattern recognition of hydrogen-bonded and geometrical features. *Biopolymers* **22,** 2577–2637.

Kayed, R., Head, E., Thompson, J. L., McIntire, T. M., Milton, S. C., Cotman, C. W., and Glabe, C. G. (2003). Common structure of soluble amyloid oligomers implies common mechanism of pathogenesis. *Science* **300,** 486–489.

Kheterpal, I., Williams, A., Murphy, C., Bledsoe, B., and Wetzel, R. (2001). Structural features of the Abeta amyloid fibril elucidated by limited proteolysis. *Biochemistry* **40,** 11757–11767.

Kheterpal, I., Zhou, S., Cook, K. D., and Wetzel, R. (2000). Abeta amyloid fibrils possess a core structure highly resistant to hydrogen exchange. *Proc. Natl. Acad. Sci. USA* **97,** 13597–13601.

Kim, D., Xu, D., Guo, J. T., Ellrott, K., and Xu, Y. (2003). PROSPECT II: Protein structure prediction program for genome-scale applications. *Protein Eng.* **16,** 641–650.

Lakdawala, A. S., Morgan, D. M., Liotta, D. C., Lynn, D. G., and Snyder, J. P. (2002). Dynamics and fluidity of amyloid fibrils: A model of fibrous protein aggregates. *J. Am. Chem. Soc.* **124,** 15150–15151.

Laskowski, R. A., MacArthur, M. W., Moss, D. S., and Thornton, J. M. (1993). PROCHECK: A program to check the stereochemical quality of protein structures. *J. Appl. Cryst.* **26,** 283–291.

Li, L., Darden, T. A., Bartolotti, L., Kominos, D., and Pedersen, L. G. (1999). An atomic model for the pleated beta-sheet structure of Abeta amyloid protofilaments. *Biophys. J.* **76**, 2871–2878.

Lindahl, E., Hess, B., and van der Spoel, D. (2001). GROMACS 3.0: A package for molecular simulation and trajectory analysis. *J. Mol. Mod.* **7**, 306–317.

Lu, K., Jacob, J., Thiyagarajan, P., Conticello, V. P., and Lynn, D. G. (2003). Exploiting amyloid fibril lamination for nanotube self-assembly. *J. Am. Chem. Soc.* **125**, 6391–6393.

Ma, B., and Nussinov, R. (2002a). Molecular dynamics simulations of alanine rich beta-sheet oligomers: Insight into amyloid formation. *Protein Sci.* **11**, 2335–2350.

Ma, B., and Nussinov, R. (2002b). Stabilities and conformations of Alzheimer's beta-amyloid peptide oligomers (Abeta 16–22, Abeta 16–35, and Abeta 10–35): Sequence effects. *Proc. Natl. Acad. Sci. USA* **99**, 14126–14131.

Murzin, A. G., Brenner, S. E., Hubbard, T., and Chothia, C. (1995). SCOP: A structural classification of proteins database for the investigation of sequences and structures. *J. Mol. Biol.* **247**, 536–540.

Perutz, M. F., Finch, J. T., Berriman, J., and Lesk, A. (2002). Amyloid fibers are water-filled nanotubes. *Proc. Natl. Acad. Sci. USA* **99**, 5591–5595.

Petkova, A. T., Ishii, Y., Balbach, J. J., Antzutkin, O. N., Leapman, R. D., Delaglio, F., and Tycko, R. (2002). A structural model for Alzheimer's beta-amyloid fibrils based on experimental constraints from solid state NMR. *Proc. Natl. Acad. Sci. USA* **99**, 16742–16747.

Petkova, A. T., Leapman, R. D., Guo, Z., Yau, W. M., Mattson, M. P., and Tycko, R. (2005). Self-propagating, molecular-level polymorphism in Alzheimer's beta-amyloid fibrils. *Science* **307**, 262–265.

Pickersgill, R. W. (2003). A primordial structure underlying amyloid. *Structure (Camb)* **11**, 137–138.

Sali, A., and Blundell, T. L. (1993). Comparative protein modelling by satisfaction of spatial restraints. *J. Mol. Biol.* **234**, 779–815.

Serpell, L. C. (2000). Alzheimer's amyloid fibrils: Structure and assembly. *Biochim. Biophys. Acta.* **1502**, 16–30.

Serpell, L. C., and Smith, J. M. (2000). Direct visualisation of the beta-sheet structure of synthetic Alzheimer's amyloid. *J. Mol. Biol.* **299**, 225–231.

Shivaprasad, S., and Wetzel, R. (2006). Scanning cysteine mutagenesis analysis of Abeta (1–40) amyloid fibrils. *J. Biol. Chem.* **281**, 993–1000.

Sipe, J. D. (1992). Amyloidosis. *Annu. Rev. Biochem.* **61**, 947–975.

Sippl, M. J. (1993). Recognition of errors in three-dimensional structures of proteins. *Proteins* **17**, 355–362.

Sunde, M., and Blake, C. (1997). The structure of amyloid fibrils by electron microscopy and X-ray diffraction. *Adv. Protein Chem.* **50**, 123–159.

Sunde, M., and Blake, C. C. (1998). From the globular to the fibrous state: Protein structure and structural conversion in amyloid formation. *Q. Rev. Biophys.* **31**, 1–39.

Torok, M., Milton, S., Kayed, R., Wu, P., McIntire, T., Glabe, C. G., and Langen, R. (2002). Structural and dynamic features of Alzheimer's Abeta peptide in amyloid fibrils studied by site-directed spin labeling. *J. Biol. Chem.* **277**, 40810–40811.

Wetzel, R. (2002). Ideas of order for amyloid fibril structure. *Structure (Camb)* **10**, 1031–1036.

Wille, H., Michelitsch, M. D., Guenebaut, V., Supattapone, S., Serban, A., Cohen, F. E., Agard, D. A., and Prusiner, S. B. (2002). Structural studies of the scrapie prion protein by electron crystallography. *Proc. Natl. Acad. Sci. USA* **99**, 3563–3568.

Williams, A., Portelius, E., Kheterpal, I., Guo, J.-T., Cook, K., Xu, Y., and Wetzel, R. (2004). Mapping Abeta amyloid fibril secondary structure using scanning proline mutagenesis. *J. Mol. Biol.* **335,** 833–842.

Xu, Y., and Xu, D. (2000). Protein threading using PROSPECT: Design and evaluation. *Proteins* **40,** 343–354.

Yankner, B. A. (1996). Mechanism of neuronal degeneration in Alzheimer's disease. *Neuron* **16,** 921–932.

Zanuy, D., Gunasekaran, K., Ma, B., Tsai, H. H., Tsai, C. J., and Nussinov, R. (2004). Insights into amyloid structural formation and assembly through computational approaches. *Amyloid* **11,** 143–161.

[19] *Ab initio* Discrete Molecular Dynamics Approach to Protein Folding and Aggregation

By Brigita Urbanc, Jose M. Borreguero,
Luis Cruz, and H. Eugene Stanley

Abstract

Understanding the toxicity of amyloidogenic protein aggregates and designing therapeutic approaches require the knowledge of their structure at atomic resolution. Although solid-state NMR, X-ray diffraction, and other experimental techniques are capable of discerning the protein fibrillar structure, determining the structures of early aggregates, called oligomers, is a challenging experimental task. Computational studies by all-atom molecular dynamics, which provides a complete description of a protein in the solvent, are typically limited to study folding of smaller protein or aggregation of a small number of short protein fragments.

We review an efficient *ab initio* computer simulation approach to protein folding and aggregation using discrete molecular dynamics (DMD) in combination with several coarse-grained protein models and implicit solvent. This approach involves different complexity levels in both the protein model and the interparticle interactions. Starting from the simplest protein model with minimal interactions, and gradually increasing its complexity, while guided by *in vitro* findings, we can systematically select the key features of the protein model and interactions that drive protein folding and aggregation. Because the method used in this DMD approach does not require any knowledge of the native or any other state of the protein, it can be applied to study degenerative disorders associated with protein misfolding and aberrant protein aggregation.

The choice of the coarse-grained model depends on the complexity of the protein and specific questions to be addressed, which are mostly suggested by *in vitro* findings. Thus, we illustrate our approach on amyloid β-protein (Aβ)

METHODS IN ENZYMOLOGY, VOL. 412
0076-6879/06 $35.00
DOI: 10.1016/S0076-6879(06)12019-4

associated with Alzheimer's disease (AD). Despite the simplifications introduced in the DMD approach, the predicted Aβ conformations are in agreement with existing experimental data. The *in silico* findings also provide further insights into the structure and dynamics of Aβ folding and oligomer formation that are amenable to *in vitro* testing.

Introduction

An increasing number of neuropathological disorders, such as Alzheimer's, Parkinson's, Creutzfeldt-Jakob, motor neuron diseases, and polyglutamine disorders, are known to be associated with protein misfolding, followed by deposition of toxic protein aggregates in tissue (Dobson, 2004; Koo *et al.*, 1999). Alzheimer's disease (AD) is a progressive, neurodegenerative disorder, pathologically characterized by senile plaques and neurofibrillary tangles (Selkoe, 2001). The primary component of senile plaques is amyloid β-protein (Aβ), which has been strongly linked to the etiology and pathogenesis of AD. Aβ aggregates into small assemblies (oligomers), protofibrils, and fibrils rich in β-sheet content. In the past decade, compelling evidence has emerged indicating that soluble oligomeric assemblies and protofibrillar intermediates that form before senile plaque deposition may be determinant pathogenetic factors (Klein *et al.*, 2004).

Determination of oligomer conformation at the atomic level and tracking pathways of assembly from monomers to oligomers requires efficient computational approaches. With the dramatic increase of computer power in recent decades, it has become possible to study the behavior of large biological molecular systems by computer simulations (Ash *et al.*, 2004; Feig and Brooks, 2004; Fersht and Daggett, 2002; Karplus and McCammon, 2002). Traditional, all-atom molecular dynamics (MD) with atomic-detail force fields in a physiological solution (which would be ideal for studying Aβ oligomerization) is not computationally accessible with current technology. An aggregation process amenable to study by all-atom MD should occur on time scales of $\leq 10^{-7}$ sec and would require the use of advanced technologies such as worldwide distributed computing (Snow *et al.*, 2002; Zagrovic and Pande, 2003; Zagrovic *et al.*, 2002). However, *in vivo* and *in vitro* studies suggest that the initial stages of oligomerization occur on time scale of hours (Bitan *et al.*, 2003a; Kayed *et al.*, 2003).

The idea of applying a fast and efficient discrete molecular dynamics method (DMD) (Rapaport, 1997) to study protein folding was proposed in 1996 (Zhou *et al.*, 1996). Soon after, the method was combined with a one-bead protein Gō model to study folding of a model three-helix bundle protein (Dokholyan *et al.*, 1998, 2000; Zhou and Karplus, 1997, 1999; Zhou *et al.*, 1997). The interparticle interactions in the Gō model are assigned on the basis of the knowledge of a native state of a protein (Taketomi *et al.*, 1975). Thus, Gō models are not *ab initio*, because they require the knowledge of a native state of a protein. Despite this drawback, they are the simplest thermodynamic models that yield a unique native state of a protein and describe folding reminiscent of a first-order phase transition (Dokholyan *et al.*, 2003). Recently, a two-bead Gō model was applied by Peng *et al.* to study aggregation of an ensemble of 28 Aβ40 peptides into a fibrillar structure (Peng *et al.*, 2004).

In 2001, a four-bead protein model in combination with the DMD method was first introduced by Smith and Hall (Smith and Hall, 2001a) inspired by earlier studies (Takada *et al.*, 1999). This model accounts for a rather accurate backbone description and is able to describe a cooperative transition of a polyalanine chain into an α-helical conformation without any *a priori* knowledge of the native state (Smith and Hall, 2001b). With the four-bead model with hydrogen bond interactions on a single 16-residue polyalanine chain, Ding *et al.* (2003) demonstrated a temperature-induced conformational change from the α-helix to the β-hairpin conformation. Because of these properties, the four-bead model with hydrogen bond interactions represents a base on which the *ab initio* modeling can be realized. The *ab initio* DMD computational approach introduces simplifications to the protein description, interparticle interactions, and treatment of the solvent. These simplifications make the DMD approach at least six orders of magnitude faster than all-atom MD with explicit solvent. To ensure biological relevance of the approach that targets different aspects of Aβ folding, oligomer, and fibril formation, up-to-date experimental findings need to be integrated into the development of the protein model and interactions, creating a much-needed partnership between computation and experiment as recognized by others [e.g., the review by Ma and Nussinov (2004)].

This review is organized in two main sections. In the first section, we describe the applications of the DMD approach to model Aβ folding and aggregation. The goal of this section is to give an idea of the kind of information we can obtain using the proposed DMD approach. In addition, we present the main hypotheses on the structure and dynamics of folding and assembly that emerged from these applications. The second section introduces in detail the implementation of the DMD method, coarse-grained protein models, interparticle interactions, and limitations that originate in simplifications associated with the approach.

Applications to Aβ Folding and Aggregation

We describe the applications of the DMD approach with the four-bead and the united-atom model to Aβ folding and aggregation to demonstrate the variety of information that can be obtained. We also review selected *in vitro* findings that shed light on different structural aspects of Aβ folding and aggregation and help guide the development of the DMD approach.

In Vitro *Findings*

The sequence DAEFRHDSGYEVHHQKLVFFAEDVGSNKGAII-GLMVGGVVIA defines the primary structure of Aβ42. Aβ40 lacks the last two amino acids, I41 and A42. The secondary structure of Aβ monomer conformations depends strongly on the environment. In an apolar membrane-like environment, Aβ40 and Aβ42 monomers adopt predominantly an α-helical conformation (Coles *et al.*, 1998; Crescenzi *et al.*, 2002), whereas in an aqueous solution Aβ prefers a collapsed coil monomer structure with a bend in the V24-K28 region (Zhang *et al.*, 2000).

Recent limited proteolysis experiments on Aβ40 and Aβ42 have shown that the region A21–A30 is highly resistant to proteolytic attack under conditions favoring oligomerization, suggesting the presence of a folded structure (Lazo *et al.*, 2005). Similar results were observed for the Aβ fragment Aβ(21–30) in monomeric solution. Lazo *et al.* postulated that this decapeptide adopts a structure that nucleates the intramolecular folding of the full-length Aβ monomer. The solution dynamics of Aβ(21–30), as determined by NMR studies, yielded two families of folded Aβ(21–30) structures both containing a turn-like motif centered at G25–S26 (Lazo *et al.*, 2005). These *in vitro* results raise questions that can be addressed *in silico*: (1) what is the driving force of folding, and (2) how does the folded structure affect the pathway of Aβ assembly?

Aβ40 and Aβ42 both have high tendencies to aggregate into fibrils, which makes studies of oligomeric intermediates difficult. To study Aβ oligomerization *in vitro*, the technique photo-induced cross-linking of unmodified proteins (PICUP) has been applied to covalently stabilize oligomers (see Chapter 12 by Bitan, 2006 in Volume 413). Using PICUP coupled with size-exclusion chromatography, Bitan *et al.* (2003a) showed that Aβ40 and Aβ42 display distinct oligomer size distributions. Solutions of Aβ40 display a rapid equilibrium among monomers, dimer, trimers, and tetramers, whereas Aβ42 preferentially forms pentamer/hexamer units (paranuclei), which further assemble into beaded superstructures similar to early protofibrils (Bitan *et al.*, 2003a). Further studies of primary structure elements controlling early oligomerization demonstrate that I41 is critical for paranucleus formation in solutions of Aβ42 and that A42 is necessary for further assembly of Aβ42 into larger oligomers (Bitan *et al.*, 2003b). In addition, oxidation of M35 blocks paranucleus formation in Aβ42 but does not alter the Aβ40 oligomer size distribution (Bitan *et al.*, 2003c). Mass spectroscopy and ion mobility measurements of Aβ42, which was subjected to filtration to remove large assemblies and immediately electrosprayed, indicated the presence of dimers, tetramers, paranuclei, and pairs of paranuclei in agreement with PICUP results (Bernstein *et al.*, 2005).

Solid-state NMR studies yielded high-resolution information on the Aβ40 fibrillar structure, in which each individual peptide displays a bend, stabilized by a salt-bridge between D23 and K28 (Petkova *et al.*, 2002; see Chapter 6 by Tycko, 2006 in Volume 413). The kinetics of Aβ40 fibril formation is typically preceded by a lag phase that is not present in a recently synthesized Aβ40-lactam (D23/K28) that contains a lactam bridge between D23 and K28 (Sciarretta *et al.*, 2005). This experimental finding explains the importance of the bend in the V24-K28 region and the associated salt-bridge D23-K28 in Aβ40 fibrillogenesis and suggests that Aβ40-lactam (D23/K28) bypasses an unfavorable folding step, leading to ~1000-fold greater rate of fibril formation (Sciarretta *et al.*, 2005). The role of the salt-bridge D23-K28 formation at different stages of Aβ folding and assembly can be addressed in the DMD approach by systematically varying the effective electrostatic potential.

Four-Bead Model with Hydrogen Bonding: Planar β-sheet Assemblies and the Role of Glycines

Urbanc *et al.* (2004a) applied a four-bead protein model with backbone hydrogen bond interactions to study Aβ40 versus Aβ42 dimer formation. The Aβ42 sequence was simplified to a polyalanine chain with glycines at positions 9, 25, 29, 33, 37, and 38. This model exhibited conformational changes with increasing temperature. The monomer adopted an α-helical conformation at low temperatures, several types of β-strand conformations including β-hairpin conformation at intermediate temperatures, and random coil-like conformation at high temperatures. A turn between G25 and G29 was consistently observed at intermediate temperatures and was shown to be induced by the presence of glycines, in particular G25. The importance of glycines was recently confirmed by an all-atom MD study of Aβ42 folding in explicit aqueous solution, which demonstrated that glycines induced local turns in the peptide and consequently caused the α-helical to β-strand conformational change (Xu *et al.*, 2005).

The turn between G25 and G29 occurred in the same protein region as the bend in the model of Aβ fibrils by Petkova *et al.* (2002). The local structure of a typical peptide within the fibril is quite different from the four-bead model prediction. Hydrogen bonds in the fibril are oriented along the fibrillar axis and link neighboring peptides with no significant intramolecular hydrogen bonding. In the simplified four-bead model, intramolecular hydrogen bonds first give rise to β-hairpin monomer

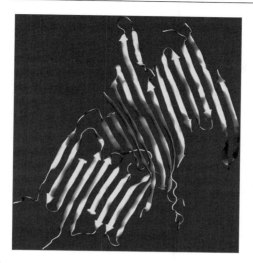

FIG. 1. An Aβ42 octamer as found within the four-bead model with hydrogen bond, but no amino acid–specific interactions. The octamer is an extended planar β-sheet with several domains that are slightly rotated with respect to one another.

conformations, which then further assemble into extended planar β-sheets. These planar β-sheet aggregates are held together exclusively by intramolecular and intermolecular hydrogen bonding (Fig. 1.). A critical observation was the lack of stacking among the β-sheets, in contradiction with the model of Aβ fibril formation (Petkova *et al.*, 2002). This result suggests that amino acid–specific interactions between pairs of side-chains are responsible for a correct description of the stacked β-sheet structure.

Four-Bead Model with Amino Acid–Specific Hydropathic Interactions:
Aβ40 versus Aβ42 Oligomer Formation

Ding *et al.* (2003) studied the effect of hydrophobic side chain interactions on the α-helix and β-hairpin monomer conformations in a 16-residue polyalanine. They found that above a certain strength of effective hydrophobic interactions ($E_{HP}/E_{HB} > 0.20$), the β-hairpin monomer conformation disappears, and it is replaced by a globular monomer conformation (Ding *et al.*, 2003). Nguyen and Hall demonstrated that the presence of a weak effective hydrophobic attraction ($E_{HP}/E_{HB} < 1/6$) between the side chains of 16-residue polyalanine peptides leads to formation of a stacked β-sheet structure, consistent with the basic structural features of the fibril formation (Nguyen and Hall, 2004a,b, 2005; see Chapter 20 by Hall and Wagoner, 2006 in this volume).

Urbanc *et al.* introduced a four-bead Aβ model with hydrogen bond interactions and effective hydrophobic and hydrophilic interactions that were amino acid-specific (Urbanc *et al.*, 2004b). They showed that such a model with strong amino acid–specific hydrophobic and hydrophilic interactions ($E_{HP}/E_{HB} = 0.3$) leads to the formation of globular oligomer structures (Urbanc *et al.*, 2004b). Urbanc *et al.* (2004b) demonstrated that this model is able to capture significant oligomerization differences between Aβ40 and Aβ42 that are consistent with *in vitro* results (Bitan *et al.*, 2003a,b,c). The effective hydrophobic attraction, as well as the effective hydrophilic repulsion, are critical features of the model that yields a steady-state distribution of Aβ oligomers of different sizes. If only the hydrophobic attraction was present in the model, the steady state would be a single globular oligomer because of the lack of forces opposing aggregation. In the presence of both hydrophobic attraction and hydrophilic repulsion, globular oligomers of various sizes coexist in a quasi-steady state. A typical globular oligomer consists of a core containing the hydrophobic parts of Aβ and a surface containing the hydrophilic N-terminal residues (Fig. 2). In agreement with experimental findings (Bitan *et al.*, 2003a), Urbanc *et al.* found that Aβ42 had an increased tendency to form pentamers, whereas dimers dominated in Aβ40. Detailed structural analysis of these *in silico* results provided new structural insights and offered a plausible explanation of the role of M35 in Aβ40 versus Aβ42 oligomerization, indicating that oxidation of M35 disrupts Aβ42 paranuclei formation but does not affect Aβ40 oligomerization (Bitan *et al.*, 2003c). Statistical analysis of the tertiary structure of *in silico* pentamers showed important differences between the two alloforms in terms of contact formation involving C-terminal residues. In Aβ42, the intramolecular contacts between V39-A42 on one side and I31, I32, L34, M35, and V36 on the other side dominated, whereas in Aβ40, the C-terminal fragment V39-A40 did not form any significant intramolecular contacts (Urbanc *et al.*, 2004b). Thus, on the basis of this structural information, Urbanc *et al.* suggested that disrupting the hydrophobic nature of M35 by oxidation would cause a disruption of important hydrophobic contacts between oxidized M35 and C-terminal fragment in the Aβ42 pentamer. Because these contacts were not present in Aβ40 pentamers, oxidation of M35 would not make much of a difference in Aβ40 oligomer formation.

United-Atom Model: Aβ(21–30) Folding Initiated by a Hydrophobic Packing between V24 and K28

Borreguero *et al.* (2005) developed a united-atom protein model in which all atoms except hydrogens are explicitly present. They applied the DMD approach with the united-atom model to study folding and unfolding transitions of Aβ(21–30) under different electrostatic interaction strengths

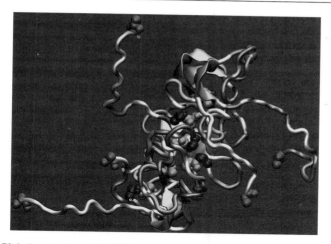

FIG. 2. Globular structure of Aβ42 hexamer as found within the four-bead peptide model with amino acid–specific interactions caused by hydropathy. D1 is represented by four red spheres to illustrate the hydrophilic N-termini at the surface of the hexamer. I41 (four green spheres) and A42 (four blue spheres), as part of the C-terminal region, are at the hydrophobic core of the hexamer. Yellow ribbons represent a β-strand, cyan tube a turn, and silver tube a random coil-like secondary structure. The image was generated within the VMD software package (Humphrey *et al.*, 1996), which includes the STRIDE algorithm for calculating the secondary structure-propensity per residue (Heinig and Frishman, 2004). (See color insert.)

(EIS) (Borreguero *et al.*, 2005). Hydrophobicity was shown to be the driving force of folding in Aβ(21–30), inducing packing between V24 and the butyl portion of K28 (Fig. 3). In addition to hydrophobicity, intermediate EIS (~1.5 kcal/mol) predominantly between E22 and K28 contributed to an optimal stability of the folded structure (Borreguero *et al.*, 2005). At higher EIS (~2.5 kcal/mol)—typically occurring in the interior of proteins— Aβ(21–30) was found to be partially unfolded because of a salt-bridge between D23 and K28. This observed prevalence of the D23-K28 interaction at highest EIS is in agreement with molecular models of protofibrils formed by full-length Aβ (Petkova *et al.*, 2002) and Aβ(16–35) (Ma and Nussinov, 2002) that show stabilization through D23-K28 salt-bridge and no E22-K28 interaction or V24-K28 packing. The study of Borreguero *et al.* exposed the binary nature of salt-bridge interactions between K28 and E22/D23 and provided a mechanistic explanation for the linkage of amino acid substitutions at E22 with AD and cerebral amyloid angiopathy (Borreguero *et al.*, 2005). Recently, Cruz *et al.* studied Aβ(21–30) and its Dutch mutant (E22Q) by all-atom MD in water, reduced-density water, and in water with salt ions. They confirmed that in water Aβ(21–30) folding is driven by hydrophobic

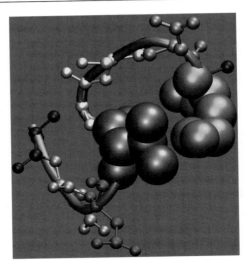

FIG. 3. Folded Aβ(21–30) decapeptide conformation as found within the united-atom model with amino acid–specific interactions caused by hydropathy and charge. All atoms except hydrogens are drawn as small spheres: A21 and A30 (blue), E22 (pink), D23 (red), V24 (tan), G25 and G29 (white), S26 (yellow), N27 (orange), and K28 (cyan). V24 and K28 are presented by large opaque spheres to illustrate their packing, a critical event in the decapeptide folding. The image was generated within the VMD software package (Humphrey *et al.*, 1996). (See color insert.)

forces involving V24 and K28. In addition, Cruz *et al.* showed that the Aβ(21–30) folded structure is very sensitive to changes in environment and that in the Dutch mutant folding events are rare (Cruz *et al.*, 2005).

Structural Hypotheses Derived from the DMD Studies

Next we summarize the main hypotheses regarding the structure and dynamics of oligomer and fibril formation that are derived from the results of the DMD studies using either the four-bead model with amino acid-specific hydropathies (Urbanc *et al.*, 2004b) or the united-atom model with atomic hydropathies and effective electrostatic interactions (Borreguero *et al.*, 2005). These hypotheses are amenable to both *in silico* and *in vitro* testing.

1. Full-length monomers of Aβ40 and Aβ42 fold from the C-terminus toward the N-terminus. First intramolecular contacts during Aβ monomer folding are formed between V36 and V39 and their neighbors (Urbanc *et al.*, 2004b).

2. Aβ40 and Aβ42 monomers fold in different ways. The Aβ42 monomer folding is associated with a turnlike element centered at G37–G38, which is not present in the Aβ40 monomer. The Aβ40 monomer has an additional parallel β-strand between A2-F4 and the central hydrophobic cluster (L17-A21), not present in the Aβ42 monomer (Urbanc et al., 2004b). The prediction of the turn at G37-G38 in Aβ42 (but not in Aβ40) is consistent with in vitro results of limited proteolysis, which shows that the region V39-A42 in Aβ42 is protease resistant, whereas the region V39-A40 in Aβ40 is not (Lazo et al., 2005).

3. In Aβ40 oligomers, the most significant intermolecular contacts exist between pairs of central hydrophobic clusters (L17-V18-F19-F20-A21), whereas in Aβ42 oligomers, contacts between pairs of C-terminal regions (V39-A42) are the most important (Urbanc et al., 2004b).

4. Despite similar globular structure with hydrophobic C-terminal residues in the core and hydrophilic N-terminal residues at the surface, the structure of Aβ40 and Aβ42 pentamers differs. The parallel β-strand structure at the N-termini of Aβ40 (as described in 2) persists in all assembly states and is completely absent from Aβ42 oligomers. Consequently, the N-termini of Aβ42 are spatially less restricted and can be found on average further away from the core of the oligomer. This difference in the N-termini properties might contribute to a more exposed hydrophobic core of Aβ42 oligomers, rendering Aβ42 more prone to further aggregate (Urbanc et al., 2004b).

5. Hydrophobic attraction between V24 and the butyl portion of K28 drives the folding of Aβ(21–30), whereas the salt-bridge E22-K28 contributes to the stability of the folded structure (Borreguero et al., 2005).

6. Because experiments show that the same region V24-G25-S26-N27-K28 is protease resistant in the full-length Aβ (Lazo et al., 2005), fibril formation should be preceded by an event that disrupts the folded loop V24-G25-S26-N27-K28. Because the E22-K28 salt-bridge contributes to the loop stability, a substitution of E22 by a non-negatively charged amino acid should enhance the fibril formation through: (1) decrease of the loop stability, and (2) increase of the rate of D23-K28 salt-bridge formation due to the absence of competition between E22 and D23 (Borreguero et al., 2005).

Based on the above predictions, one can introduce selected amino acid substitutions in Aβ40 and Aβ42 that would hypothetically disrupt or change monomer and oligomer conformations. Should in vitro and other experimental findings that target the structure of Aβ folded monomers and oligomers determine that any of the above hypotheses is not valid, the DMD approach can be refined in two ways: (a) by introducing more detail into the protein model; and (b) by refining the interactions between the side-chain atoms and possibly introducing locally modified interactions.

Methods

Discrete Molecular Dynamics Method

MD is a computer simulation method in which particles move according to specific interparticle forces on the basis of classical dynamics. Newton's equations of motion must be numerically integrated at each time step for all particles to update instantaneous velocities and positions. DMD is a simplified version of MD and is applicable whenever the interparticle potentials can be represented by one or more square wells (Rapaport, 1997). Within each well, the potential is constant, because the force between the two particles is zero, and thus the particles move with constant velocities until they reach a distance at which the potential is discontinuous. At that moment, an elastic or inelastic collision occurs, and the two particles change their velocities instantaneously while conserving the total energy, momentum, and angular momentum. No numerical integration is needed. The only events are two-particle collisions, and the main challenge is to keep track of collision times. Consequently, DMD simulations are considerably faster than continuous MD simulations.

During a DMD simulation, the number of particles, volume, and temperature are held constant. Periodic boundary conditions are implemented to avoid interactions with the walls of the simulation box. The size of the box is chosen to be larger than the stretched protein under study. We implement temperature control in our model using the method proposed by Berendsen *et al.* (1984). In this method, a heating rate coefficient, α, is introduced. The temperature is rescaled at regular intervals Δt: $T(t + \Delta t) = T(t) + \alpha \Delta t \, [T_\infty - T(t)]$ where $T(t)$ is the instantaneous temperature, $T(t + \Delta t)$ is the rescaled temperature, and T_∞ is the target temperature of the heat bath; α^{-1} is a characteristic time, in which the temperature equilibrates. The time interval Δt corresponds to about N collisions, where N is the number of particles. Temperature is defined by the total kinetic energy of particles as follows $\frac{3}{2}\kappa_B T = \frac{1}{N}\sum_{i=1}^{N}\frac{m_i v_i^2}{2}$, where κ_B is the Boltzmann constant, v_i are the velocities of each of the N particles, and m_i their masses. Rescaling temperature requires rescaling all N velocities v_i by a factor $\sqrt{T(t + \Delta t)/T(t)}$, which is followed by recalculation of the collision times. To avoid the time-consuming task of recalculating collision times, we introduce a rescaled time variable and rescaled potentials, keeping the velocities and collision times intact. This transformation does not alter the trajectory in any way. We keep track of the original simulation time by keeping track of the rescaling factors, so that results are expressed in original units. A more detailed description has been given elsewhere (Borreguero, 2004).

Four-Bead Model Implementation

We use the four-bead protein model introduced by Ding *et al.* (2003). In the four-bead model, the backbone is represented by three beads, corresponding to the amide (N), the α-carbon (C_α), and the carbonyl (C) groups. Each side-chain (except G, which lacks the side-chain group) is represented by one side-chain bead (C_β). Each bead (atom) is characterized by its mass and hard-core radius. In the simplest version of the model, all atoms have equal mass and their hard-core radii are set to their van der Waals radii (Creighton, 1993). Each side-chain atom is characterized by a type, which determines its interactions with other atoms. Any two atoms can only be at a distance $d > d_{min}$, where d_{min} is the sum of their hard-core radii. Thus, the potential is set to an "infinitely" large value for $d < d_{min}$. Pairs of atoms can be linked by a covalent bond or an angular constraint to account for the protein geometry as shown in Fig. 4A. If two atoms are linked in this way, there is a distance d_{max} such that for $d > d_{max}$ the potential is infinite to prevent the two atoms from breaking the bond. The lengths of bonds and angular constraints are determined phenomenologically by calculating their distributions using known folded protein structures of ~7700 proteins from the Protein Data Bank (PDB) (http://www.rcsb.org/pdb). The values of the lengths of covalent bonds and angular constraints, which are allowed to vary around their average values by 2%, were reported elsewhere (Ding *et al.*, 2003).

Backbone Hydrogen Bond. In proteins, the most ubiquitous hydrogen bond interaction involves the carbonyl oxygen and the amide hydrogen of two amino acids. In the four-bead model, because the carbonyl oxygen and the amide hydrogen are not explicitly present, an effective backbone

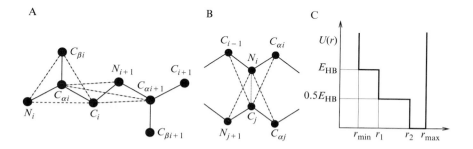

FIG. 4. (A) Covalent bonds (solid lines) and constraints (dashed lines) that define the four-bead peptide model. (B) Hydrogen bond between C_j and N_i (dotted line) and the corresponding auxiliary bonds (dashed lines) that define the geometry of hydrogen bonding in the four-bead model. (C) Interparticle potential $U(r)$ for the four auxiliary bonds shown in (B).

hydrogen bond is introduced between the nitrogen atom N_i of the i–th amino acid and the carbon atom C_j of the j–th amino acid (Ding $et\ al.$, 2003) (Fig. 4B–C). On formation of the hydrogen bond, atoms N_i and C_j in the model change types to prevent their involvement in additional hydrogen bond formation. When amino acids i and j belong to the same protein, we allow them to form a hydrogen bond only if they are at least three amino acids apart, $|i - j| \geq 4$. This constraint is a consequence of angular restrictions of the hydrogen bond that requires that the CO and NH bonds are approximately collinear. These same angular restrictions are enforced by introducing four auxiliary bonds involving the left and the right neighboring beads of N_i and C_j (Fig. 4B). The hydrogen bond between N_i and C_j forms only if all six beads are at energetically favorable distances. Each of the four auxiliary interactions is modeled by a double-step potential (Fig. 4C), and the particular values of the hydrogen bond parameters are chosen phenomenologically to best match the distribution of the corresponding distance in real proteins (Ding $et\ al.$, 2003). The additional auxiliary interactions take place only in the presence of the hydrogen bond interaction. During the hydrogen bond formation or deletion, the other interactions involving the N_i and C_j beads remain intact. When a new hydrogen bond is formed, the new hard-core collision distance between N_i and C_j is assigned to be 4.0 Å, such that at the lowest energy state of a hydrogen bond, the optimal distances of the four auxiliary pairs allow for approximately linear alignment of the CO and the NH bonds.

Amino Acid–Specific Interactions Caused by Hydropathy. Because the solvent is not explicitly present in our DMD approach, effective interactions between the side-chain atoms are introduced to mimic the solvent effects. We introduce hydrophobic attraction and hydrophilic repulsion between pairs of side-chains, depending on the hydropathic nature of individual side-chains. In our model, the potential energy decreases when two hydrophobic residues interact, thus minimizing their solvent accessible surface area (SASA). Conversely, the potential energy increases when two hydrophilic residues interact. This potential thus favors noninteracting hydrophilic residues, which maximizes their SASA.

There are different ways of implementing amino acid–specific hydropathic interactions. We chose the empirical amino acid hydropathy scale derived by Kyte and Doolittle (1982) as previously described (Urbanc $et\ al.$, 2004b). We consider the following amino acids: I, V, L, F, C, M, and A hydrophobic; N, Q, and H non-charged hydrophilic; and D, E, K, and R charged hydrophilic. The remaining amino acids with absolute values of hydropathies below threshold values are considered neutral, so that two neutral side-chain atoms only interact through their hard-core interaction. The hydropathic interactions are of two types: (1) an attractive interaction between two hydrophobic side-chains; and (2) a repulsive interaction between two noncharged hydrophilic or a charged hydrophilic and a noncharged hydrophilic side-chain. Interactions are implemented using a square-well potential between the pairs of side-chain beads $C_{\beta,i}$ and $C_{\beta,j}$, so that they interact if the distance between their centers is less than the interaction range distance 7.5 Å (Fig. 5A,B). The potential energy of the effective attractive hydrophobic interaction E_{HP} is proportional to the mean of the relative hydrophobic strengths, I (-1.0), V (-0.93), L (-0.84), F (-0.62), M (-0.42), and A (-0.40), where the negative sign reflects the attractive nature of the interaction. The potential energy of the effective repulsive hydrophilic interaction is proportional to the mean of the relative hydrophilic strengths, R (1.0), K (0.87), E (0.78), D (0.78), N (0.78), Q (0.78), and H (0.71), where the positive sign reflects the repulsive nature of the interaction. H, with a pK_a value ~ 6.0, is considered a noncharged hydrophilic amino acid because at physiological conditions (pH = 7.4) only about 4% of H is charged.

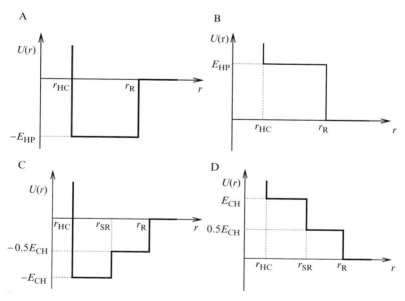

FIG. 5. (A) Effective hydrophobic interaction with the potential energy $-E_{HP}$ between two hydrophobic side-chain atoms as implemented in the four-bead model. (B) Effective hydrophilic interaction with the potential energy E_{HP} between two hydrophilic side-chain atoms as implemented in the four-bead model. The range of the hydrophobic and hydrophilic interactions is set to $r_R = 7.5$ Å. (C) Effective electrostatic interaction with the electrostatic potential energy $-E_{CH}$ between two oppositely charged atoms as implemented in the four-bead model. (D) Effective electrostatic interaction with the electrostatic potential energy E_{CH} between two atoms of the same charge as implemented in the four-bead model. The range of the interaction is $r_R = 7.5$ Å and the soft range is $r_{SR} = 6$ Å. The hard-core repulsion distance is denoted by r_{HC}.

Amino Acid–Specific Interactions Caused by Charge. The effective electrostatic interaction between two charged side-chain atoms is implemented using a double attractive/repulsive square well potential with the interaction range $r_R = 7.5$ Å and a "soft" interaction range $r_{SR} = 6$ Å (Fig. 5C,D). The energy of the effective electrostatic interaction, E_{CH}, is tunable and is typically set in the range $E_{CH}/E_{HB} \in [0, 1]$, with different E_{CH} values corresponding to different solvent conditions.

United-Atom Model Implementation

In the united-atom model, all protein atoms except hydrogens are explicitly represented. The backbone of the protein is represented by four atoms, corresponding to the amide group (N), the α carbon (C_α), the prime carbon (C), and the oxygen (O). On formation of the backbone hydrogen bond, N and O change their types. A special atom type is introduced for the amide group (N) of P to describe its characteristic covalent bond to the P C_δ side-chain atom. We assign a different type for each side-chain atom of the 19 amino acids. For each atom, we assign an individual atomic mass, a phenomenologically estimated radius (Tsai *et al.*, 1999) and a nominal charge of ($+1$) to the amino groups of K and R, and a (-1) charge to the carboxy groups of D and E.

To achieve the correct description of the flexibility of the protein, we assign three types of bonding between protein atoms to account for the backbone and side-chain geometries: (1) covalent bonding; (2) angular constraints; and (3) rotameric constraints (Fig. 6). The rotameric constraints were first introduced in the context of the six-bead model (Ding *et al.*, 2005) and later expanded to account for the united-atom model (Borreguero, 2004). These lengths and their variances are derived from the library of potentials using a statistical analysis of a specific database of protein structures (PDB40), which is a subset of the Structural Classification of Proteins (SCOP) database (http://scop.berkeley.edu) of protein structures (Chandonia *et al.*, 2004). The exact values were given by Borreguero (2004). Typically, the covalent bonds are allowed to vary by $\sim$4%, angular constraints by $\sim$6%, and the rotameric constraints by 4–28%.

We implement covalent and angular constrains as square-well potentials with two "infinite" walls representing the limits of typical interatomic distances for the particular bond or constraint under consideration (Fig. 8A) As an example, we describe a rotameric constraint by implementing the χ_1 rotamer angle (e.g., for Val). χ_1 is the angle between the two planes generated by atoms $N–C_\alpha–C_\beta$ and $C_\alpha–C_\beta–C_{\gamma 2}$, respectively (Fig. 7A). Thus, χ_1 is determined by the positions of the four atoms $N–C_\alpha–C_\beta–C_{\gamma 2}$. Other four-atom sets are equally valid, namely $N–C_\alpha–C_\beta–C_{\gamma 1}$, $C–C_\alpha–C_\beta–C_{\gamma 1}$,

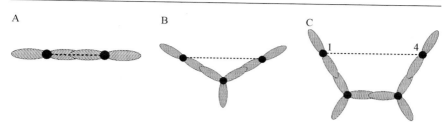

FIG. 6. Bond types between two atoms. (A) Covalent bond. (B) Angular constraint to model the central hybridized molecular orbital. (C) Rotameric constraint 1–4 to reproduce the statistically observed preference of atom 4 to orient itself with respect to the position of atom 1.

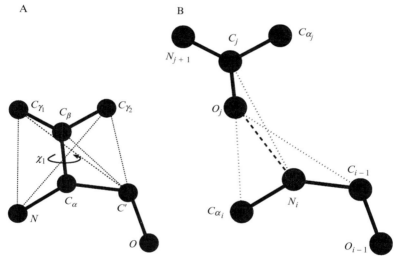

FIG. 7. (A) Schematic diagram of the χ_1 rotamer of valine. Dashed lines represent distances between atoms involved in rotameric-constraint interactions. Any two of these interactions uniquely determine χ_1. (B) Schematic diagram of the backbone hydrogen bond. The dashed line represents the N—O bond, and the dotted lines represent auxiliary constraints maintaining the correct N—O orientation.

or $C–C_\alpha–C_\beta–C_{\gamma 2}$. χ_1 determines the distance between the first and last atom for each set (i.e., $\overline{NC_{\gamma 2}}$ (Fig. 8B–C), $\overline{NC_{\gamma 1}}$, $\overline{CC_{\gamma 1}}$, and $\overline{CC_{\gamma 2}}$). Conversely, any two of these four distances are sufficient to uniquely determine χ_1. The distance distributions for each of the four constraints defines a respective potential energy function (Fig. 8D).

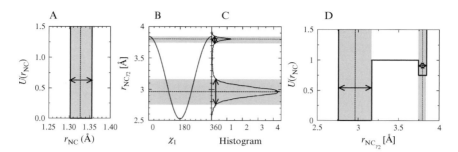

FIG. 8. (A) Potential energy associated with the bond between N and C groups in the protein backbone. The most probable distance (dashed line) and a typical range of distances (arrow) derived from the PDB40 structural database. (B) The distance $r_{NC_{\gamma 2}}$ as a function of the rotamer χ_1. (C) Histogram of distances $r_{NC_{\gamma 2}}$ as derived from the PDB40 dataset. (D) Associated rotameric potential between N and $C_{\gamma 2}$. We estimate the width and depth of the two potential wells as the width and the area of the two respective probability peaks. We normalize the potential units in this plot with respect to the deepest well.

Soft-Core Interactions. Two atoms can approach to distances smaller than the sum of their van der Waals radii. However, these distances are not energetically favorable, and the two atoms have to overcome an energy barrier (Tsai *et al.*, 1999). From the occurrence probability of these events at room temperature, we estimate the energy barrier to be threefold the thermal energy. We set the lower limit of the soft-core region to be 85% of the sum of their respective van der Waals radii (Fig. 9A). Distances smaller than this lower limit, which correspond to less than 1% of the observed distances in protein structures, are not allowed.

Backbone Hydrogen Bond. We adopt the hydrogen bonding first implemented into the four-bead model (between the amide N and carbonyl C groups) (Ding *et al.*, 2003) and further developed to account for the explicit backbone O group (Ding *et al.*, 2005). The hydrogen bond between N and O atoms is implemented into the model as a "reaction"-type interaction. On formation of the backbone hydrogen bond, N and O change their types into new "bonded" types to prevent additional hydrogen bonding to the third atom. The geometry of the backbone hydrogen bond is modeled by a four-body interaction. In addition to the N–O bond, three additional constraints between neighboring N and O atoms (Fig. 7B) reproduce the orientation of the hydrogen bond. Details of this backbone hydrogen bond model were described elsewhere (Borreguero, 2004; Ding *et al.*, 2005).

Amino Acid–Specific Interactions Caused by Hydropathy. As in the four-bead model, we introduce an implicit treatment of the solvent with an attractive/repulsive potential energy between pairs of side-chain atoms

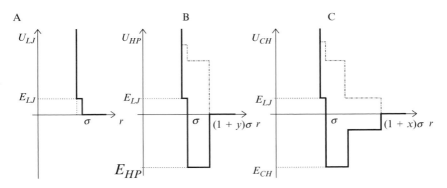

FIG. 9. Three types of potentials, U_{LJ}, U_{HP}, and U_{CH}, in dependence on the interparticle distance r. (A) Lenard-Jones potential U_{LJ}. σ is the sum of the Van der Waals radii of the two interacting atoms, and $E_{LJ} = 3\kappa_B T$ is a finite repulsive potential energy of two atoms at distances smaller than σ. (B) Effective hydrophobic potential U_{HP} with $y = 0.2$ between two noncharged atoms with the potential energy E_{HP}. The dotted-dashed line represents the effective repulsion between a charged and a noncharged atom. (C) Effective electrostatic potential U_{CH} with $x = 1.3$. The first potential well at distances $\sigma < r < 1.4\sigma$ corresponds to the potential energy E_{CH}. The second potential well at distances $1.4\sigma < r < 2.3\sigma$ corresponds to the potential energy $0.3E_{CH}$.

(Fig. 9B). We define an atomic hydropathy scale on the basis of an experimental estimation of the gain/loss of the free energy on transferring a particular amino acid from an aqueous solution to a gas phase (Wesson and Eisenberg, 1992). Knowing the gain/loss of the free energy for each amino acid, we then estimate the atomic solvation energies (Zhang *et al.*, 1997). When the distance between two atoms becomes smaller than 120% of the sum of their Van der Waals radii, we consider the interfacial volume as solvent excluded, and the two atoms interact with a potential energy equal to the sum of their hydropathy values. For a particular atom of type t, we define hydropathy values $HP_t = -\sigma_t SASA_t/n_t$, where σ_t is the atomic solvation parameter (free energy gain/loss per unit of solvent-exposed area of the atom type t), $SASA_t$ is the solvent accessible surface area (Wesson and Eisenberg, 1992), and n_t is an estimated number of the neighboring nonbonded atoms (usually $1 \leq n_t \leq 3$).

Amino Acid–Specific Interactions Caused by Charge. We implement an effective electrostatic interaction between two charged atoms using a double attractive/repulsive square well potential (Fig. 9C). The cutoff between two charged atoms is set to a value 2.33-fold larger than the sum of their Van der Waals radii ($\sim$7 Å). A tunable potential energy of the charged interaction within the range 0–2.5 kcal/mol allows us to perform simulations in a wide range of solvent conditions. This energy range is within and above the experimentally measured values for the free energy gain on salt-bridge formation on the surface of proteins, 0.24–1.26 kcal/mol (Horovitz and Fersht, 1992; Searle et al., 1999). A more detailed description was given elsewhere (Borreguero et al., 2005).

Limitations of the DMD Approach

As described previously, the peptide model parameters (the peptide bonds and constraints) and the parameters of the hydrogen bonding are defined phenomenologically using the known crystalline structure of proteins from the Protein Data Bank. Such a phenomenological approach to modeling (Ding et al., 2002) has been discussed extensively by Zhang et al. (2004). The phenomenologically derived force-field was shown to be essential for successful folding of the Trp-cage protein (Ding et al., 2005). As the PDB expands, however, these parameters could change. Different protein databases (α-protein database, β-protein database, α/β-protein database, etc.) could yield different parameters as well. It has been shown that knowledge-based potentials yield different results when trained on either NMR or X-ray resolved structures (Godzik et al., 1995). We considered the filtered protein structural database PDB40, which contains representatives of all known protein folds. The filter ensures that no two proteins have more than 40% sequence identity, preventing any bias in the statistical analysis toward overrepresented homologous sequences. For our purposes, the peptide model parameters obtained from this database are "fixed" and represent the definition of the model peptide.

In our approach, the hydrogen bond interaction is not amino acid-specific. The hydrogen-bond potential energy, E_{HB}, is the energy unit. This choice does not imply that we treat the hydrogen bond interaction as a fixed interaction independent of the environment. In fact, the free energy cost of breaking a hydrogen bond strongly depends on the local environment (Honig and Yang, 1995). This cost may be small in aqueous solutions but is typically large in organic solvents. In aqueous solution, the effective hydrogen bond is strong at the hydrophobic core and weak at the surface of a protein or a protein assembly. These local variations of the hydrogen bonding caused by variations in the dielectric constant within a solvent are neglected in our approach. However, our simulation approach allows for different environments by assigning different nonvariable dielectric constants to different solvent conditions and correspondingly renormalizing interactions in the model to the reference energy given by the hydrogen bond energy, E_{HB}.

Amino acid–specific interactions caused by hydropathy and charge also depend on the particular environment under study. The strength of the effective hydropathic interactions depends on the solvent and other global variables, such as temperature, molar concentration, and pH. Our model accounts for these global variables by varying the amino acid–specific hydropathies and charge, temperature, number of peptides, and the volume of the simulation box. In an explicit solvent, however, the hydrophobic/hydrophilic effect and electrostatic interactions depend strongly on the local variations of SASA and the dielectric constant. As in the case of the hydrogen bond, in water the hydrophobic/hydrophilic effect is strongest at the surface of the protein or protein assembly. The opposite is true for the electrostatic interactions. The free energy change associated with breaking a salt-bridge within the hydrophobic core of a protein is much higher than at its surface, where polar water molecules shield the charged atoms and thus effectively weaken the electrostatic interactions between the side-chains.

The DMD approach described in this review neglects local variations in the dielectric constant and in the SASA of each side-chain. The effective hydrophobic/hydrophilic interactions in coarse-grained models are based on hydropathy scales. A number of different hydropathy scales exist, some of them phenomenological (for example, Kyte and Doolittle [1982]); others are based on the *in vitro* gain/loss of the free energy when a particular atom is transferred from an aqueous solvent to a gas phase (for example, Wesson and Eisenberg [1992]). A question that needs to be addressed in the future is how robust the results of the DMD approach are with respect to different hydropathy scales.

The effective electrostatic interaction is modeled by a two-step square-well potential. We neglect the long-range nature of the Coulombic interaction between two charged particles (and/or two dipoles). Implementing a true long-range electrostatic interaction would require a considerable computational effort with a potential approximated by a multistep square-well of an "infinite" range. The interaction range problem is addressed in all-atom MD either by using Ewald sums in combination with periodic boundary conditions, multipole expansions, or a field-reaction method (Rapaport, 1997). However, even these sophisticated algorithms neglect the electrostatic forces above a certain cutoff distance. In the DMD approach, the solvent is implicit. When implementing the effective interaction between charged atoms, one needs to consider effects of the aqueous solution. The charged groups of a peptide, surrounded by water molecules, are effectively shielded because of the polar nature of water molecules. Because of this shielding, we can approximate the effective electrostatic potential by a two-step square-well potential with a finite distance range as a first-order approximation. In addition, one can assign an effective charge to a given side-chain. As the charge of a particular side-chain depends on pH, we can model different pH environments by reassigning the charge of a particular amino acid. For example, H, which is considered neutral at pH = 7, would be considered positively charged at low pH. Again, our approach neglects the fact that in the core of a peptide assembly the electrostatic interactions are stronger than at the surface of the assembly.

In principle, it is possible to model local variations of interparticle interactions that depend on SASA of individual side-chain atoms in the DMD approach by keeping track of the neighborhood of each atom. All interparticle interactions except the hard-core and soft-core interactions would need to be rescaled at regular simulation time intervals to account for atom-to-atom variability of SASA. However, this would require computational effort and could significantly slow down the simulations.

Conclusion

In this chapter, we described in detail the DMD method, coarse-grained protein models, and interparticle interactions that were developed to study Aβ folding and aggregation. We also addressed weaknesses of the DMD approach that originate in simplifications of the peptide description, interparticle interactions, and the absence of explicit solvent. The strengths of the DMD approach are: (1) its *ab initio* nature that does not require any experimental parameters specific to Aβ as input parameters; (2) its efficiency that allows for study of not only folding but also of oligomerization and fibril formation of full-length Aβ; and (3) its biological relevance that can be achieved through structural *in vitro* $\leftrightarrow$ *in silico* feedback-guided development of the model and interactions. Because of these advantages, this approach is also applicable to studies of proteins associated with other neurodegenerative diseases.

Soluble oligomers are a common feature of amyloid assembly, but their significance in the pathway of fibril assembly is not clear (Glabe, 2004). Are oligomers obligatory intermediates on a single pathway from misfolded monomers to fibrils or are oligomers reversible off-pathway intermediates that only buffer the concentration of misfolded monomers? Another fundamental question is how the assembly state of Aβ correlates with its function and toxicity (Klein *et al.*, 2004). The DMD approach can provide detailed structural information on different assembly states and explain the pathways of Aβ oligomer and fibril formation. Assuming that the assembly structure is directly correlated to toxicity, *in silico* findings could yield mechanistic hypotheses about why certain assemblies are more toxic than others and thus provide directions for further *in vitro* testing and identifying drug targets that would disrupt formation of these assemblies.

Acknowledgments

We thank S. V. Buldyrev, F. Ding, and N. V. Dokholyan for helpful discussions on the implementation of coarse-grained models and for continued intellectual support. We greatly appreciate many valuable discussions with G. Bitan, N. D. Lazo, and D. B. Teplow that guided us in our approach. We are thankful to the NIH for support on the grant AG023661 and Alzheimer's Association for the Zenith Fellows award. We are grateful to Stephen Bechtel, Jr. for a private donation.

References

Ash, W. L., Zlomislic, M. R., Oloo, E. O., and Tieleman, D. P. (2004). Computer simulations of membrane proteins. *Biochim. Biophys. Acta* **1666,** 158–189.

Berendsen, H. J. C., Postma, J. P. M., Vangunsteren, W. F., Dlinola, A., and Haak, J. R. (1984). Molecular dynamics with coupling to an external bath. *J. Chem. Phys.* **81,** 3684–3690.

Bernstein, S. L., Wyttenbach, T., Baumketner, A., Shea, J.-E., Bitan, G., Teplow, D. B., and Bowers, M. T. (2005). Amyloid β-protein: Monomer structure and early aggregation states of Aβ42 and its Pro[19] alloform. *J. Am. Chem. Soc.* **127,** 2075–2084.

Bitan, G. (2006). Structural study of metastable amyloidogenic protien oligomers by photo-induced cross-linking of unmodified protiens. *Methods Enzymol.* **413,** in press.

Bitan, G., Kirkitadze, M. D., Lomakin, A., Vollers, S. S., Benedek, G. B., and Teplow, D. B. (2003a). Amyloid β-protein (Aβ) assembly: Aβ–40 and Aβ–42 oligomerize through distinct pathways. *Proc. Natl. Acad. Sci. USA* **100,** 330–335.

Bitan, G., Vollers, S. S., and Teplow, D. B. (2003b). Elucidation of primary structure elements controlling early amyloid β-protein oligomerization. *J. Biol. Chem.* **278,** 34882–34889.

Bitan, G., Tarus, B., Vollers, S. S., Lashuel, H. A., Condron, M. M., Straub, J. E., and Teplow, D. B. (2003c). A molecular switch in amyloid assembly: Met[35] and amyloid β-protein oligomerization. *J. Am. Chem. Soc.* **125,** 15359–15365.

Borreguero, J. M. (2004). Computational studies of protein stability and folding kinetics. PhD. Thesis, Boston University.

Borreguero, J. M., Urbanc, B., Lazo, N. D., Buldyrev, S. V., Teplow, D. B., and Stanley, H. E. (2005). Discrete molecular dynamics study of the amyloid β-protein decapeptide Aβ(21–30). *Proc. Natl. Acad. Sci. USA* **102,** 6015–6020.

Chandonia, J. M., Hon, G., Walker, N. S., Lo Conte, L., Koehl, P., Levitt, M., and Brenner, S. E. (2004). The ASTRAL compendium in 2004. *Nucleic Acids Res.* **32,** D189–D192.

Coles, M., Bicknell, W., Watson, A. A., Fairlie, D. P., and Craik, D. J. (1998). Solution structure of amyloid-β peptide (1–40) in a water-micelle environment. Is the membrane spanning domain where we think it is? *Biochemistry* **37,** 11064–11077.

Creighton, T. E. (1993). "Proteins: Structures and Molecular Properties," 2nd Ed. Freeman and Company.

Crescenzi, O., Tomaselli, S., Guerrini, R., Salvatori, S., D'Ursi, A. M., Temussi, P. A., and Picone, D. (2002). Solution structure of the Alzheimer amyloid β-peptide (1–42) in an apolar microenvironment. Similarity with a virus fusion domain. *Eur. J. Biochem.* **269,** 5642–5648.

Cruz, L., Urbanc, B., Borreguero, J. M., Lazo, N. D., Teplow, D. B., and Stanley, H. E. (2005). Solvent and mutation effects on the nucleation of amyloid β-protein folding. *Proc. Natl. Acad. Sci. USA* **102,** 18258–18263.

Ding, F., Dokholyan, N. V., Buldyrev, S. V., Stanley, H. E., and Shakhnovich, E. I. (2002). Direct molecular dynamics observation of protein folding transition state ensemble. *Biophys. J.* **83,** 3525–3532.

Ding, F., Borreguero, J. M., Buldyrev, S. V., Stanley, H. E., and Dokholyan, N. V. (2003). A mechanism for the α-helix to β-hairpin transition. *Proteins: Struct. Func. Genet.* **53**, 220–228.

Ding, F., Buldyrev, S. V., and Dokholyan, N. V. (2005). Folding Trp-cage to NMR resolution native structure using a coarse-grained protein model. *Biophys. J.* **88**, 147–155.

Dobson, C. M. (2004). Principles of protein folding, misfolding, and aggregation. *Cell Dev. Biol.* **15**, 3–16.

Dokholyan, N. V., Buldyrev, S. V., Stanley, H. E., and Shakhnovich, E. I. (1998). Discrete molecular dynamics studies of folding of a protein-like model. *Folding Design* **3**, 577–587.

Dokholyan, N. V., Buldyrev, S. V., Stanley, H. E., and Shakhnovich, E. I. (2000). Identifying the protein folding nucleus using molecular dynamics. *J. Mol. Biol.* **296**, 1183–1188.

Dokholyan, N. V., Borreguero, J. M., Buldyrev, S. V., Ding, F., Stanley, H. E., and Shakhnovich, E. I. (2003). Identifying importance of amino acids for protein folding from crystal structures. *Methods Enzymol.* **374**, 616–638.

Feig, M., and Brooks, C. L., III (2004). Recent advances in the development and application of implicit solvent models in biomolecule simulations. *Curr. Opin. Struct. Biol.* **14**, 217–224.

Fersht, A. R., and Daggett, V. (2002). Protein folding and unfolding at atomic resolution. *Cell* **108**, 573–582.

Glabe, C. G. (2004). Conformation-dependent antibodies target diseases of protein misfolding. *Trends Biochem. Sci.* **29**, 542–547.

Godzik, A., Kolinski, A., and Skolnick, J. (1995). Are proteins ideal mixtures of amino acids? Analysis of energy parameter sets. *Prot. Sci.* **4**, 2107–2117.

Hall, C. K., and Wagoner, V. A. (2006). Computational approaches to fibril structure and formation. *Methods Enzymol.* **412**, 338–365.

Heinig, M., and Frishman, D. (2004). STRIDE: A web server for secondary structure assignment from known atomic coordinates of proteins. *Nucl. Aci. Res.* **32**, W500–W502.

Honig, B., and Yang, A. S. (1995). Free energy balance in protein folding. *Adv. Protein Chem.* **46**, 27–58.

Horovitz, A., and Fersht, A. R. (1992). Co-operative interactions during protein folding. *J. Mol. Biol.* **224**, 733–740.

Humphrey, W., Dalke, A., and Schulten, K. (1996). VMD: Visual molecular dynamics. *J. Mol. Graphics* **14**, 33–38.

Karplus, M., and McCammon, J. A. (2002). Molecular dynamics simulations of biomolecules. *Nat. Struct. Biol.* **9**, 646–652.

Kayed, R., Head, E., Thompson, J. L., McIntire, T. M., Milton, S. C., Cotman, C. W., and Glabe, C. G. (2003). Common structure of soluble amyloid oligomers implies common mechanisms of pathogenesis. *Science* **300**, 486–489.

Klein, W. L., Stine, W. B., Jr., and Teplow, D. B. (2004). Small assemblies of unmodified amyloid β-protein are the proximate neurotoxin in Alzheimer's disease. *Neurobiol. Aging* **25**, 569–580.

Koo, E. H., Lansbury, P. T., Jr., and Kelly, J. W. (1999). Amyloid diseases: Abnormal protein aggregation in neurodegeneration. *Proc. Natl. Acad. Sci. USA* **96**, 9989–9990.

Kyte, J., and Doolittle, R. F. (1982). A simple method for displaying the hydropathic character of a protein. *J. Mol. Biol.* **157**, 105–132.

Lazo, N. D., Grant, M. A., Condron, M. C., Rigby, A. C., and Teplow, D. B. (2005). On nucleation of amyloid β-protein monomer folding. *Prot. Sci.* **14**, 1581–1596.

Ma, B., and Nussinov, R. (2002). Stabilities and confomations Alzheimer's β-amyloid peptide oligomers ($A\beta_{16-22}$, $A\beta_{16-35}$, and $A\beta_{10-35}$): Sequence effect. *Proc. Natl. Sci. USA* **99**, 14126–14131.

Ma, B., and Nussinov, R. (2004). From computational quantum chemistry to computational biology: Experiments and computations are (full) partners. *Phys. Biol.* **1**, P23–P26.

Nguyen, H. D., and Hall, C. K. (2004a). Molecular dynamics simulations of spontaneous fibril formation by random-coil peptides. *Proc. Natl. Acad. Sci. USA* **101**, 16180–16185.

Nguyen, H. D., and Hall, C. K. (2004b). Phase diagrams describing fibrillization by polyalanine peptides. *Biophys. J.* **87**, 4122–4134.

Nguyen, H. D., and Hall, C. K. (2005). Kinetics of fibril formation by polyalanine peptides. *J. Biol. Chem.* **280**, 9074–9082.

Peng, S., Ding, F., Urbanc, B., Buldyrev, S. V., Cruz, L., Stanley, H. E., and Dokholyan, N. V. (2004). Discrete molecular dynamics simulations of peptide aggregation. *Phys. Rev.* **E69**, 041908.

Petkova, A., Ishii, T. Y., Balbach, J. J., Antzutkin, O. N., Leapman, R. D., Delaglio, F., and Tycko, R. (2002). A structural model for Alzheimer's β-amyloid fibrils based on experimental constraints from solid state NMR. *Proc. Natl. Acad. Sci. USA* **99**, 16742–16747.

Rapaport, D. C. (1997). "The Art of Molecular Dynamics Simulation." Cambridge University Press.

Sciarretta, K. L., Gordon, D. J., Petkova, A. T., Tycko, R., and Meredith, S. C. (2005). Aβ40-Lactam (D23/K28) models a conformation highly favorable for nucleation of amyloid. *Biochemistry* **44**, 6003–6014.

Searle, M. S., Griffiths-Jones, S. R., and Skinner-Smith, H. (1999). Energetics of weak interactions in a β-hairpin peptide: Electrostatic and hydrophobic contributions to stability from lysine salt bridges. *J. Acad. Chem. Soc.* **121**, 11615–11620.

Selkoe, D. J. (2001). Alzheimer's disease: Genes, proteins, and therapy. *Physiol. Rev.* **81**, 741–766.

Smith, A. V., and Hall, C. K. (2001a). α-Helix formation: Discontinuous molecular dynamics on an intermediate-resolution protein model. *Proteins* **44**, 344–360.

Smith, A. V., and Hall, C. K. (2001b). Assembly of a tetrameric α-helical bundle: Computer simulations on an intermediate-resolution protein model. *Proteins* **44**, 376–391.

Snow, C. D., Nguyen, N., Pande, V. S., and Gruebele, M. (2002). Absolute comparison of simulated and experimental protein-folding dynamics. *Nature* **420**, 102–106.

Takada, S., Luthey-Schulten, Z., and Wolynes, P. G. (1999). Folding dynamics with nonadditive forces: A simulation study of a designed helical protein and a random heteropolymer. *J. Chem. Phys.* **110**, 11616–11629.

Taketomi, H., Ueda, Y., and Gō, N. (1975). Studies on protein folding, unfolding and fluctuations by computer simulations. *Int. J. Peptide Protein Res.* **7**, 445–459.

Tsai, J., Taylor, R., Chothia, C., and Gerstein, M. (1999). The packing density in proteins: Standard radii and volumes. *J. Mol. Biol.* **290**, 253–266.

Tycko, R. (2006). Characterization of amyloid structures at the molecular level by solid state nuclear magnetic resonance spectroscopy. *Methods Enzymol.* **413**, in press.

Urbanc, B., Cruz, L., Ding, F., Sammond, D., Khare, S., Buldyrev, S. V., Stanley, H. E., and Dokholyan, N. V. (2004a). Molecular dynamics simulation of amyloid-β dimer formation. *Biophys. J.* **87**, 2310–2321.

Urbanc, B., Cruz, L., Yun, S., Buldyrev, S. V., Bitan, G., Teplow, D. B., and Stanley, H. E. (2004b). *In silico* study of amyloid β-protein (Aβ) folding and oligomerization. *Proc. Natl. Acad. Sci. USA* **101**, 17345–17350.

Wesson, L., and Eisenberg, D. (1992). Atomic solvation parameters applied to molecular dynamics of proteins in solution. *Protein Sci.* **1**, 227–235.

Xu, Y., Shen, J., Luo, X., Zhu, W., Chen, K., Ma, J., and Jiang, H. (2005). Conformational transition of amyloid β-peptide. *Proc. Natl. Acad. Sci. USA* **102**, 5403–5407.

Zagrovic, B., Snow, C. D., Shirts, M. R., and Pande, V. S. (2002). Simulation of folding of a small α-helical protein in atomistic detail using worldwide-distributed computing. *J. Mol. Biol.* **323**, 927–937.

Zagrovic, B., and Pande, V. S. (2003). Solvent viscosity dependence of the folding rate of a small protein: Distributed computing study. *J. Comput. Chem.* **24**, 1432–1436.

Zhang, C., Vasmatzis, G., Cornette, J. L., and DeLisi, C. (1997). Determination of atomic desolvation energies from the structures of crystallized proteins. *J. Mol. Biol.* **267,** 707–726.

Zhang, C., Liu, S., Zhou, H., and Zhou, Y. (2004). The dependence of all-atom statistical potentials on structural training database. *Biophys. J.* **86,** 3349–3358.

Zhang, S., Iwata, K., Lachenmann, M. J., Peng, J. W., Li, S., Stimson, E. R., Lu, Y., Felix, A. M., Maggio, J. E., and Lee, J. P. (2000). The Alzheimer's peptide Aβ adopts a collapsed coil structure in water. *J. Struct. Biol.* **130,** 130–141.

Zhou, Y., Hall, C. K., and Karplus, M. (1996). First-order disorder-to-order transition in an isolated homopolymer model. *Phys. Rev. Lett.* **77,** 2822–2825.

Zhou, Y., Karplus, M., Wichert, J. M., and Hall, C. K. (1997). Equilibrium thermodynamics of homopolymers and clusters: Molecular dynamics and Monte-Carlo simulations of system with square-well interactions. *J. Chem. Phys.* **107,** 10691–10708.

Zhou, Y., and Karplus, M. (1997). Folding thermodynamics of a three-helix-bundle protein. *Proc. Natl. Acad. Sci. USA* **94,** 14429–14432.

Zhou, Y., and Karplus, M. (1999). Folding of a model three-helix bundle protein: A thermodynamic and kinetic analysis. *J. Mol. Biol.* **293,** 917–951.

[20] Computational Approaches to Fibril Structure and Formation

By Carol K. Hall and Victoria A. Wagoner

Abstract

Assembly of normally soluble proteins into amyloid fibrils is a cause or associated symptom of numerous human disorders. Although some progress toward understanding the molecular-level details of fibril structure has been made through *in vitro* experiments, the insoluble nature of fibrils make them difficult to study experimentally. We describe two computational approaches used to investigate fibril formation and structure: intermediate-resolution discontinuous molecular dynamics simulations and atomistic molecular dynamics simulations. Each method has its strengths and weaknesses, but taken together the two approaches provide a useful molecular-level picture of fibril structure and formation.

Introduction

The hallmark of many neurodegenerative diseases, including Alzheimer's disease, is the accumulation and deposition of protein plaques in specific tissues within various organs in the body (Koo, 2002). These plaques are composed of ordered protein aggregates known as amyloid fibrils that form when normally soluble disease-specific proteins undergo a conformational change that leads to their aberrant assembly. Many of the 24 known so-called

METHODS IN ENZYMOLOGY, VOL. 412   0076-6879/06 $35.00
DOI: 10.1016/S0076-6879(06)12020-0

amyloid diseases are fatal. (Bucciantini *et al.*, 2002; Dobson, 2001; Kelly, 1998; Prusiner, 1997).

In vitro experiments on the different proteins associated with amyloid diseases (e.g., β-amyloid [Alzheimer's], prion protein [transmissible spongiform encephalopathies], and huntingtin [Huntington's]) indicate that even though these proteins have no obvious sequence homology, the amyloid fibrils formed share similar morphological properties (Blake and Serpell, 1996; Serpell, 2000; Serpell *et al.*, 2000; Sunde and Blake, 1997; Sunde *et al.*, 1997). X-ray diffraction studies of amyloid fibrils show that they are straight, rigid structures of varying length that consist of two or more smaller fibrils, called protofilaments. The protofilament itself is a "cross-β" structure with β strands perpendicular to the fibril axis connected by backbone hydrogen bonds to form β-sheets parallel to the fibril axis. Interestingly, proteins other than those known to cause one of the amyloid diseases have been found to form fibrils when their protein's native state is disrupted, such as in the presence of denaturant or under concentrated conditions (Chiti *et al.*, 1999, 2001). As a consequence, a number of researchers have begun to think that it is the basic interactions experienced by all proteins (e.g., the hydrophobic interactions and hydrogen bonding) rather than the chemistry associated with specific sequences that drive amyloid formation (Chiti *et al.*, 1999; Dobson, 2001). This implies that fibril or protofibril (Caughey and Lansbury, 2003) formation may be an intrinsic property of many different proteins (Hou and Zagorski, 2004; Kelly, 1997, 1998; Rochet and Lansbury, 2000) not just those associated with the amyloid diseases.

Although experimental research into the formation of amyloid is vital, a number of investigators have turned to computer simulation to gain a better understanding of the molecular-level details associated with amyloid formation. For example, Ma and Nussinov recently used all-atom molecular dynamics simulation to study preformed model fibrils of short alanine-rich fragments (5–40 residues) of proteins associated with disease. These model fibrils remained stable over the length of the simulation (Ma and Nussinov, 2002a), indicating that their postulated fibril structure may, indeed, be that of the real fibril. In addition, our group has developed an intermediate-resolution protein model that, when combined with discontinuous molecular dynamics, allows the simulation of protofilament formation starting from a completely random initial configuration of 48 chains of Ac-KA$_{14}$K-NH$_2$ (Nguyen and Hall, 2004b). These studies indicate that we may be able to shed light on the complex problem of protein aggregation by studying simple sequences with computer simulations. Important lessons about the essential physical features necessary for the formation and structure of disease-related amyloid fibrils could be learned with the use of computer simulations, provided that the models used are reasonable representations of the real molecule.

In this chapter, we focus on the two methods mentioned in the previous paragraph, molecular dynamics simulations based on all-atom protein models and discontinuous molecular dynamics simulation based on intermediate resolution protein models, and their application to protein aggregation. Both methods are rooted in molecular dynamics simulation. In molecular dynamics, the trajectories of studied atoms are computed by solving Newton's equation of motion at regularly spaced time intervals, called time steps. At the beginning of each time step, the net force acting on each molecule because of all the other molecules is calculated. Knowledge of this force allows the determination of each molecule's acceleration (F = ma), which, in turn, allows the prediction of the position and velocity of each molecule at the beginning of the next time step. At each time step, the instantaneous values of thermodynamic properties are computed and recorded. Thermodynamic properties are obtained by averaging the instantaneous properties over time (Allen and Tildesley, 1987; Frenkel and Smit, 1996; Leach, 2001).

Discontinuous molecular dynamics (DMD) is a variant on standard molecular dynamics that is applicable to systems of molecules interacting by means of discontinuous potentials (e.g., hard sphere and square-well potentials). Unlike soft potentials such as the Lennard–Jones potential, discontinuous potentials exert forces only when particles collide, enabling the exact (as opposed to numerical) solution of the collision dynamics. This imparts great speed to the algorithm, allowing sampling of longer time scales and larger systems than traditional molecular dynamics. The particle trajectories are followed by analytically integrating Newton's equations of motion, locating the time between collisions and then advancing the simulation to the next collision (event) (Alder and Wainwright, 1959; Smith *et al.*, 1997). DMD on chain-like molecules is generally implemented using the "bead string" algorithm introduced by Rapaport (Rapaport, 1978, 1979) and later modified by Bellemans *et al.* (1980). Chains of square-well spheres can be accommodated in this algorithm by introducing well-capture, well-bounce, and well-dissociation "collisions" when a sphere enters, attempts to leave, or leaves the square well of another sphere. In canonical ensemble DMD simulations (constant system size, volume and temperature), the temperature is maintained constant by implementing the Andersen thermostat method (Andersen, 1980); all spheres are subjected to random, infrequent collisions with ghost particles whose velocities are chosen randomly from a Maxwell Boltzmann distribution centered at the system temperature.

Perhaps the most important aspect of molecular simulations, whether we are speaking of continuous or discontinuous molecular dynamics, is what the simulation can teach us about the proteins we are studying. Simulations provide molecular snapshots of a protein that currently cannot be obtained from experiments. The snapshots tell us how proteins (or a system of proteins) arrange themselves, which residues are interacting through a hydrogen bond, the distance between two key amino acids, etc. Structural properties can also be determined such as the radius of gyration, which is a measure of the distance of particle i from the center of mass of the entire molecule:

$$R_g^* = \frac{< 1/N \sum_{i=1}^{N} [(x_i - x_c)^2 + (y_i - y_c)^2 + (z_i - z_c)^2] >}{\sigma^2 N} \tag{1}$$

In addition, knowing the internal energy (E) of the system allows us to calculate many other properties such as the melting temperature, free energy, and the heat capacity:

$$C_v^* = \frac{C_v}{k_B} = \frac{[< E^2 > - (< E >)^2]}{(\varepsilon T^*)^2} \tag{2}$$

by using statistical mechanical relationships. These properties can be compared with experimentally determined properties.

Protein Models

Protein folding models range from very simple lattice models or single-sphere-per-amino-acid continuum models to very complex all-atom continuum models in explicit solvent. All-atom simulation packages such as CHARMM (Brooks et al., 1983), AMBER (Weiner et al., 1986), DISCOVER (Dauber-Osguthorpe et al., 1988), ENCAD (Levitt, 1983), and ECEPP (Zimmerman et al., 1977) provide explicit detail on the protein's geometry and its interactions with itself and solvent molecules, thereby accounting for all of the interactions involved in protein folding. The atomistic detail provided by these software packages comes at a cost in terms of the time scale that can be examined during a simulation. The longest all-atom simulation to-date was a 1 μsec simulation of the 36-residue villin headpiece starting from an initial configuration with some native state characteristics performed by Duan and Kollman (Duan and Kollman, 1998). [In comparison, the time scale required for a protein to fold from a random coil is 10–100 μsec (Leach, 2001)]. This simulation took 2 full months on 256 dedicated parallel supercomputers, yet they only achieved partial folding of the protein. Therefore, to simulate the complete folding process from a random coil configuration to the protein's native (folded) state, simplified representations (models) of proteins must be used. The big question is: How much detail is enough?

All-Atom Molecular Dynamics

All-atom molecular dynamics simulations represent the protein at atomic resolution, generally accounting for every atom on the protein and on the solvent molecules. These simulations can provide crucial information regarding the stability of a final fibril structure. They can also provide a detailed description of the events occurring during the initial stages of aggregation, most commonly the formation of β-sheets, which is believed to be a key step in amyloid fibril formation (Khurana *et al.*, 2001). As mentioned earlier, in traditional molecular dynamics simulations, Newton's equations of motion are numerically integrated to obtain the trajectories of each participating molecule. A necessary input into these calculations is a description of the molecular geometry and of the forces acting on each molecule. The latter is expressed in terms of an energy function that is usually empirically based as opposed to being derived directly from quantum calculations. The energy is then the summation of the intramolecular and intermolecular interactions between all of the molecules. There are many all-atom molecular dynamics computer programs (CHARMM, AMBER, and ECEPP to name a few). The programs differ primarily in terms of the description of the forces acting on each atom (the force field). We will focus on the CHARMM energy function for illustrative purposes.

CHARMM

One of the most popular computer programs used to perform all-atom molecular dynamics simulations is CHARMM. In any of the all-atom MD packages, a molecular model (physical description of the protein) is needed before dynamic calculations can be made. In CHARMM, the molecular model includes information such as the protein sequence and the residue topology that are stored in the protein structure file (PSF). By residue topology, we mean a file containing data on the 20 amino acid residues including the identities of the constituent atoms, the values of the bond and torsional angles, the charge on each residue, and the identities of the atoms that are capable of hydrogen bonding. The purpose of the topology file is to enable the computer to generate a representation of the protein in question from its sequence alone (Molecular Simulations, 1999). CHARMM also incorporates another kind of file called a patch, which contains for example information about which residues can form disulfide bridges. In addition to the protein structure file, CHARMM requires a coordinate file, which contains the relative positions of all atoms (protein and solvent) in the input configuration, and a parameter file, which contains information necessary for calculating the energy of the molecule such as the force constants, the equilibrium bond distance, bond angle, and dihedral angles, and the van der Waals radii. The Cartesian coordinates of a particular protein can also be generated from X-ray crystallography data or NMR data provided by the protein databank. CHARMM uses periodic boundary conditions to limit the number of particles necessary to model the protein in solution. A summary of the files and information required in CHARMM to calculate the energy of a molecule in a particular conformation is shown in Fig. 1 (Brooks *et al.*, 1983; Molecular Simulations, 1999). These features are common to all CHARMM codes regardless of the type of simulation being performed.

The next component of CHARMM is the empirical energy function. There are many different forms for the empirical energy function, but all can be expressed as a sum of the following types of terms:

$$E = E_b + E_\theta + E_\varphi + E_\omega + E_{vdW} + E_{el} + E_{hb} + E_{cr} + E_{c\varphi}. \tag{3}$$

The terms in the energy function include potential energies for the bond length, E_b, bond angle, E_θ, dihedral angle, E_φ, improper torsions, E_ω, electrostatic interactions, E_{el}, hydrogen bonding interactions, E_{hb}, and structural constraints, E_{cr} and $E_{c\varphi}$. Each term in Eq. (3) is a continuous potential that represents either a bond length, bond angle, torsional angle, or nonbonded pairwise interaction term. One example of a nonbonded interaction is the van der Waals energy, E_{vdW}, which is often represented with a Lennard–Jones potential. Another nonbonded interaction is the

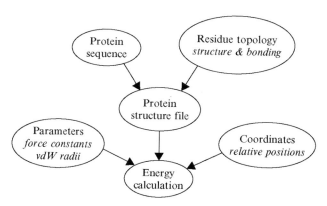

FIG. 1. Summary of the information required by CHARMM at the beginning of a simulation (Molecular Simulations, 1999).

electrostatic potential, E_{el}, which can be represented in a number of different ways including a constant dielectric (Coulomb's law), distance-dependent dielectric, shifted dielectric, dielectric that depends on the distance between the center of geometry of the two groups containing electrostatic atoms i and j, or extended dielectric that is the sum of a constant dielectric term and a second term that accounts for the approximate potential and field at atom i because of all other atoms not connected to atom i by angles and bonds outside a predetermined cutoff radius. (Brooks et al., 1983).

An important consideration in computer simulations of protein folding and aggregation is the treatment of the solvent. There are many different options for representing solvent and/or its impact (e.g., the hydrophobic effect) in CHARMM. One of the most common potentials used in CHARMM to describe water–water interaction is the ST2 potential (Brooks et al., 1983), but other models have been used successfully. The detailed description of water–water interactions that characterizes explicit-solvent all-atom molecular dynamics simulations severely limits the time scale that can be accessed. For example, in simulating a single protein molecule solvated with 3000 water molecules, most of the computational time is dedicated to keeping track of the water–water interactions. The point is that all-atom explicit-solvent calculations are very expensive computationally and significantly reduce the accessible time scales even for a single protein.

In an effort to access longer time scales and to study more complex systems, several energy functions have been developed that model the solvent implicitly as a potential of mean force. A potential of mean force accounts for both the hydrophobic effect and electrostatic screening effects. One popular implicit-solvent model is the CHARMM PARAM19 model (Ferrara et al., 2002; Gsponer et al., 2003). In the CHARMM PARAM19 model, the potential of mean force is described as the sum of both the intrasolute interactions and a mean solvation term. The solvation term is the sum over all the atoms of the product of an atomic solvation parameter and the solvent-accessible surface area (Ferrara et al., 2002). This model assumes that the major contributor to the solvation energy is the interaction between the protein and its first shell of solvent molecules (Roux and Simonson, 1999). In an effort to increase computational speed, Ferrara et al. used an approximate method developed by Hasel and co-workers (Hasel et al., 1988) to obtain an analytical expression for the solvent accessible surface area. In addition, this model enforces a free energy penalty when a charged residue is buried in the protein interior and accounts for electrostatic screening effects with distance-dependent dielectric. The CHARMM PARAM19 implicit-solvent model is much faster computationally than the explicit-solvent models and requires a mere 50% increase in computational time compared with *in vacuo* simulations (Ferrara *et al.*, 2002). However, even with the use of an implicit-solvent model, the time scales accessible in all-atom protein simulations are limited because of the molecular detail of the model.

Once the energy function is fully defined and the protein structure file is generated, the starting initial configuration is subjected to an energy minimization step to find the nearest local free energy minimum configuration. The initial structure configuration is cooled to a temperature at or close to 0 K (absolute zero) in which case the entropy is zero and so the internal energy is equivalent to the free energy. CHARMM has five different energy minimization techniques: steepest descent (SD), conjugated gradient (CONJ), Powell (POWE), Newton–Raphson (NRAP), and truncated-Newton minimization package (TNPACK). These methods have the common goal of finding a set of coordinates that corresponds to a molecular conformation whose potential energy is at a minimum (Molecular Simulations, 1999). Energy minimization can also be used to eliminate structural defects caused by overlapping atoms and distorted bond and torsional angles. Energy minimization is performed not only in the case of a single molecule but also in the case in which the initial configuration is a preformed fibril or aggregate structure.

The next step after energy minimization is the calculation of the dynamics during which Newton's second law is numerically integrated, most often with the Verlet algorithm (Allen and Tildesley, 1987; Frenkel and Smit, 1996). There are six basic steps in a typical dynamics run: the preliminary step (discussed previously), which involves generating the protein structure file (PSF), the energy minimization step (also discussed previously), the heating step, the equilibration step, the production step, and the quenching step. In the heating step, the velocities, which are randomly assigned to the atoms in the system from a Gaussian distribution, are increased at preset time intervals during dynamics calculations until the temperature of interest is reached. The heating step is followed by a period of equilibration until a stable temperature and structure are reached. The production step starts with the equilibrated structure; the particle trajectories (the dynamics) are followed by numerically integrating Newton's laws of motion with a time step on the order of 1–2 fs and for time scales of 10 ns (Molecular Simulations, 1999). Although the overall procedure seems quite straightforward, it can take months to reach some biologically relevant time scales, like β-sheet formation, on a system of parallel supercomputers (Duan and Kollman, 1998). All-atom molecular dynamics simulations are further confounded by the complex potential energy surface associated with protein complexes, making it difficult to determine whether a global energy minimum has been obtained or whether the structure has become kinetically trapped in local minima. Significant effort has been put forth to develop free energy optimization methods to address the issue of kinetic trapping, but this is outside the scope of this chapter (Mortenson and Wales, 2001).

Simulation Results on Fibril Structure Using All-Atom MD

Nussinov and coworkers have used the CHARMM polar hydrogen force field package in the presence of solvent to monitor the motions of 24 chains of Aβ(16–22), KLVFFAE, initially arranged in a double-layered antiparallel β-sheet configuration (Ma and Nussinov, 2002b). At the end of a 2.5 ns simulation, these short highly hydrophobic sequences formed a double-layered β-sheet conformation. The strands within each β-sheet were antiparallel to each other with a 15° twist as shown in Fig. 2. A more recent study by Haspel and coworkers on systems containing six or nine copies of a short sequence of the amyloidogenic human calcitonin hormone peptide (15–19) (Haspel et al., 2004), DFNKF, led to the observation of a stable parallel β-sheet conformation in a single sheet layer.

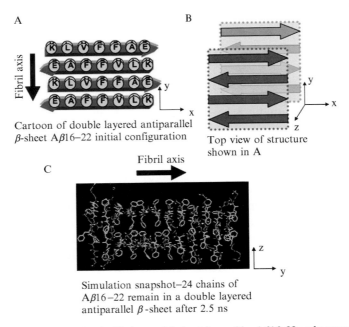

A

Fibril axis

Cartoon of double layered antiparallel
β-sheet Aβ16–22 initial configuration

B

Top view of structure
shown in A

C

Fibril axis

Simulation snapshot–24 chains of
Aβ16–22 remain in a double layered
antiparallel β-sheet after 2.5 ns

FIG. 2. Illustration of the double-layered β-sheet formed by Aβ16–22 and a snapshot from the CHARMM simulation (Ma and Nussinov, 2002a). (See color insert.)

The atomic-resolution simulations approach discussed thus far provides valuable information on the orientation of the peptides within a fibril but no information on the misfolding events that precede the formation of the fibril. In other words, the model fibril structure is postulated and is not based on known or experimental information. In contrast, the initial configuration in a molecular dynamics simulation of $A\beta_{1-40}$ by Petkova et al. (2002) was taken from their group's solid-state NMR data on phi–psi angles. When experimental data were not available, the phi–psi angles were set to $-140°$ and $140°$, respectively, the dihedral angles ω were set to $180°$, and the signs for the dihedral angles were chosen to allow for intermolecular backbone hydrogen bonding. The CHARMM forcefield was used to model five copies of $A\beta_{1-40}$ starting from the in-register parallel (within each β-sheet) cross-β arrangement obtained from the NMR data as described previously and shown in Fig. 3A. After energy minimization, they observed a fibril structure with the following characteristics: a salt bridge between residues K28 and D23, an offset horseshoe structure for each strand that was held together by intrachain association of hydrophobic side chains, and interchain backbone hydrogen bonding along the fibril axis as shown in Fig. 3B.

Information regarding fibril structure can also be gleaned from atomic resolution simulations of the initial stages of aggregation. Gsponer et al. (2003) explored the initial steps of yeast prion aggregation using CHARMM PARAM19, an implicit solvent model. They performed 20 MD simulations

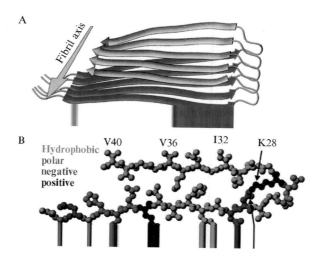

Fig. 3. (A) Ribbon diagram depicting the arrangement of the five copies of $A\beta(10–40)$. (B) $A\beta(10–40)$ as depicted by CHARMM all-atom simulations package with residues colored according to type. (Petkova et al., 2002). (See color insert.)

on a system containing three copies of the heptapeptide sequence, GNNQQNY, residues 7–13 of the yeast protein Sup35, starting from random conformations at 330K for a total of 20 μsec. They found that the three peptides arranged themselves in a parallel orientation with predominantly β-structure. Santini et al. (2004) studied a system of three Aβ_{16-22} peptides arranged in six different starting configurations, including random unfolded chains, and an antiparallel β-sheet, to explore the early stages of aggregation. They used the ART-OPEP model in conjunction with the GROMACS simulation package with implicit solvent for a total simulation time of 20 ns. They found that the system preferred to be in an ordered β-sheet configuration. The system size was too small to make any conclusive remarks about parallel versus antiparallel arrangements.

Atomic resolution simulations have also been used to study destabilizing events that are believed to happen early on during the fibrillization process, such as the α-helix to β-strand conversion or the partial unfolding of the native state into a conformation likely to aggregate into ordered structures. Recently, Tarus et al. (2005) explored the initial stage of aggregation of Aβ_{10-35}, which is thought to be a dimerization based on experimental observations. They used the GRAMM energy minimization program to form two types of dimers, the φ-dimer, which is stabilized by hydrophobic contacts, and the ε-dimer, which is stabilized by electrostatic interactions. Each of the two dimers was then simulated for 10 ns using CHARMM PARAM22 with explicit solvent to determine its structural stability. The φ-dimer did not dissociate, whereas the ε-dimer did, indicating that the forcing out of water molecules because of the interaction between hydrophobic residues on the φ-dimer may play a significant role in its structural stability.

Intermediate-Resolution Protein Model with Discontinuous Molecular Dynamics

Given the computational difficulties and limitations associated with simulating the folding of just a single protein, one would imagine that these problems would be even greater when the number of chains in the system is sizable as in aggregation simulations. There is also the added difficulty of the long time scales involved. For example, it can take hours for proteins to aggregate in vitro (Blondelle et al., 1997; Forood et al., 1996). Certain aspects of protein aggregation (for example the cross-β structure in protofilaments) seem to be less dependent on the protein sequence than protein folding (Chiti et al., 1999, 2001). In fact, Wetzel and coworkers have shown experimentally that mutations in the hydrophobic core of Aβ1–40, residues 17–20, destabilizes the fibril structure, but this effect is often counterbalanced by the formation of additional hydrogen bonds (Williams *et al.*, 2004, 2006). These observations suggest that protein aggregation may be less sensitive to the details of the intramolecular and intermolecular interactions than protein folding. Therefore, simplified approaches have gained in popularity for studies of aggregation in multipeptide systems.

There are two avenues by which all-atom molecular dynamics can be simplified. One approach is to change the manner in which the dynamics is calculated, and the second approach is to reduce the detail involved in the description of the molecule. These two simplifications can either be used together or separately. In our group's work, we do both. We simplify the dynamics calculation by using discontinuous molecular dynamics, and we reduce the detail in the protein's representation by using an intermediate resolution protein model.

Discontinuous Molecular Dynamics

DMD simulations begin by placing the model molecules into a virtual simulation box that is replicated in all dimensions to eliminate wall effects; this is known as periodic boundary conditions. The volume fraction of molecules in the simulation box for the case in which the molecules are spherical is given by $\varphi = \pi N \sigma^3 / 6V$, where N is the number of molecules in the box, σ is the particle diameter, and V is the volume of the simulation box. In protein simulations, one more commonly refers to the concentration of the system that is given by $c = N/L^3$, where L is the simulation box length. The temperature is set according to the equipartition theorem which relates the system kinetic energy to the temperature:

$$\frac{1}{2}(m_i v_i)^2 = \frac{3}{2} N k_B T \qquad (4)$$

where m is the particle mass, v is the particle velocity, and k_B is the Boltzmann constant. As mentioned previously, the velocities are assigned randomly from a Maxwell–Boltzmann distribution about the system temperature. Protein aggregation simulations are often performed in the canonical ensemble, which means that the number of particles, volume, and temperature are held constant. Constant temperature is maintained by implementing the Andersen thermostat method (Andersen, 1980).

The main steps in a DMD simulation are the calculation of collision times, t_{ij}, between particles i and j, and the calculation of the post-collision velocities for the colliding pair. DMD proceeds by calculating the collision times for all possible pairs of particles, determining which collision pair, i and j, has the smallest collision time, advancing the system to that event, and computing the system dynamics. A simple procedure for DMD is shown in Fig. 4.

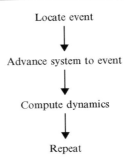

Locate event

Advance system to event

Compute dynamics

Repeat

FIG. 4. A general procedure for DMD.

For two particles, i and j, with diameter σ undergoing a hard sphere collision, the collision will occur at time t_{ij}, which satisfies the condition:

$$|r_{ij}(t + t_{ij})| = |r_{ij} + v_{ij}t_{ij}| = \sigma \qquad (5)$$

where $r_{ij} = r_i - r_j$ and $v_{ij} = v_i - v_j$. The collision time t_{ij} is obtained by squaring Eq. (5) and solving the resulting quadratic equation:

$$t_{ij} = \frac{-b_{ij} \pm \sqrt{b_{ij}^2 - v_{ij}^2(r_{ij}^2 - \sigma^2)}}{v_{ij}^2} \qquad (6)$$

where b_{ij} is defined as $b_{ij} = r_{ij} v_{ij}$. This equation must meet certain criteria for the calculated t_{ij} to be correct. If b_{ij} is greater than 0, then the spheres are moving away from each other, and no collision will occur. If b_{ij} is less than 0 and the square-root discriminant is positive (to give a real solution), the spheres will collide (Alder and Wainwright, 1959; Allen and Tildesley, 1987; Smith et al., 1995a, 1997). For a hard sphere collision we are interested in the smallest collision time; this means the discriminant in Eq. (6) should be subtracted. Eq. (6) can be used to generate a list of all collision times for all possible colliding pairs, i and j. The next collision time t_c is the minimum collision time in this list. The system is advanced by t_c to the new position $r_i(t + t_c)$, at which a single pair of particles is in contact preparing to undergo a collision:

$$r_i(t + t_c) = r_i(t) + v_i t_c. \qquad (7)$$

After the collision occurs, the post-collision velocities are calculated. The new velocities for interacting particles i and j are determined by imposing conservation of kinetic energy and conservation of linear momentum. The new velocities are given by:

$$v_i \, (after) = v_i \, (before) + \Delta v_i \qquad (8)$$

$$v_j \left(after\right) = v_j \left(before\right) - \Delta v_i \tag{9}$$

where v_i represents each component of the velocity vector and the velocity change, Δv_i, for a hard core collision between collision partners i and j separated by a distance σ is:

$$\Delta v_i = -\Delta v_j = -\left(\frac{b_{ij}}{\sigma^2}\right) r_{ij}. \tag{10}$$

After each event between partners i and j, the collision time list is updated for particles i and j and any other particles that would have interacted with either i or j. (Alder and Wainwright, 1959; Allen and Tildesley, 1987; Smith, 1997; Smith *et al.*, 1995b).

The equations described thus far are applicable to hard sphere particles undergoing a core collision. Real intermolecular potentials, however, include not just repulsive interactions (like the hard sphere case) but also attractive interactions. The square-well potential is one such potential that has a repulsive interaction at short distances and an attractive interaction at intermediate distances. The square-well potential is mathematically described as:

$$U(r_{ij}) = \begin{cases} \infty & (r < \sigma_1) \\ -\varepsilon & (\sigma_1 \leq r \leq \sigma_2) \\ 0 & (r > \sigma_2) \end{cases}. \tag{11}$$

The equations described previously for hard sphere particles can easily be extended to the case of the square-well potential (Alder and Wainwright, 1959).

In the square-well potential model, two basic types of collisions or events can occur (Alder and Wainwright, 1959). Two particles either undergo (1) a core collision as described for the hard sphere system at $r = \sigma_1$, or (2) an attractive collision at $r = \sigma_2$, which can either be a capture, dissociation, or bounce. A capture event occurs when the attractive wells of particles i and j collide, a dissociation event occurs when two particles already inside the well become separated, and a bounce event occurs when two particles inside a square well fail to separate because of insufficient kinetic energy. The general procedure is the same as that shown in Fig. 4. The collision times are functions of the separation σ_1 or σ_2, resulting in slightly different formulas for calculating the collision times and the velocity changes. Alder and Wainwright provide a detailed flowchart that can be used to determine which type of collision (event) will take place between two interacting particles, i and j. This is reproduced in Fig. 5 (Alder and Wainwright, 1959). In addition, in DMD we have pseudo-events for bookkeeping purposes. These pseudo-events include implementing the thermostat, implementing efficiency techniques, and data collection.

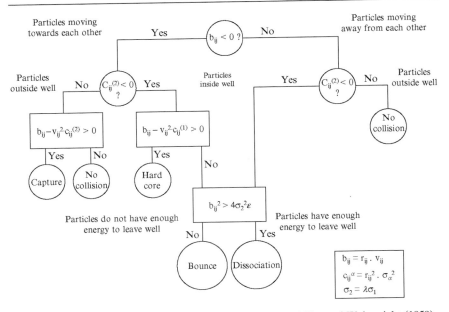

FIG. 5. Flowchart for square-well dynamics. Adapted from Alder and Wainwright (1959).

Although we have only discussed systems of spheres thus far, the equations that we have introduced can easily be adapted to systems of chain-like molecules (peptides or polymers). DMD on chain-like molecules is implemented using the "bead-string" algorithm of Rapaport (Rapaport, 1978, 1979) and later modified by Bellemans (Bellemans et al., 1980). In Bellemans's algorithm, adjacent spheres along a chain are bonded together by short, invisible strings whose length ensures that the bond length between spheres varies freely between $(1 + \delta)l$ (bond extension) and $(1 - \delta)l$ (bond contraction), where l is the ideal bond length and δ is a tolerance for bond length fluctuations ($\delta << l$). This means that bonded spheres along the chain are partially decoupled from each other, moving freely along linear trajectories between bond stretch (extension) collisions and core (contraction) collisions. The collision times and post-collision velocities for a bond contraction are the same as for the hard sphere case. However, the bond extension time is given by:

$$t_{ij}^{Bond} = \frac{-b_{ij} + \sqrt{b_{ij}^2 - v_{ij}^2(r_{ij}^2 - (1 + \delta)^2 l^2)}}{v_{ij}^2} \qquad (12)$$

where $(1 + \delta)l$ represents the maximum bond length.

The execution speed of DMD is proportional to N^2, but there are several efficiency techniques developed by Smith *et al.* that can be used to reduce the execution speed to be proportional to N. See Smith *et al.* (1997) for a more detailed discussion.

Intermediate Resolution Protein Models–PRIME

Inspired by the early reduced representation model of Takada (Takada *et al.*, 1999), our group developed an intermediate-resolution protein model for simulations of protein folding and aggregation (Smith and Hall, 2001a, b,c). In this model, which we now call PRIME (for *P*rotein *I*ntermediate-Resolution *Mo*del), the protein backbone is represented by three united atom spheres, one for the amide group (NH), one for the carbonyl group (CO), and one for the alpha-carbon and its hydrogen ($C_\alpha H$). The side chains are modeled with a single sphere of variable size. All backbone bond lengths and angles are set to their ideal values. As mentioned earlier for DMD on chain-like molecules, the covalent bonds are maintained with a hard sphere interaction occurring when the bond lengths move outside of the range $(1 + \delta)l$ to $(1 - \delta)l$, where l is the ideal bond length and δ is the tolerance for acceptable fluctuation in bond lengths that is set at 2.375% (Bellemans *et al.*, 1980). The covalent bond lengths in this model are given in Table I. Ideal backbone bond angles, C_α–C_α distances and residue L-isomerization are fixed through a series of pseudobonds that are also allowed to fluctuate within 2.375% of their given length. A depiction of the protein model for alanine showing the backbone united atoms, NH, $C_\alpha H$, and C=O, and side chain united atom, CH_3, along with the covalent bonds and pseudobonds is given in Fig. 6. The values of bond angles and pseudobond lengths are also given in Table I. Local interactions between united atoms separated along the protein backbone by three or fewer bonds are modeled in a similar manner to nonlocal interactions but with different bead diameters. Takada *et al.* found that it was more appropriate to describe the local interactions with the real atomic diameters (N, C_α, C) rather than the united atom diameters given in Table I. To account for the interactions between atoms connected by three or fewer bonds, we allow 25% overlap of their united atom bead diameters. This treatment of local interactions successfully limits the motion of the phi and psi dihedral angles, yielding reasonable Ramachandran plots (Smith and Hall, 2001b; Takada *et al.*, 1999). In addition, in PRIME, the side chain can either be represented with a single sphere, as in the case of alanine, or by several spheres, as in the case of glutamine; glycine is incorporated into the alpha carbon united atom and therefore not represented as a sphere.

TABLE I
DMD SIMULATION PARAMETERS

United atom diameters, σ	(Å)
NH	3.300
$C_\alpha H$	3.700
CO	4.000
Well diameters, $\lambda\sigma$	(Å)
NH	4.200
CO	4.200
Bond lengths, l	(Å)
$N_i H - C_{\alpha,i} H$	1.460
$C_{\alpha,i} H - C_i O$	1.510
$C_i O - N_{i+1} H$	1.330
Pseudobond lengths, l	(Å)
$N_i H - C_i O$	2.45
$C_{\alpha,i} H - N_{i+1} H$	2.41
$C_i O - C_{\alpha,i+1} H$	2.45
$C_{\alpha,i} H - C_{\alpha,i+1} H$	3.80
Bond angles	(°)
$<N_i H - C_{\alpha,i} H - C_i O$	111.0
$<C_{\alpha,i} H - C_i O - N_{i+1} H$	116.0
$<C_i O - N_{i+1} H - C_{\alpha,i+1} H$	122.0

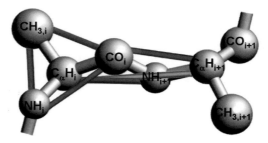

FIG. 6. Geometry of intermediate resolution protein model, PRIME, for alanine. (See color insert.)

Hydrogen bonding between the peptide backbone amide groups and carbonyl groups on the same or neighboring chain is represented by a directionally dependent square well attraction between the NH and C=O united atoms of depth ε_{HB} whenever: (1) the virtual hydrogen and oxygen atoms (whose location can be calculated at any time) are separated by 4.2Å (the sum of the NH and C=O well widths), (2) the nitrogen–hydrogen and carbon–oxygen vectors point toward each other within a fairly generous tolerance, and (3) neither the NH nor the C=O are already involved in a hydrogen bond with a different partner. To ensure that criteria 1 and 2 are satisfied, we require that the four atom pairs N_i–$C_{\alpha j}$, N_i–N_{j+1}, C_j–$C_{\alpha i}$, C_j–C_{i-1} shown connected by thick dashed lines in Fig. 7 (hereafter referred to as auxiliary pairs), be separated by a distance greater than d_{ij}, which is chosen to maintain the hydrogen bond angle constraints (Ding *et al.*, 2003; Nguyen *et al.*, 2004); their values are given in Table II. On the formation of a bond between N_i and C_j, these auxiliary pairs temporarily interact by means of a square-shoulder potential:

$$U(r_{ij}) = \begin{cases} \infty & (r < \sigma) \\ \varepsilon_{HB} & (\sigma_1 \leq r \leq d) \\ 0 & (r > \sigma) \end{cases} \qquad (13)$$

where r is the distance between spheres i and j; σ is the sphere diameter; ε_{HB} is the shoulder height (equal to the well depth of the hydrogen bond between N_i and C_j), and d is the square-shoulder width. These auxiliary pairs return to their original interactions when the hydrogen bond is broken.

The solvent molecules are modeled implicitly by means of a square well attraction (potential of mean force) between two hydrophobic residues and a hard sphere repulsion between two polar residues or between a polar and a hydrophobic residue. The depth of the square well attraction, ε_{HP},

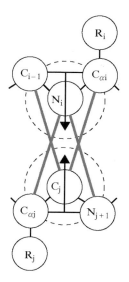

FIG. 7. Backbone hydrogen bonding where the dashed circle represents the attractive square well of N_i and C_j (Nguyen *et al.*, 2004).

TABLE II
PARAMETER D_{IJ}

Pairs	d_{ij} (Å)
N_i–$C_{\alpha j}$	5.00
N_i–N_{j+1}	4.74
C_j–$C_{\alpha i}$	4.86
C_j–C_{i-1}	4.83

between two hydrophobic residues is scaled relative to ε_{HB} by a factor R, which describes the solvent characteristics of the system (Nguyen *et al.*, 2004); R is defined as $\varepsilon_{HP}/\varepsilon_{HB}$. All system parameters are scaled by ε_{HB}, so that the system reduced temperature is $T^* = k_B T/\varepsilon_{HB}$.

Simulation Results on Fibril Formation and Structure using DMD

We performed simulations on a single model polyalanine sequence, Ac-KA$_{14}$K-NH$_2$ (Nguyen *et al.*, 2004). Polyalanine was chosen for study because Ac-KA$_{14}$K-NH$_2$ forms fibrils *in vitro* as shown by Blondelle and coworkers (1997). We began by exploring how the temperature and hydrophobic interaction strength, as modeled through the parameter R, affected the conformational conversion of the isolated chain. At low temperatures and low hydrophobicity (0 < R < 1/10), we observed a transition from an α-helix to random coils. However, as the hydrophobic interaction strength was increased (1/4 < R < 1/2), a third conformational transition from an α-helix to a β-sheet structure was observed. In that case, as the temperature increased, the isolated peptide adopted an α-helix, then a β-hairpin or β-sheet, and finally a random coil configuration. Finally, at high hydrophobic interaction strength (R > 1/2), the model polyalanine formed only random coils. Because there is little evidence to support a three state transition (α-helix → β-hairpin → random coils), a value of R was chosen low enough to avoid the three state transition but yet high enough so that the most stable state at a low temperature is an α-helix.

We next used PRIME and DMD to explore the phenomena of protein aggregation with large systems comprised of up to 96 16-mers with the KA$_{14}$K sequence. Starting from a random conformation, we were able to simulate the spontaneous formation of a protofilament or small fibril (Nguyen and Hall, 2004b). Snapshots of the simulation are shown in Fig. 8. The protofilament was composed of β-sheets with residues positioned in an in-register parallel arrangement. By performing many simulations at different temperatures, concentrations, and solvent strengths, we determined that as temperature and concentration were increased, the number of α-helices

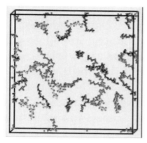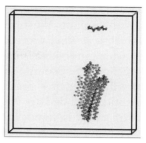

FIG. 8. Snapshots of a 48-peptide system at various reduced times, t*. The simulation proceeds from a random initial configuration at concentration c = 10 mM and temperature T* = 0.14 until the formation of a protofilament at t* = 205.9 (Nguyen and Hall, 2004). (See color insert.)

decreased and the number of extended ordered structures increased. The protofilament or fibril was found to be stable at temperatures higher than the folding temperature.

We went on to describe the dependence of peptide aggregation on peptide concentration and temperature by conducting equilibrium simulations using the replica-exchange technique (Sugita and Okamota, 1999) on a system containing 96 16-mers of Ac-KA14K-NH$_2$. A phase diagram in the temperature–concentration space was mapped out, illustrating which structures were stable at each condition (Nguyen and Hall, 2004c). The α-helical region was stable at low temperatures and low concentration. The nonfibrillar β-sheet region was stable at intermediate temperatures and relatively low concentrations and expanded to higher temperatures as concentration was increased. The fibril region was primarily stable at intermediate temperatures and intermediate concentrations and expanded to lower temperatures as the peptide concentration was increased. Finally, the random coil region was stable at high temperatures at all concentrations. Interestingly, we were able to observe the formation of small fibrils (protofilaments) for systems containing 96 peptides within 160 h on an AMD Athlon MP2200+ workstation.

We also investigated the kinetics of fibrillization by simulating systems of 48–96 peptides of sequence KA$_{14}$K (Nguyen and Hall, 2004a) using the intermediate resolution protein model, PRIME. We performed both seeded and unseeded simulations to determine whether fibril formation was a nucleation-dependent event. In the presence of a seed (a preformed aggregate), the lag time, which is the time to form a fibril or fibril component, disappeared. The lag time decreased with increasing temperature and concentration. Fibril formation proceeded in the following way: small amorphous aggregates associated, rearranged into ordered β-sheet structures, and ultimately formed a "nucleus," which rapidly grew into a small fibril or protofilament. We observed two growth mechanisms: lateral addition in which a β-sheet was added to the side of the fibril, and end-to-end growth in which individual peptides were attached to the end of each β-sheet (this mechanism accounts for the indeterminate length of the fibril). Once the fibrillar structure reached a critical number of β-sheets, the monomeric peptides tended to attach to an already formed β-sheet rather than to form a new isolated β-sheet. The number of critical β-sheets was a function of system size. A 12-peptide system formed a fibril with 2–3 β-sheets, a 24-peptide system formed a fibril with 3–4 β-sheets, a 48-peptide system formed a fibril with 3–6 β-sheets, and a 96-peptide system formed a fibril with 4–6 β-sheets.

We have extended the PRIME model to polyglutamine to study the aggregation of polyglutamine-containing proteins (Marchut and Hall, 2006). The backbone of a polyglutamine residue is modeled with the same level of detail as a polyalanine residue (a single sphere for the carbonyl group, amide group, and alpha-carbon group). The difference between the polyglutamine and polyalanine models is in the description of the side chain. The polyglutamine side chain is represented with four united atom residues, two for the hydrophobic methyl groups (CH_2, in blue), one for the carbonyl group (CO, in red) and one for the amine group (NH, in green) as shown in Fig. 9. The carbonyl and amine groups along the side chain allow the side chain to participate in hydrogen bonding either with the backbone amide or carbonyl groups or with other side chains. Simulations were conducted on a system of 24 16-mers (Q_{16}) at a concentration of 5 mM over a range of temperatures starting from a random initial configuration. Amorphous aggregates were formed at low temperatures, annular structures composed of β-sheets were formed at intermediate temperatures, and random coil configurations were found at high temperatures. Figure 10 shows a snapshot of one of the annular structures. Similar annular structures were observed by Wacker and coworkers on two cleavage products of the huntingtin protein with 20 and 53 polyglutamine repeats, respectively (Wacker *et al.*, 2004) using atomic force microscopy and were also predicted by Perutz from X-ray scattering data on $D_2Q_{15}K_2$ (Perutz *et al.*, 2002).

The PRIME model is now being extended to the description of heteroproteins. We have performed DMD simulations on the Mouse prion peptide (111–120) of sequence, VAGAAAGAV (Wagoner and Hall, unpublished). Preliminary results from simulations of the isolated peptide reveal that it is too short to adopt a discernible secondary structure. It forms one to two beta-hydrogen bonds at temperatures between T* = 0.07 and 0.11, and it exists as a random coil at temperatures greater than T* = 0.11. Aggregation

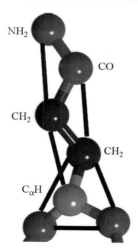

FIG. 9. Geometry of intermediate resolution protein model, PRIME, for glutamine. (See color insert.)

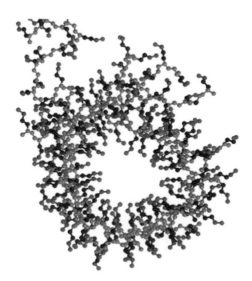

FIG. 10. Tube formed during simulation of 24 polyglutamine 16mers (Marchut and Hall, 2006). (See color insert.)

studies have been performed on a system of 48 peptides of VAGAAAA-GAV at a concentration, c = 5 mM, a range of reduced temperatures, T* = 0.07–0.14, and a hydrophobic interaction strength, R = 1/10, starting from a configuration of random coils. Amorphous aggregation is observed at or below T* = 0.11. Ordered aggregates are formed at temperatures greater than T* = 0.11 but less than T* = 0.13. The ordered aggregates are composed of stacks of four to six β-sheets. The peptides remain in random coil configurations at temperatures greater than or equal to T* = 0.13. We have also conducted simulations at concentrations, c = 1 mM, 2.5 mM, and 10 mM at reduced temperature, T* = 0.12. The peptides formed β-sheets at concentrations greater than c = 1 mM but less than c = 5 mM and formed ordered aggregates (a protofilament) at concentration, c = 10 mM. Figure 11 is a snapshot of an ordered aggregate formed by a 48-peptide system of VAGAAAAGAV at concentration, c = 1 mM and reduced temperature, T* = 0.12, where peptides that form a β-sheet are colored the same. For example, in Fig. 11, there are two β-sheets formed, green and blue, respectively, which then associate to form an ordered aggregate.

The great speed of DMD simulations with reduced representation protein models has inspired other groups to develop protein models based on hard sphere and square well interactions. Dokholyan and coworkers have applied DMD to the study of a two-sphere per residue coarse-grained protein model (Ding *et al.*, 2002a,b) with a Go-type potential used to describe the intermolecular interactions between the side chains. Ding *et al.* (2002b) used this two-sphere model to study the aggregation of a system containing eight copies of

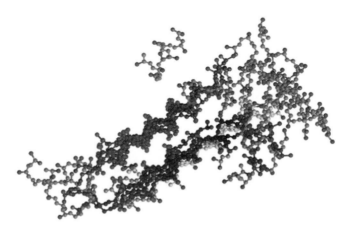

FIG. 11. Snapshot of a 48-peptide ordered aggregate obtained from the c = 1 mM simulation at T* = 0.12 (Wagoner and Hall, unpublished). (See color insert.)

the Src SH3 protein. Starting from a system of random coils, they observed a possible pathway for amyloidogenesis that included the formation of four dimers → two bundles of four proteins → one aggregate. Peng *et al.* (2004) applied the two-sphere model to study a large system of β-amyloid proteins, $A\beta_{1-40}$. They started with 28 copies of the peptide arranged randomly, with each peptide in a α-helical conformation as determined by the NMR measurements of Coles *et al.* (1998), at concentration c = 6 mM. They observed the formation of amorphous aggregates at temperatures below the melting temperature of a single peptide, T *= 0.4, and dissociation of all structures at temperatures greater than T* = 1.10. Multilayer β-sheet structures formed over a range of simulation temperatures between T* = 0.55 and T* = 1.10.

Recently, Urbanc *et al.* (2004a,b) used a model similar to the four-sphere intermediate resolution model introduced by Smith and Hall (2001a,b) along with DMD to study the oligomerization of amyloid β-protein, $A\beta_{1-40}$ and $A\beta_{1-42}$. They observed an α-helix to β-strand structural transition at intermediate simulation temperatures followed by another transition from β-strand to random coil at relatively high simulation temperatures. They also observed a β-turn between residues D23 and K28. Although a turn is observed experimentally between residues D23 and K28, it is not a true β-turn (which would involve hydrogen bonding between residues D23 and K28) but rather an electrostatically driven salt-bridge that leaves the backbone amide and carbonyl groups free to hydrogen bond with another $A\beta_{1-42}$ peptide (Petkova *et al.*, 2002). Urbanc *et al.* (2004b) went on to study the aggregation of 32 copies of $A\beta_{1-40}$ and $A\beta_{1-42}$ starting from a mostly α-helical configuration at concentration 3.4 mM and simulation temperature T* = 0.15. They found that $A\beta_{1-40}$ preferred to form a dimer, and $A\beta_{1-42}$ preferred to form a trimer and a pentamer. The cores of both oligomers were composed primarily of hydrophobic residues. The observation of the pentamer is interesting, because experiments suggest that the composition of protofibrils or paranuclei is predominantly pentameric and hexameric (Bitan *et al.*, 2003a,b). Although they were able to see the formation of oligomers, their simulation time was too short to observe the formation of the fibril structure reported by Petkova *et al.* (2005).

Conclusion

Great progress has been made over the past 5 years in advancing the study of fibril structure and formation through the use of computer simulations. We have discussed two simulation approaches: atomic resolution models and intermediate-resolution protein models. Each gives us an insightful, but incomplete, view of the fundamentals associated with fibrillization. Which approach is the most fruitful avenue to pursue? In trying to answer this question, the authors usually wind up thinking about the fable of the six blind men who went to "see" an elephant. Each touches a different part of the elephant—side, tusk, trunk, knee, ear, tail—and concludes that an elephant is like—a wall, a spear, a snake, a tree, a fan, a rope. Mistaking the parts for the whole, the blind men then argue passionately about the nature of the elephant, each convinced that his view is correct. Our community is, of course, wiser than the blind men (at least when it comes to science), because we know that the different computational tools (as well as experimental tools) that we use afford us views of different aspects of fibril formation. It is only by sharing this information and piecing together our various observations that we will be able to assemble a good comprehensive picture of the nature of fibril formation and structure.

Acknowledgment

We thank Dr. Hung D. Nguyen (Scripps Research Institute, La Jolla, CA) and Dr. Alexander J. Marchut (Bristol-Myers Squibb, New Brunswick, NJ) for insightful discussions. This work was supported by National Institutes of Health Grant GM-56766.

References

Alder, B. J., and Wainwright, T. E. (1959). Studies in molecular dynamics, I: General method. *J. Chem. Phys.* **31,** 459–466.

Allen, M. P., and Tildesley, D. J. (1987). "Computer Simulation of Liquids." Oxford University Press, New York.

Andersen, H. C. (1980). Molecular dynamics simulation at constant temperature and/or pressure. *J. Chem. Phys.* **72,** 2384.

Bellemans, A., Orbans, J., and Belle, D. V. (1980). Molecular dynamics of rigid and non-rigid necklaces of hard disks. *Mol. Phys.* **39,** 781–782.

Bitan, G., Kirkitadze, M. D., Lomakin, A., Vollers, S. S., Benedek, G. B., and Teplow, D. B. (2003a). Amyloid beta-protein (Abeta) assembly: Abeta 40 and Abeta 42 oligomerize through distinct pathways. *Proc. Natl. Acad. Sci. USA* **100,** 330–335.

Bitan, G., Vollers, S. S., and Teplow, D. B. (2003b). Elucidation of primary structure elements controlling early amyloid beta-protein oligomerization. *J. Biol. Chem.* **278,** 34882–34889.

Blake, C., and Serpell, L. C. (1996). Synchotron X-ray studies suggest that the core of transthyretin amyloid fibril is a continuous b-sheet helix. *Structure* **4,** 989–998.

Blondelle, S. E., Forood, B., Houghten, R. A., and Perez-Paya, E. (1997). Polyalanine-based peptides as models for self-associated b-pleated-sheet complexes. *Biochemistry* **36,** 8393–8400.

Brooks, B. R., Bruccoleri, R. E., Olafson, B. D., Stales, D.J, Swaminathan, S., and Karplus, M. (1983). CHARMM: A program for macromolecular energy minimization and dynamics calculation. *J. Comp. Chem.* **4,** 187.

Bucciantini, M., Giannoni, E., Chiti, F., Baroni, F., Fornigli, L., Zurdo, J., Takkei, N., Ramponi, G., Dobson, C., and Stefani, M. (2002). Inherent toxicity of aggregates implies a common mechanism for protein misfolding diseases. *Nature* **416,** 507.

Caughey, B., and Lansbury, P. T. (2003). Protofibrils, pores, fibrils, and neurodegeneration: Separating the responsible protein aggregates from the innocent bystanders. *Annu. Rev. Neurosci.* **26,** 267–298.

Chiti, F., Bucciantini, M., Capanni, C., Taddei, N., Dobson, C. M., and Stefan, M. (2001). Solution conditions can promote formation of either amyloid protofilaments or native fibrils from the HypF N-terminal domain. *Prot. Sci.* **10,** 2542.

Chiti, F., Webster, P., Toddei, N., Clark, A., Stefan, M., Ramponi, G., and Dobson, C. (1999). Designing conditions for *in vitro* formation of amyloid protofilaments and fibrils. *Proc. Natl. Acad. Sci. USA* **96,** 3590–3594.

Coles, M., Bicknell, W., Watson, A. A., Fairlie, D. P., and Craik, D. J. (1998). *Biochemistry* **37,** 11064.

Dauber-Osguthorpe, P., Roberts, V. A., Osguthorpe, D. J., Wolff, J., Genest, M., and Hagler, A. T. (1988). Structure and energetics of ligand binding to proteins: *Escherichia coli* dihydrofolate reductase trimethoprim, a drug receptor system. *Proteins: Struct. Funct. Genet.* **4,** 31.

Ding, F., Borreguero, J. M., Buldyrev, S. V., Stanley, H. E., and Dokholyan, N. V. (2003). Mechanism for the alpha-helix to beta-hairpin transition. *Proteins: Struct. Funct. and Genet.* **53,** 220–228.

Ding, F., Dokholyan, N. V., Buldyrev, S. V., Stanley, H. E., and Shakhnovich, E. (2002a). Direct molecular dynamics observation of protein folding transition state ensemble. *Biophys. J.* **83,** 3525.

Ding, F., Dokholyan, N. V., Buldyrev, S. V., Stanley, H. E., and Shakhnovich, E. (2002b). Molecular dynamics simulation of the SH3 domain aggregation suggests a generic amyloidogenesis mechanism. *J. Mol. Biol.* **324,** 851.

Dobson, C. M. (2001). The structural basis of protein folding and its links with human disease. *Phil. Trans. R. Soc. Lond. B.* **356,** 133.

Duan, Y., and Kollman, P. A. (1998). Pathways to a protein folding intermediate observed in a 1-microsecond simulation in aqueous solution. *Science.* **282,** 740.

Ferrara, P., Apostolakis, J., and Caflisch, A. (2002). Evaluations of a fast implicit solvent model for molecular dynamics simulations. *Proteins: Struct. Funct. Genet.* **46,** 24–33.

Forood, B., Perez-Paya, E., Houghten, R. A., and Blondelle, S. E. (1996). Structural characterization and 5'-mononucleotide binding of polyalanine b-sheet complexes. *J. Mol. Recognit.* **9,** 488.

Frenkel, D., and Smit, B. (1996). "Understanding Molecular Simulation: From Algorithms to Applications." Academic Press, Chestnut Hill, MA.

Gsponer, J., Haberthur, U., and Caflisch, A. (2003). The role of side-chain interactions in the early steps of aggregation: Molecular dynamics simulations of an amyloid-forming peptide from the yeast prion Sup35. *Proc. Natl. Acad. Sci. USA* **100,** 5154–5159.

Hasel, W., Hendrickson, T., and Still, W. (1988). A rapid approximation to the solvent accessible surface areas of atoms. *Tetrahedron Comput. Methodol.* **1,** 103–116.

Haspel, N., Zanuy, D., Ma, B., Wolfson, H., and Nussinov, R. (2004). A comparative study of amyloid fibril formation by residues 15–19 of the human calcitonin hormone: A single b-sheet model with a small hydrophobic core. *J. Mol. Biol.* **345,** 1213–1227.

Hou, L., and Zagorski, M. G. (2004). Sorting out the driving forces for parallel and antiparallel alignment in the Ab peptide fibril structure. *Biophys. J.* **86,** 1–2.

Kelly, J. (1997). Amyloid fibril formation and protein misassembly: A structural quest for insights into amyloid and prion diseases. *Structure.* **5,** 595–600.

Kelly, J. W. (1998). The alternative conformations of amyloidogenic proteins and their multi-step assembly pathways. *Curr. Opin. Struct. Biol.* **8,** 101–106.

Khurana, R., Gillespie, J. R., Talapatra, A., Minert, L. J., Ionescu-Zanetti, C., Millett, I., and Fink, A. L. (2001). Partially-folded intermediates as critical precursors of light chain amyloid fibrils and amorphous aggregates. *Biochemistry.* **40,** 3525.

Koo, E. H. (2002). The b-Amyloid precursor protein (APP) and Alzheimer's disease: Does the tail wag the dog? *Traffic.* **3**, 763–770.

Leach, A. R. (2001). "Molecular Modelling: Principles and Applications." Pearson Education Limited.

Levitt, M. (1983). Molecular dynamics of native protein: I. Computer simulation trajectories. *J. Mol. Biol.* **168**, 595.

Ma, B., and Nussinov, R. (2002a). Molecular dynamics simulations of alanine rich b-sheet oligomers: Insight into amyloid formation. *Prot. Sci.* **11**, 2335–2350.

Ma, B., and Nussinov, R. (2002b). Stabilities and conformations of Alzheimer's b-amyloid peptide oligomers (Ab16-22, Ab16-35, and Ab10-35): Sequence effects. *Proc. Natl. Acad. Sci. USA* **99**, 14126–14131.

Marchut, A. J., and Hall, C. K. (2006). Spontaneous formation of annular structures observed in molecular dynamics simulations of polyglutamine peptides. *Comput. Biol. Chem.* **30**, 215–218.

Molecular Simulations Inc. (1999). "CHARMm Principles." Inc., Molecular Simulations, San Diego.

Mortenson, P. N., and Wales, D. J. (2001). Energy landscapes, global optimization and dynamics of polyalanine Ac(ala)$_8$NHMe. *J. Chem. Phys.* **114**, 6443–6454.

Nguyen, H. D., and Hall, C. K. (2004a). Kinetics of fibril formation by polyalanine peptides. *J. Biol. Chem.* **280**, 9074–9082.

Nguyen, H. D., and Hall, C. K. (2004b). Molecular dynamics simulations of spontaneous fibril formation by random-coil peptides. *Proc. Natl. Acad. Sci. USA* **101**, 16180–16185.

Nguyen, H. D., and Hall, C. K. (2004c). Phase diagrams describing fibrillation by polyalanine peptides. *Biophys. J.* **87**, 4122–4134.

Nguyen, H. D., Marchut, A. J., and Hall, C. K. (2004). Solvent effects on the conformational transition of a model polyalanine peptide. *Prot. Sci.* **13**, 2909–2924.

Peng, S., Ding, F., Urbanc, B., Buldyrev, S. V., Cruz, L., Stanley, H. E., and Dokholyan, N. V. (2004). Discrete molecular dynamics simulations of peptide aggregation. *Phys. Rev. E.* **69**, 041908-1–041908-7.

Perutz, M. F., Finch, J. T., Berriman, J., and Lesk, A. (2002). Amyloid fibers are water-filled nanotubes. *Proc. Natl. Acad. Sci. USA* **99**, 5591.

Petkova, A. T., Ishii, Y., Balbach, J. J., Antzutkin, O. N., Leapman, R. D., Delaglio, F., and Tycko, R. (2002). A structural model for Alzheimer's b-amyloid fibrils based on experimental constraints from solid state NMR. *Proc. Natl. Acad. Sci. USA* **99**, 16742–16747.

Petkova, A. T., Leapman, R. D., Guo, Z., Yau, W.-M., Mattson, M. P., and Tycko, R. (2005). Self-propagating, molecular-level polymorphism in Alzheimer's b-amyloid fibrils. *Science.* **307**, 262–265.

Prusiner, S. B. (1997). Prion diseases and the BSE crisis. *Science.* **278**, 245–251.

Rapaport, D. C. (1978). Molecular dynamics simulation of polymer chains with excluded volume. *J. Phys. A.* **11**, L213.

Rapaport, D. C. (1979). Molecular dynamics study of polymer chains. *J. Chem. Phys.* **71**, 3299.

Rochet, J. C., and Lansbury, P. T. (2000). Amyloid fibrillogenesis: Themes and variations. *Curr. Opin. Struct. Biol.* **10**, 60–68.

Roux, B., and Simonson, T. (1999). Implicit solvent models. *Biophys. Chem.* **78**, 1–20.

Santini, S., Mousseau, N., and Derreumaux, P. (2004). In Silico Assembly of Alzheimer's Ab16–22 Peptide into b-sheets. *J. Am. Chem. Soc.* **126**, 11509–11516.

Serpell, L., Sunde, M., Benson, M., Tennent, G., Pepys, M., and Fraser, P. (2000). The protofilament substructure of amyloid fibrils. *J. Mol. Biol.* **300**, 1033–1039.

Serpell, L. C. (2000). Alzheimer's amyloid fibrils: Structure and assembly. *Biochim. Biophys. Acta* **1502**, 16–30.

Smith, A. V., and Hall, C. K. (2001a). Assembly of a tetrameric a-helical bundle: Computer simulations on an intermediate-resolution protein model. *Proteins: Struct. Func. Genet.* **44**, 376.

Smith, A. V., and Hall, C. K. (2001b). a-Helix formation: Discontinuous molecular dynamics on an intermediate resolution model. *Protein: Struct. Func. Genet.* **44**, 344–360.

Smith, A. V., and Hall, C. K. (2001c). Protein refolding versus aggregation: Computer simulations on an intermediate resolution model. *J. Mol. Biol.* **312**, 187.

Smith, S. W., Hall, C. K., and Freeman, B. D. (1995a). Large scale molecular dynamics study of entangled hard-chain fluids. *Phys. Rev. Lett.* **75**, 1316.

Smith, S. W., Hall, C. K., and Freeman, B. D. (1995b). Molecular dynamics study of transport coefficients for hard-chain fluids. *J. Chem. Phys.* **102**, 1057–1073.

Smith, S. W., Hall, C. K., and Freeman, B. D. (1997). Molecular dynamics for polymeric fluids using discontinuous potentials. *J. Comput. Phys.* **134**, 16.

Sugita, Y., and Okamota, Y. (1999). Replica exchange molecular dynamics method for protein folding. *Chem. Phys. Lett.* **314**, 141.

Sunde, M., and Blake, C. C. F. (1997). The structure of amyloid fibrils by electron microscopy and X-ray diffraction. *Adv. Protein Chem.* **50**, 123.

Sunde, M., Serpell, L. C., Bartlam, M., Fraser, P. E., Pepys, M. B., and Blake, C. C. F. (1997). Common core structure of amyloid fibrils by synchrotron X-ray diffraction. *J. Mol. Biol.* **273**, 729–739.

Takada, S., Luthey-Schulten, Z., and Wolynes, P. G. (1999). Folding dynamics with non-additive forces: A simulation study of a designed helical protein and a random heteropolymer. *J. Chem. Phys.* **110**, 11616.

Tarus, B., Straub, J. E., and Thirumalai, D. (2005). Probing the initial stage of aggregation of the Ab10–35-protein: Assessing the propensity for peptide dimerization. *J. Mol. Biol.* **345**, 1141–1156.

Urbanc, B., Cruz, L., Ding, F., Sammond, D., Khare, S., Buldyrev, S. V., Stanley, H. E., and Dokholyan, N. V. (2004a). Molecular dynamics simulation of amyloid b-dimer formation. *Biophys. J.* **87**, 1–12.

Urbanc, B., Cruz, L., Yun, S., Buldyrev, S. V., Bitan, G., Teplow, D. B., and Stanley, H. E. (2004b). In silico study of amyloid b-protein folding and oligomerization. *Proc. Natl. Acad. Sci. USA* **101**, 17345–17350.

Wacker, J. L., Zareie, M. H., Fong, H., Sarikaya, M., and Muchowski, P. J. (2004). Hsp70 and Hsp40 attenuate formation of spherical and annular polyglutamine oligomers by partitioning monomer. *Nat. Struct. Mol. Biol.* **11**, 1215.

Weiner, S. J., Kollman, P. A., Nguyen, D. T., and Case, D. A. (1986). An all atom force field for simulations of proteins and nucleic acids. *J. Comp. Chem.* **7**, 230.

Williams, A. D., Portelius, E., Kheterpal, I., Guo, J. T., Cook, K. D., Xu, Y., and Wetzel, R. (2004). Mapping abeta amyloid fibril secondary structure using scanning proline mutagenesis. *J. Mol. Biol.* **335**, 833–842.

Williams, A. D., Shivaprasad, S., and Wetzel, R. (2006). Alanine scanning mutagenesis of Abeta(1–40) amyloid fibril stability. *J. Mol. Biol.* **357**, 1283–1294.

Zimmerman, S., Pottle, M. S., Nemethy, G., and Scheraga, H. A. (1977). Conformational analysis of the 20 naturally occurring amino acid residues using ECEPP. *Macromolecules.* **10**, 1–9.

Author Index

A

Abe, M., 164
Abid, K., 3
Abraham, C., 146, 218
Abraham, F. F., 288
Abrahamson, E. E., 123, 129, 133, 136
Abramowski, D., 141
Acevedo, S. F., 276
Acevedo-Cruz, A., 258
Acosta, L., 258
Adachi, K., 297
Adam, S., 259
Adjou, K. T., 224
Adlard, P. A., 145
Adler, C., 203
Aebersold, R., 79, 90
Aebi, M., 34, 186, 188
Aerni, H. R., 94
Affeck, D., 165
Agdeppa, E. D., 148, 149, 150, 151, 152
Aggarwal, N. T., 128
Agid, Y., 257
Aguet, M., 224
Aguzzi, A., 4, 188, 224, 229
Ahram, M., 80, 81
Ahringer, J., 273, 275, 277
Aigaki, T., 248
Aiken, J. M., 80
Aitken, A., 276
Aizawa, M., 257
Akagi, T., 64, 66
Akamatsu, K., 162
Akatsu, H., 148
Akbari, Y., 141
Akiguchi, M., 257
Alavi, A., 147
Albin, R. L., 258
Alder, B. J., 340, 350, 351
Alex, S., 165
Alexander, G. E., 147
Alexandrov, I. M., 34, 35, 36, 37, 40, 41
Alexoff, D., 153, 154

Aliev, G., 78
Allen, K. D., 34, 37, 40, 41
Allen, M. P., 340, 345
Allen, W. C., 163
Allibone, R., 50
Almqvist, E., 259
Alonso, M. E., 257
Alperovitch, A., 3
Alter, J. R., 258
Altschul, S. F., 304
Alvir, J., 258
Ambros, V., 262
Ancsin, J. B., 162
Anderes, L., 5, 16
Anderle, M., 87
Andersen, J. S., 87
Anderson, H. C., 340, 349
Anderson, K., 208
Anderson, L., 257
Anderson, R. G., 14
Ando, Y., 225, 232
Andrade, R. S., 248
Andre, B., 208
Andren, P. E., 94
Andreoletti, O., 225, 227
Andresen, J. M., 258
Andrew, S. E., 259
Andrews, B., 218
Andrews, N. J., 224
Andriola, I., 257
Antalffy, B., 107, 257, 274
Antoni, G., 134, 136, 140, 149
Antzutkin, O. N., 301, 302, 311, 318, 319, 321, 347, 361
Anwyl, R., 235
Apostolakis, J., 344, 345
Arai, H., 148
Arai, K., 146
Arbustini, E., 50
Archer, F., 225, 227
Arias, H. R., 78
Arkin, A. P., 208
Arnheim, N., 235, 258

S

Subject Index

A

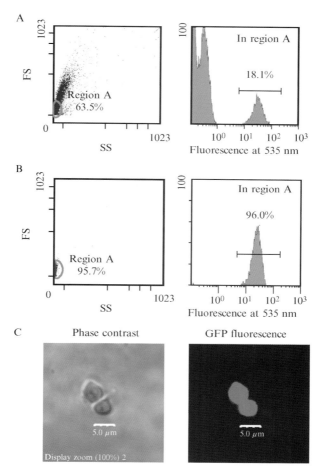

M<small>ITSUI</small> *ET AL.*, C<small>HAPTER</small> 5, F<small>IG.</small> 1. Result of purification of tNhtt-150Q-EGFP aggregates with a cell sorter. (A) Sorting profiles of aggregates from lysates of tNhtt-150Q-EGFP-expressing neuro 2A are shown in a scattergram (*left*; forward scattering [FS] vs side scattering [SS]) and in a histogram for the fluorescence at 535 nm (*right*). Aggregates with bright fluorescence are observed mostly in the particles within region A of the *left panel*. Numbers in the *left* and *right panels* represent the relative frequency of particles in region A against the total particles and the frequency of bright particles in region A, respectively. (B) Results of the reanalysis of the sorted aggregates. Note that sorted particles are homogeneous with respect to the forward scattering, side scattering, and the fluorescence at 535 nm. (C) Microscopic observation of purified aggregates presented by phase contrast image (*left*) and the corresponding fluorescence image (*right*). Scale bars, 5 μm.

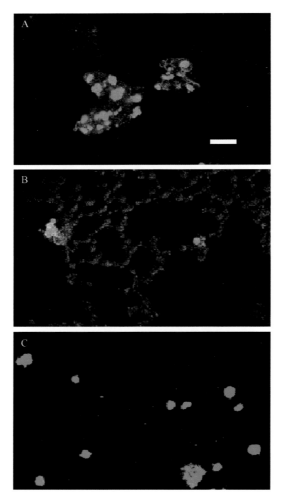

Mitsui ET AL., Chapter 5, Fig. 2. Result of purification of tNhtt-150Q-EGFP-NLS aggregates. (A) Immunofluorescence of aggregates treated with nucleases. Red fluorescence indicates the localization of lamin B as a marker of the nuclear matrix. (B) Nuclear aggregates after 2 *M* NaCl treatment. (C) Nuclear aggregates after 4% sarcosyl treatment. Note that aggregate particles are clearly removed from nuclear structures. Scale bars, 20 μm.

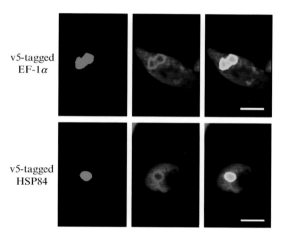

v5-tagged
EF-1α

v5-tagged
HSP84

MITSUI *ET AL.*, CHAPTER 5, FIG. 4. Immunochemical analysis of AIP candidates in tNhtt-150Q-EGFP–expressing Neuro 2A cells. Cells were transfected with v5-tagged EF-1α or v5-tagged HSP84 in pcDNA3.1 expression vector, then differentiation and tNhtt-150Q-EGFP expression were induced with dbcAMP and ecdysone, respectively. Two days later, the cells were fixed and incubated with antibody against the v5-epitope. *Left column,* fluorescence of tNhtt-150Q-EGFP aggregates. *Middle column,* v5-tagged proteins detected with anti-v5 antibody and Alexa Fluor 546-labeled secondary antibody. *Right column,* merged images of the two signals. Scale bars 20 μm.

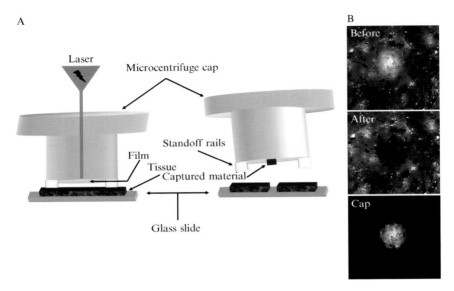

GOZAL *ET AL.*, CHAPTER 6, FIG. 1. Isolation of amyloid plaques by laser capture microdissection. (A) When a polymer-coated cap is placed on the top of fixed tissue, standoff rails prevent the direct contact of the coat with tissue to reduce contamination. Under a microscope, a laser pulse generates enough energy to melt the coated film and to glue the area of interest to the cap. Lifting the cap leads to the removal of targeted area of tissue. (B) The "before" and "after" images indicate the removal of a plaque region from a thioflavin-S–stained AD brain section. The isolated plaque was attached to the cap as shown. Moreover, the surrounding nonplaque regions were also captured as control on a different cap (not shown). This figure is adopted from our previously published sources (Liao *et al.*, 2004).

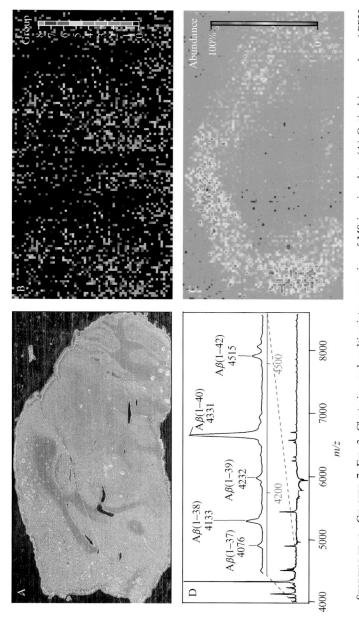

STOECKLI *ET AL.*, CHAPTER 7, FIG. 3. Clustering and multivariate processing of MS imaging data. (A) Optical image of an APP23 mouse brain section. PCA followed by ISODATA clustering groups pixels with similar mass spectra (B). PCA-DA was performed on clusters four (gray) and eight (pink), the resulting image is shown (C). Contrast is based on the complex peak pattern shown in the spectrum of (D), in which various Aβ peptides play a significant role.

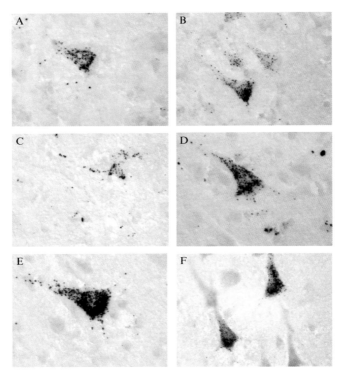

OSMAND *ET AL.*, CHAPTER 8, FIG. 2. Polyglutamine recruitment in several cortical pyramidal neurons in Huntington's disease (HD) brain (A–E) and in presymptomatic HD (F). Aggregation foci are seen widely distributed in the cytoplasm of affected cells, with the absence of nuclear involvement (A, C, E) and occasionally appearing ordered along presumably cytoskeletal elements (C). There is a noticeable accumulation of foci toward the axonal pole (B, E, F), although foci are frequently observed in proximal dendrites (D, E); bPEGQ30, 10 n*M* (A–E) and 20 n*M* (F) without tyramide enhancement.

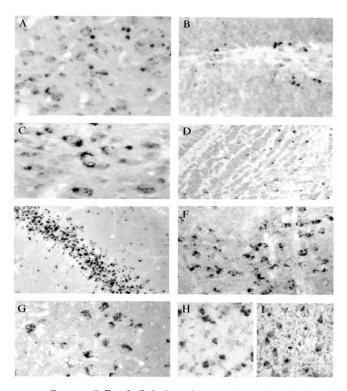

OSMAND *ET AL.*, CHAPTER 8, FIG. 3. Polyglutamine recruitment in animal models of CAG–repeat diseases; recruitment sites in these models are only revealed after biotin tyramide amplification. In the YAC 128 mouse (A–D), recruitment is widely distributed, shown here in thalamus (A), in the neurogenic layer of the hippocampal dentate gyrus (B), in magnocellular neurons of the red nucleus (C), and in the granule cell layer of the olfactory bulb (D). In the CAG140 mouse, recruitment foci were also widely distributed and are shown in the CA3 region of hippocampus (E), in neurons in the ventral tegmental area (F), and in the motor trigeminal nucleus (G). Recruitment sites were less widely distributed in the transgenic rat being observed in the basal ganglia and thalamus (not shown) and cortex (H). In the SCA3 mouse, nuclear and cytoplasmic recruitment were seen most strikingly in the vestibular nucleus (I); bPEGQ30 20 n*M* with tyramide recruitment (A–H) and mbPEGQ30 10 n*M* with tyramide recruitment (I).

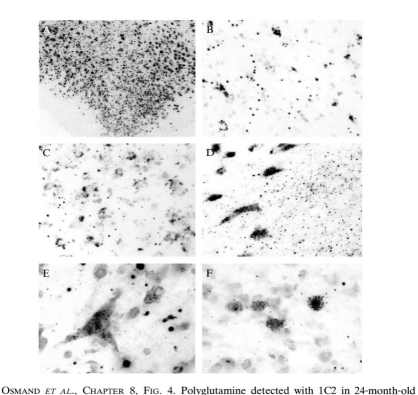

OSMAND *ET AL.*, CHAPTER 8, FIG. 4. Polyglutamine detected with 1C2 in 24-month-old heterozygous transgenic rat (A–D). Intranuclear accumulation and intranuclear inclusions, as well as neuropil aggregates are present densely in ventral striatum, shown here, in the olfactory tubercle (A). In cortex both cytoplasmic punctae, presumably corresponding to aggregation foci, and neuropil aggregates, occasionally appearing as "chains," are seen (B). Both weakly staining cytoplasmic sites and neuropil aggregates are shown in thalamus (C). In substantia nigra (D), neurons in the pars compacta show numerous minute cytoplasmic polyglutamine aggregates, whereas many neuropil aggregates are seen in the pars reticulata. In HD cortex (E, F), 1C2 staining demonstrates both punctate cytoplasmic staining in pyramidal neurons, as well as the presence of neuropil aggregates; weak peripheral staining of glial nuclei is also seen. 1C2 1:30,000 after formic acid treatment, with (A–D) and without (E, F) tyramide amplification.

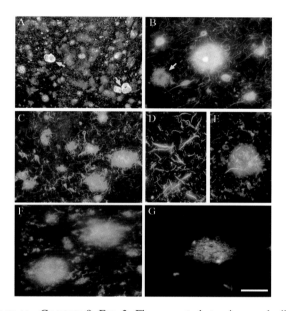

IKONOMOVIC *ET AL.*, CHAPTER 9, FIG. 2. Fluorescent photomicrographs illustrate region-specific morphological details of X-34 labeled structures and the spectrum of X-34 fluorescence color (Note: for color image please refer to the end of this volume). (A) Low-power photomicrograph of X-34 fluorescence in the middle temporal cortex of an AD case reveals prominent staining of large cyan-labeled blood vessels (arrows) and numerous SP and NFT. (B) Higher magnification of a classic, cored, SP in the medial temporal cortex shows bright cyan fluorescence in the central core region (bright white in the grayscale image) composed of highly aggregated amyloid. The surrounding halo of X-34 fluorescence is less prominently cyan-stained, whereas an adjacent diffuse-appearing SP (arrow) shows more greenish brown (gray in the gray scale image). Fibrils are easily observed in NT surrounding SP and in several NFT that also show bright cyan fluorescence and display a range of NFT morphology. (C) In the frontal cortex, numerous neuritic SP and NT are observed. (D, E) Higher power magnification of NT (D) and DN associated with one neuritic SP (E). (F) SP in the caudate nucleus stain with greenish brown fluorescence of the diffuse-appearing type, without compact amyloid cores. Only sporadic neuritic elements are seen in the surrounding neuropil. (G) An isolated diffuse-appearing SP in the molecular layer of cerebellar cortex shows cyan fluorescence of X-34 stained fibrils that appear loosely dispersed inside the amorphous SP formation. There are no associated neuritic elements in the neuropil. Scale bar = 250 μm (A), 50 μm (B, C, E–G), and 25 μm (D).

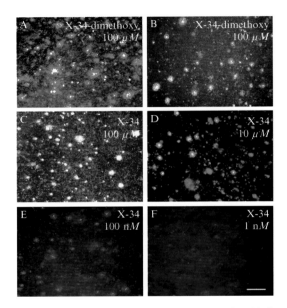

IKONOMOVIC *ET AL.*, CHAPTER 9, FIG. 4. Photomicrographs of X-34-dimethoxy and X-34 stainings in AD temporal cortex show that at comparable concentrations (100 μM and 10 μM), X-34-dimethoxy produces brighter and more cyan fluorescence compared with X-34 (A–D; Note: for color image please refer to the end of this volume). Dilution effect on X-34-dimethoxy (A, B) and X-34 (C–F) labeling in the temporal cortex from an AD patient is demonstrated. At the highest concentration, which shows optimal signal to background contrast (100 μM; C), X-34 labels the four main pathological structures in AD that contain the β-pleated sheet structure: SP, NFT, DN, and NT. At 10 μM (D), NT staining is significantly reduced; NFT and DN are less prominent, whereas SP remains intensely fluorescent. At 100 nM (E), NT are no longer detectable, NFT and DN are barely visible, and SP show reduced fluorescence. At the highest dilution tested (1 nM; F), compact SP are the only remaining structures exhibiting X-34 fluorescence. Scale bar = 100 μm.

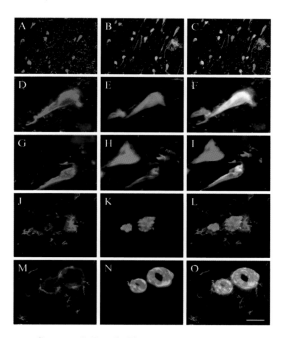

IKONOMOVIC *ET AL.*, CHAPTER 9, FIG. 5. Fluorescence photomicrographs of tissue sections from an AD hippocampus and transgenic mouse cortex processed for red fluorescence IHC (left) and counterstained with blue fluorescent X-34 histochemistry (center) shown with merged red/blue images (right column). For color image please refer to the end of this volume. In human hippocampus (A–C, D–F, and G–I), a subset of X-34 (blue)–stained NFT contains AT-8 immunostaining. A large extracellular NFT ("ghost tangle" in H) lacks AT-8 immunofluorescence. In transgenic mouse cortex, red GFAP immunostaining of astrocytes is observed in association with blue X-34 staining of amyloid material in plaques (J–L) and walls of large blood vessels (M–O). Scale bar = 90 μm (A–C), 15 μm (D–I), 35 μm (J–O).

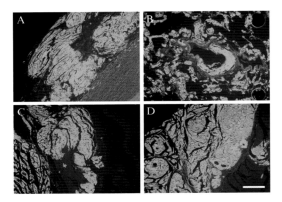

IKONOMOVIC *ET AL.*, CHAPTER 9, FIG. 6. Photomicrographs of X-34 staining in formalin-fixed, paraffin embedded histological sections of autopsy tissue from a patient with amyloid light-chain (AL) amyloidosis. The bright X-34 fluorescence marks amyloid aggregates in the heart (A), lung (B), bladder (C), and bone (D). For color image please refer to the end of this volume. Scale bar = 200 μm.

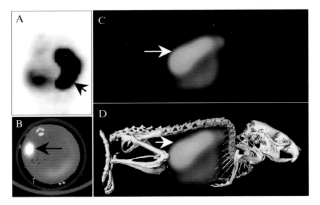

WALL *ET AL.*, CHAPTER 11, FIG. 3. Standard representations of microSPECT and microCT images obtained from ^{125}I-SAP labeling of AA-amyloid deposits in H2/IL-6 mice. (A) Planar, 2-D, scintigraphic image of a mouse with accumulation of ^{125}I-SAP in the spleen (arrow) and to a lesser degree in the liver. (B) Axial slice of a coregistered microSPECT/CT through the upper abdomen showing significant activity in the spleen (arrow) and diffuse accumulation of the SAP in the ventral liver. (C) Three-dimensional volume–texture (voltex) rendering of the microSPECT data discriminates the high amyloid content in the spleen (arrow) from the more diffuse hepatic deposits. (D) Precise anatomical localization of the amyloid is provided when a 3-D voltex rendering of the ^{125}I-SAP distribution is displayed with a surface-rendered microCT data set.

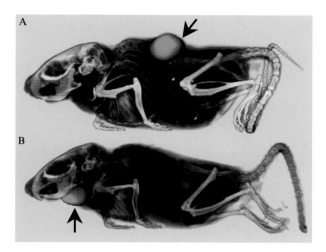

WALL *ET AL.*, CHAPTER 11, FIG. 4. Anatomical localization of a human AL amyloidoma. (A) The amyloidoma is readily visible (arrow) and anatomically pinpointed when a 3-D surface rendering of the ^{125}I-11–1F4 mAb distribution is displayed with the CT image. (B) When a ^{125}I-labeled isotype-matched control antibody (MOPC-31c IgG1κ; Sigma, St. Louis MO) is used, no activity is observed in the amyloidoma. However, free iodide is visible in the thyroid (arrow). In both (A) and (B), a threshold was applied to the SPECT data to display the 40% maximal activity.

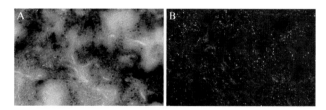

WALL *ET AL.*, CHAPTER 11, FIG. 6. Representative microautoradiographic study of splenic tissue from an AA-amyloidotic H2/IL-6 mouse administered ^{125}I-SAP for SPECT imaging. Consecutive sections were stained using photographic emulsion (A) and alkaline Congo red (B). In (A), dense areas of amyloid appear as dark punctate regions, whereas more diffuse accumulation of ^{125}I-SAP is manifest as shadowed areas. The amyloid deposits in (B) appear in cross-polarized illumination as green-blue birefringent areas. Magnification was ×40 (5× objective).

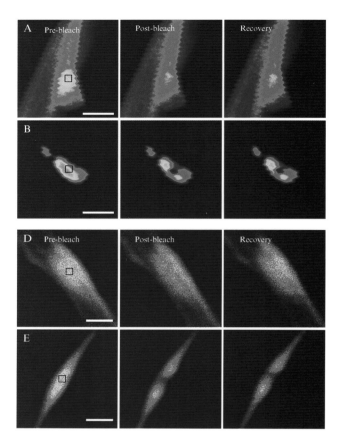

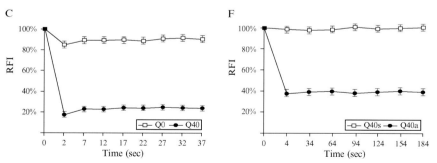

BRIGNULL *ET AL.*, CHAPTER 16, FIG. 2. FRAP analysis of polyglutamine::YFP solubility in living animals. FRAP distinguishes between Q0::YFP (A) and a Q40::YFP foci (B) in muscle cells. Image presented are representative and include an image before photobleaching (pre-bleach) of the boxed area (black box), immediately after photobleaching (post-bleach) and after a recovery period (time indicated in accompanying graph). As expected, Q0 recovers rapidly from photobleaching (A), whereas Q40 protein in foci does not recover from

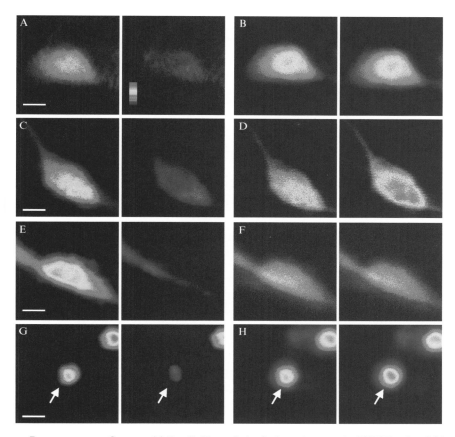

Brignull *et al.*, Chapter 16, Fig. 3. Expanded polyglutamine proteins FRET *in vivo*. Q86 protein in neuronal aggregates exhibits FRET, indicating close and roughly ordered interactions at the molecular level. YFP photobleaching is seen on the YFP channel (A, C, E, and G) and its effect on the CFP donor in the same cell (B, D, F, H). CFP and YFP coexpression (A, B) do not FRET, $E^c = 0.08$ ($\pm$ standard deviation of 0.16), $n = 5$. Animals expressing CFP::YFP do FRET (C, D), $E^c = 0.25$ (± 0.08), $n = 20$. Neurons coexpressing Q19:: CFP and Q19::YFP (E, F) do not FRET, $E^c = -0.08$ (± 0.05), $n = 15$, whereas coexpression of Q86::CFP and Q86::YFP (G, H) does produce FRET, $E^c = 0.22$ (± 0.08), $n = 17$. Cells shown are representative of FRET experiments. Intensity is by a color scale (G), where blue is least intense and red is most intense. Scale bar = 2 μm.

photobleaching (B), consistent with an insoluble protein. Bar = 2 μm. (C) Results from FRAP are quantified by determining the relative fluorescence intensity (RFI) for each time point the graph represents the average of analysis of a minimum of five independent measurements. Error bars indicate SEM. In neurons, FRAP was able to distinguish between a soluble (40s, D) and insoluble species (40a, E) of the same Q40::YFP protein, quantified in (F). Scale bars indicate 2 μm. Quantification of neuronal data is the mean of five neurons for 40s and 10 neurons for 40a. Error bars indicate SEM.

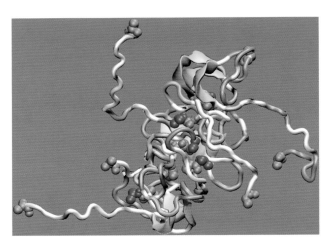

URBANC *ET AL.*, CHAPTER 19, FIG. 2. Globular structure of Aβ42 hexamer as found within the four-bead peptide model with amino acid–specific interactions caused by hydropathy. D1 is represented by four red spheres to illustrate the hydrophilic N-termini at the surface of the hexamer. I41 (four green spheres) and A42 (four blue spheres), as part of the C-terminal region, are at the hydrophobic core of the hexamer. Yellow ribbons represent a β-strand, cyan tube a turn, and silver tube a random coil-like secondary structure. The image was generated within the VMD software package (Humphrey *et al.*, 1996), which includes the STRIDE algorithm for calculating the secondary structure-propensity per residue (Heinig and Frishman, 2004).

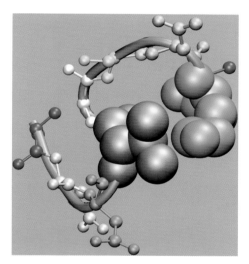

URBANC *ET AL.*, CHAPTER 19, FIG. 3. Folded Aβ(21–30) decapeptide conformation as found within the united-atom model with amino acid–specific interactions caused by hydropathy and charge. All atoms except hydrogens are drawn as small spheres: A21 and A30 (blue), E22 (pink), D23 (red), V24 (tan), G25 and G29 (white), S26 (yellow), N27 (orange), and K28 (cyan). V24 and K28 are presented by large opaque spheres to illustrate their packing, a critical event in the decapeptide folding. The image was generated within the VMD software package (Humphrey *et al.*, 1996).

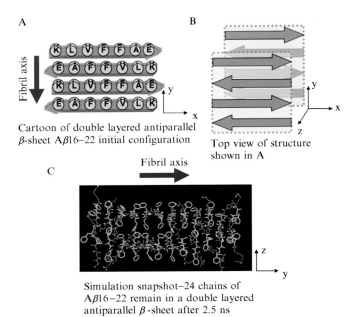

A

Fibril axis

Cartoon of double layered antiparallel
β-sheet Aβ16–22 initial configuration

B

Top view of structure
shown in A

Fibril axis

C

Simulation snapshot–24 chains of
Aβ16–22 remain in a double layered
antiparallel β-sheet after 2.5 ns

HALL AND WAGONER, CHAPTER 20, FIG. 2. Illustration of the double-layered β-sheet formed by Aβ16–22 and a snapshot from the CHARMM simulation (Ma and Nussinov, 2002a).

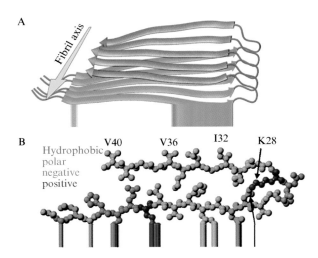

A

Fibril axis

B

Hydrophobic
polar
negative
positive

V40   V36   I32   K28

HALL AND WAGONER, CHAPTER 20, FIG. 3. (A) Ribbon diagram depicting the arrangement of the five copies of Aβ(10–40). (B) Aβ(10–40) as depicted by CHARMM all-atom simulations package with residues colored according to type. (Petkova *et al.*, 2002).

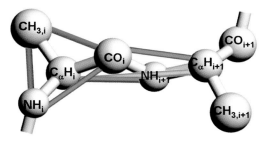

HALL AND WAGONER, CHAPTER 20, FIG. 6. Geometry of intermediate resolution protein model, PRIME, for alanine.

HALL AND WAGONER, CHAPTER 20, FIG. 8. Snapshots of a 48-peptide system at various reduced times, t^*. The simulation proceeds from a random initial configuration at concentration $c = 10$ mM and temperature $T^* = 0.14$ until the formation of a protofilament at $t^* = 205.9$ (Nguyen and Hall, 2004).

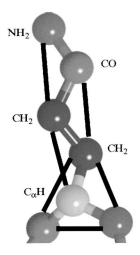

HALL AND WAGONER, CHAPTER 20, FIG. 9. Geometry of intermediate resolution protein model, PRIME, for glutamine.

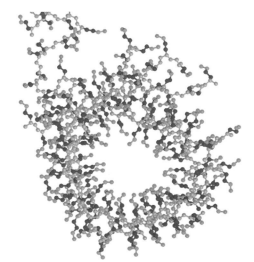

HALL AND WAGONER, CHAPTER 20, FIG. 10. Tube formed during simulation of 24 poly-glutamine 16mers (Marchut and Hall, 2006).

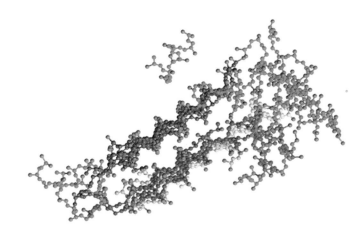

HALL AND WAGONER, CHAPTER 20, FIG. 11. Snapshot of a 48-peptide ordered aggregate obtained from the c = 1 mM simulation at T* = 0.12 (Wagoner and Hall, unpublished).